TURBO CODE APPLICATIONS

Turbo Code Applications

A Journey from a Paper to Realization

Edited by

KEATTISAK SRIPIMANWAT

National Electronics and Computer Technology Center (NECTEC),
Pathumthani, Thailand

Springer

A C.I.P. Catalogue record for this book is available from the Library of Congress.

ISBN 10 1-4020-3686-8 (HB)
ISBN 13 978-1-4020-3686-6 (HB)
ISBN 10 1-4020-3685-X (e-book)
ISBN 13 978-1-4020-3685-9 (e-book)

Published by Springer,
P.O. Box 17, 3300 AA Dordrecht, The Netherlands.

www.springeronline.com

Printed on acid-free paper

Printed in the Netherlands.

To all scientists who have dedicated
their efforts to the growth of communication
engineering and information theory societies

Preface

Turbo Code Applications: *a journey from a paper to realization* presents contemporary applications of turbo codes in thirteen technical chapters. Each chapter focuses on a particular communication technology utilizing turbo codes, and they are written by experts who have been working in related areas from around the world. This book is published to celebrate the 10th year anniversary of turbo codes invention by *Claude Berrou Alain Glavieux and Punya Thitimajshima* (1993-2003). As known for more than a decade, turbo code is the astonishing error control coding scheme which its performance closes to the *Shannon's* limit. It has been honored consequently as one of the seventeen great innovations during the first fifty years of information theory foundation. With the amazing performance compared to that of other existing codes, turbo codes have been adopted into many communication systems and incorporated with various modern industrial standards. Numerous research works have been reported from universities and advance companies worldwide. Evidently, it has successfully revolutionized the digital communications.

Turbo code and its successors have been applied in most communications starting from the ground or terrestrial systems of data storage, ADSL modem, and fiber optic communications. Subsequently, it moves up to the air channel applications by employing to wireless communication systems, and then flies up to the space by using in digital video broadcasting and satellite communications. Undoubtedly, with the excellent error correction potential, it has been selected to support data transmission in space exploring system as well.

To emphasize on its applications, the effort for editing this book is not only to focus on the technical aspect of turbo code, but also to depict its impacts and up-to-date research works. This book aims to place in courses for graduate students, to involve in research for professional scientists and engineers, and to be a reference book. These interests lie in the field of digital communications, coding theory and information technology. Principle of turbo codes can be found widely in many text books and other online materials. Thus, this book intends to provide an advance coverage of turbo code applications for readers with background experience in this topic, and targets to review up-to-date applications of turbo code and its successors. With the best effort of well-known authors in related fields including the strong support of technical committee, readers are expected of having a technical book that obtains contemporary fruitful results of turbo codes.

Acknowledgments

The organization of this editorial textbook was co-sponsored by 1) the Electrical Engineering/Electronics, Computer, Telecommunications, and Information Technology Association of Thailand (ECTI) and 2) the National Electronics and Computer Technology Center (NECTEC) of the National Science and Technology Development Agency (NSTDA). That was also with technical supports from Thailand chapters of the IEEE Communications Society, IEEE CAS Society, and IEEE MTT/AP/ED. Completing this editorial book, it was with valuable contribution from many people in various ways. To express this appreciation, first special thanks are for consulting members: *Sawasd Tantaratana* (SIIT)-Chair, *Witold Krzymien* (U of Alberta), *Johann Weinrichter* (TU-Wien), *Sirikiat Ariyavisitakul* (Texas Instruments), and *Thaweesak Koanantakool* (NECTEC).

This book is particularly indebted to technical committee and reviewers. Their suggestions for all possible improving of the manuscript and ensuring of chapters consistency, are highly appreciated. Thus, grateful thanks are for *Sorin Adrian Barbulescu, Piengpen Butkatanyoo, Ditsapon Chumchewkul, Ivan Fair, Peter Hamilton, Nguyen H. Ha, Chutima Indaraprasirt, Chaiwat Keawsai, Busaba Kramer, Pham Manh Lam, Komsak Meksamoot, Amporn Poyai, Chumnarn Punyasai, R.M.A.P. Rajatheva, Athikom Roeksabutr, Hamid Sadjadpour, Mathini Sellathurai, Christian Seyringer, Morakot Sriswasdi, Nidapan Sureeratanan, Phubate Udomsaph, Johann Weinrichter, and Yan Xin.*

It is pleased to acknowledge the administrative support for preparing the final manuscript. That was provided by an assistant editor, *Theeraputh Mekathikom*, with co-supporting from *Warapong Suwanarak*. Also, thanks to *Mark de Jongh* and *Helga Melcherts* of Springer publisher for their supportive.

Since the kick off time for organizing this book, it was strongly encouraged by the general secretary team. Grateful thank is then for *Pornchai Supnithi* (KMITL), and the profound appreciation is to *Poramate Tarasak* (U of Victoria) for the support that was provided continuously throughout the entire steps of this book.

Finally, to all scientists for the dedication to develop our communication engineering society, and to their (our) families for understanding, thank you.

ECTI & NECTEC - NSTDA
Pathumthani, Thailand

Keattisak Sripimanwat
February 2005

Contents

Part V Implementations

List of Contributors

C. Argon
Seagate Technology
Bloomington
MN 55435, USA

S. A. Barbulescu
Institute for Telecommunications
Research, University of South
Australia, Mawson Lakes
SA 5095, Australia

G. P. Calzolari
European Space Agency
D/OPS, ESOC
Robert-Bosch-Straße 5
64293 Darmstadt, Germany

S. Cheng
Lane Dept. of Computer Science
and Electrical Engineering
West Virginia University
Morgantown, WV 26506-6109, USA

F. Chiaraluce
Dipartimento di Elettronica
Intelligenza Artificiale e
Telecomunicazion, Università
Politecnica delle Marche
Via Brecce Bianche
60131 Ancona, Italy

R. Garello
Dipartimento di Elettronica
Politecnico di Torino
Corso Duca degli Abruzzi
24, 10129 Torino, Italy

Y. L. C. de Jong
Communications Research Centre
3701 Capling Ave.,Ottawa
Ontario K2H 8S2, Canada

B. V. K. Vijaya Kumar
Dept. of Electrical and
Computer Engineering
Carnegie Mellon University
Pittsburgh
PA 15213, USA

B. Lu
Silicon Laboratories
Broomfield
CO 80021, USA

G. Masera
Dipartimento di
Elettronica, Politecnico di
Torino, Corso Duca degli
Abruzzi 24-10129, Torino, Italy

S. W. McLaughlin
School of Electrical and
Computer Engineering
Geogia Institute of Technology
Atlanta GA 30332, USA

T. Mekathikom
National Electronics and
Computer Technology Center
-NSTDA
Thailand Science Park
Pathumthani 12120, Thailand

S. Ölçer
IBM Research Division
Zurich Research Laboratory
Saeumerstrasse 4
8803 Rueschlikon, Switzerland

H. Sadjadpour
School of Engineering
University of California
1556 High Street
Santa Cruz, CA 95064, USA

C. Schlegel
Dept. of Electrical
and Computer Engineering
University of Alberta
Edmenton, AB
T6G 2V4, Canada

M. Sellathurai
Communications Research Centre
3701 Capling Ave.,Ottawa
Ontario K2H 8S2, Canada

R. I. Seshadri
Lane Dept. of Computer Science
and Electrical Engineering
West Virginia University
Morgantown
WV 26506-6109, USA

H. Song
Agere Systems
1921 Corporate Center Circle
Suite 3-A Longmont
CO 80504, USA

K. Sripimanwat
National Electronics and
Computer Technology Center
-NSTDA
Thailand Science Park
Pathumthani 12120, Thailand

P. Supnithi
Dept. of Telecommunications
Engineering, Faculty of Engineering
King Mongkut's Institute
of Technology Ladkrabang
Bangkok 10520, Thailand

P. Tarasak
Dept. of Electrical and
Computer Engineering
University of Victoria
Victoria, BC
V8W3P6, Canada

D. Tujkovic
Centre for Wireless
Communications (CWC)
University of Oulu
P.O. Box 4500,
FIN-90014 Oulu, Finland

M. C. Valenti
Lane Dept. of Computer Science
and Electrical Engineering
West Virginia University
Morgantown
WV 26506-6109, USA

E. Vassallo
European Space Agency
D/OPS, ESOC
Robert-Bosch-Straße 5
64293 Darmstadt, Germany

X. Wang
Department of Electrical
Engineering, Columbia University
New York, NY 10027, USA

List of Acronyms

3GPP	3rd Generation Partnership Project
4-PSK	Quaternary Phase Shift Keying
8PSK	8-ary Phase Shift Keying
ADSL	Asymemtric Digital Subscriber Line
AGC	Automatic Gain Control
APP	*a posteriori* probability
APSK	Amplitude Phase Shift Keying
ARQ	Automatic Repeat Request
ASIC	Application Specific Integrated Circuit
ASIP	Application Specific Instruction set Processor
ASK	Amplitude Shift Keying
ASM	Attached Sync Marker
ATM	Asynchronous Transfer Mode
AWGN	Additive White Gaussian Noise
BCC	Binary Convolutional Codes
BCCC	Binary Concatenated Convolutional Codes
BCH	Bose-Chaudhuri-Hocquenghem code
BCJR	Bahl, Cocke, Jelinek, and Raviv Algorithm
BER	Bit Error Rate
BICM	Bit-Interleaved Coded Modulation
bps	bit per second
BPSK	Binary Phase Shift Keying
BSC	Binary Symmetric Channel
BTC	Block Turbo Code
BWA	Broadband Wireless Access
CCSDS	Consultative Committee for Space Data Systems
CD	Compact Disc
CDMA	Code Division Multiple Access
CD-R	Read-only CD
CD-ROM	Read-only Memory CD

CD-RW	Rewritable CD
CGA	Chase-GMD Algorithm
CIRS	Cross-Interleaved Reed-Solomon codes
CMOS	Complementary Metal Oxide Semiconductor
CNES	Centre National d'Etudes Spatiales
CPM	Continuous-Phase Modulation
CRC	Cyclic Redundancy Check
CRSC	Circular Recursive Systematic Convolutional
CSI	Channel State Information
CTC	Convolutional Turbo Code
DFE	Decision Feedback Equalizer
DFG	Data Flow Graph
DLR	Deutschen Zentrum für Luft- und Raumfahrt
DMC	Discrete Memoryless Channel
DMS	Discrete Markov Source
DMT	Discrete Multi-Tone
DOW	Direct Overwrite
DSB-SC	Double Side-Band Suppressed Carrier
DSL	Digital Subscriber Line
DSP	Digital Signal Processing
DVB	Digital Video Broadcasting
DVB-RCS	Digital Video Broadcasting-Return Channel via Satellite
DVB-S	Digital Video Broadcasting-Satellite
DVB-S2	Digital Video Broadcasting-Satellite (second generation)
DVD	Digital Video Disc
DVD-R	Read-only DVD
DVD-RW	Rewritable DVD
EFM	Eight-to-Fourteen Modulation
eIRA	extended Irregular Repeat Accumulate (code)
EM	Expectation-Maximization algorithm
ESA	European Space Agency
ETSI	European Telecommunications Standards Institute
EXIT	Extrinsic Information Transfer
FER	Frame Error Rate
FIR	Finite Impulse Response
FPGA	Field Programmable Gate-Array
FSE	Fractionally-Spaced Equalizer
FSM	Finite State Machine
FWHM	Full Width at Half Maximum density
GF	Galois Field
GMD	Generalized Minimum Distance
GMSK	Gaussian Minimum Shift Keying
GPR	Generalized Partial Response
HCCC	Hybrid Concatenated Convolutional Code
HD-DVD	High-Density DVD

HDL	Hardware Description Language
i.i.d.	Independent and Identically Distributed
IIR	Infinite Impulse Response
ISI	Inter Symbol Interference
JAXA	Japan Aerospace Exploration Agency
LAN	Local Area Network
LBC	Linear Block Code
LDPC	Low Density Parity Check code
LLR	Log-Likelihood Ratio
LMMSE	Linear Minimum Mean-Square-Error
MAC	Medium Access Control
MAN	Metropolitan Area Network
MAP	Maximum *a posteriori* Probability
MIMO	Multiple-Input Multiple-Output
MLC	Multilevel Coded modulation
MLSD	Maximum Likelihood Sequence Detector
MMF	Multimode Fiber
MMSE	Minimum-Mean Square Error
MO	Magneto-Optical
MPEG	Moving Picture Experts Group
M-PSK	M-ary Phase Shift Keying
MTF	Modulation Transfer Function
NASA	National Aeronautics and Space Administration
NPML	Noise-Predictive Maximum-Likelihood
NRC	Non Recursive Convolutional
NRZ	Nonreturn-to-Zero
NRZI	Non-Return-to-Zero-Inverted
OFDM	Orthogonal Frequency Division Multiplexing
PAM	Pulse Amplitude Modulation
PC	Phase Change
PCC	Parallel Concatenated Code
PCCC	Parallel Concatenated Convolutional Code
PCE	Parallel Concatenated Encoder
PR	Partial Response
PRML	Partial Response Maximum Likelihood
PWM	Pulse-Width Modulation
QAM	Quadrature Amplitude Modulation
QPSK	Quadrature Phase Shift Keying
RAM	Random Access Memory
RCPCC	Rate-Compatible Punctured Convolutional Code
RCST	Return Channel Satellite Terminal
RIBB	Ring Interleaver Bottleneck Breaker
RLL	Runlength-Limited
RM	Reed-Muller Code
RS	Reed-Solomon Code

RSC	Recursive Systematic Convolutional
RSPC	Reed-Solomon Product Codes
SCC	Serial Concatenated Code
SCCC	Serial Concatenation Convolutional Code
SCE	Serial Concatenated Encoder
SCTC	Serially-Concatenated Turbo Codes
SCTCM	Serial Concatenated Trellis Coded Modulation
SDR	Sigma-to-Dynamic Ratio
SER	Symbol Error Rate
SIC	Soft Interference Cancellation
SIHO	Soft-Input / Hard-Output
SIMO	Single-Input Multiple-Output
SISO	Soft-Input / Soft-Output
SMF	Single-Mode Fiber
SNR	Signal to Noise Ratio
SOVA	Soft Output Viterbi Algorithm
SPB	Sphere Packing Bound
SSPA	Solid State Power Amplifier
ST	Space-Time
ST-BICM	Space-Time Bit-Interleaved Coded Modulation
STTrCs	Space-Time Trellis Codes
ST-TTCM	Space-Time Turbo Trellis Coded Modulation
TCC	Turbo Convolutional Code
TCM	Trellis Coded Modulation
TPC	Turbo Product Code
TTCM	Turbo Trellis Coded Modulation
UEP	Unequal Error Protection
UMTS	Universal Mobile Telecommunication Service
VA	Viterbi Algorithm
VLSI	Very Large Scale Integrated circuits
VSAT	Very Small Aperture Terminal
WEF	Weight Enumerating Function
WER	Word Error Rate
WGN	White Gaussian Noise
WSSUS	Wide Sense Stationary random processes with Uncorrelated Scattering
ZF-LE	Zero-Forcing Linear Equalizer

Chapter 1

Book Introduction

Keattisak Sripimanwat

National Electronics and Computer Technology Center-NSTDA, Thailand

"The invention of turbo codes did not result from a linear limit mathematical demonstration. It was the outcome of an empirical construction of a global coding/decoding scheme, using existing bricks that had never been put together in this way before." [1]

Claude Berrou (2001)

Getting a method to control or to mitigate error for data transmission or storage in digital communication systems, error control coding is one of the main communication techniques for this purpose. Obviously for more than fifty years, in advanced communication systems error control coding has played a very important role. It has been developing and adopted successfully into many application platforms.

Briefly regarding the historical timeline of error correcting codes, it was officially started in the year 1948 with the introduction of an information theory by *Claude E. Shannon*. A prediction of *Shannon* is that arbitrarily reliable communications are achievable by redundant channel coding. Subsequently, there were many pioneer works or milestones after Shannon's discovery. Starting in early 1950s, most researches emphasized on theoretical side or on the foundation of concerned mathematics [2]. Next, greater effort on searching for good codes structure was done during 1960s. Through the 1970s, the design of families of codes with larger code lengths and better performance was focused as the main target. Then, the transformation from theoretical era to the practice was concentrated in 1980s. It is noted that new design of encoders and decoders were presented frequently to digital communication engineering community during this period of time.

In that past fifty years, intensive research efforts have been done worldwide in order to achieve coding solution for solving related communication problems. Those are, among other things, 1) to have the better coding gain, 2) to reduce decoding complexity, and 3) to support or to associate working

K. Sripmanwat (ed.), Turbo Code Applications: a journey from a paper to realization, 1–14.

with other communication techniques. As the coding target, performance of the systems has been sailing closing to that *Shannon's* limit gradually. Resulting to recognized milestones along the past five decades, development of that error control coding came up many successful results. For examples, they are block codes, Hamming codes, Convolutional codes and Viterbi algorithm, Bose and Chaudhuri and Hocquenghem codes (BCH), Reed-Solomon codes (RS), and Trellis Coded Modulation (TCM). The historical breakthrough of turbo codes then arrived at early of 1990s.

In the year 1993, an annual international conference on communications or ICC was organized in Geneva, Switzerland. In that technical event, it was recognized that a paper of *Claude Berrou Alain Glavieux* and *Punya Thitimajshima* introduced an invention of new error control coding scheme. This novel method provides virtually error-free communications or obtains much better coding gain beyond that of any other existing codes. Gradually, it became a forefront of communication research and also inspires to generate other new numerous ideas until date. Turbo codes, on the same hand, plays an important role in most modern communication systems. It stepped out from that paper and successfully entered for the commercialization in the present telecommunication market. Undoubtedly from those accomplishments, a number of awards were then honored to its inventors [3]. As known for more than ten years, the first appearance of turbo codes to the public was on a paper entitled *"Near Shannon limit error-correcting coding and decoding: turbo-codes"* [4].

In this first chapter, it is an introduction of this great coding invention with related stories to the motivation and the organization of this book. That would give readers with more basic point of view before going on to its application in the following chapters. This book emphasizes mainly on advanced turbo codes applications. For more information, readers can find more details for the concept of error control coding and the principle of turbo codes from a number of other good sources. The helpful materials are available both online and in hardcopy styles. Some suggested books are as in [2, 5–7].

To follow by Sec. 1.1, it engulfs a brief turbo codes history. That provides the explanation to its evolutions and milestones, main related publications, patents, and awards. Sec. 1.2 guides readers to the organization of the book which emphasizes on the utilization. It summarizes all further thirteen chapters which present the grasp of turbo codes applications, and were written by leading scientists in the related communication areas.

1.1 A Brief History of Turbo Codes

> *"At first, it was a great surprise to observe that the bit error rate (BER) of these reconstructed symbols after decoding was lower than that of decoded information d. We were unable to find any explanation for this strange behavior in the literature."* [8]
>
> *Claude Berrou and Alain Glavieux (1998)*

This section is giving readers with a collection of important materials along the turbo codes discovery. That begins with a group of scientists which their work based on the contemporary scheme of convolutional encoding and Viterbi algorithm decoding. The main events are also depicted in the timeline of Fig. 1.1. Its details are presented as follows.

1.1.1 Evolutions and Milestones

Refer in the "Reflections on the Prize Paper: Near optimum error-correcting coding and decoding: turbo codes" published on June 1998 in IEEE information theory society newsletter [8], *Claude Berrou, Alain Glavieux,* and *Patrick Adde* were mentioned as key persons prior to the time of turbo codes invention. At the Ecole Nationale Supérieure des Télécommunications de Bretagne of France, these scientists started their work focusing on the *Soft-Output Viterbi Algorithm (SOVA).* It was based on the literature of *G. Battail* in 1987 [9] and of *J. Hagenauer* and *P. Hoeher* in 1989 [10]. Those were certainly referred to famous papers of *A.J. Viterbi,* "Convolutional codes and their performance in communication systems" [11], and of *G.D. Forney,* "The Viterbi algorithm" [12]. Initially, their research was to transfer the SOVA algorithm into hardware platform on MOS transistors in the simplest possible way as the target.

Consequently, they observed that SOVA can be considered as a signal-to-noise (SNR) amplifier. This could be mentioned as the beginning of "turbo"-codes concept because it stimulated them to consider "feed back" techniques that commonly used with electronic amplifier circuits. To explore that concept, they cascaded that signal-to-noise (SNR) amplifier or their SOVA version in order to obtain large asymptotic gains. This connection bases on "concatenation" coding technique of the well known concept in the literature. Their experiments were done on a serial concatenation of two ordinary convolutional codes at the early step. It was later concentrated on parallel concatenation. Because the idea of two component decoders working with the same clock signal matches with that the reason of hardware implementation (in parallel) for clock signal distribution. This parallel concatenation with amplifiers was considered to be meaningful only if the code is systematic, and it was a straightforward to use recursive systematic convolutional (RSC) codes at the final.

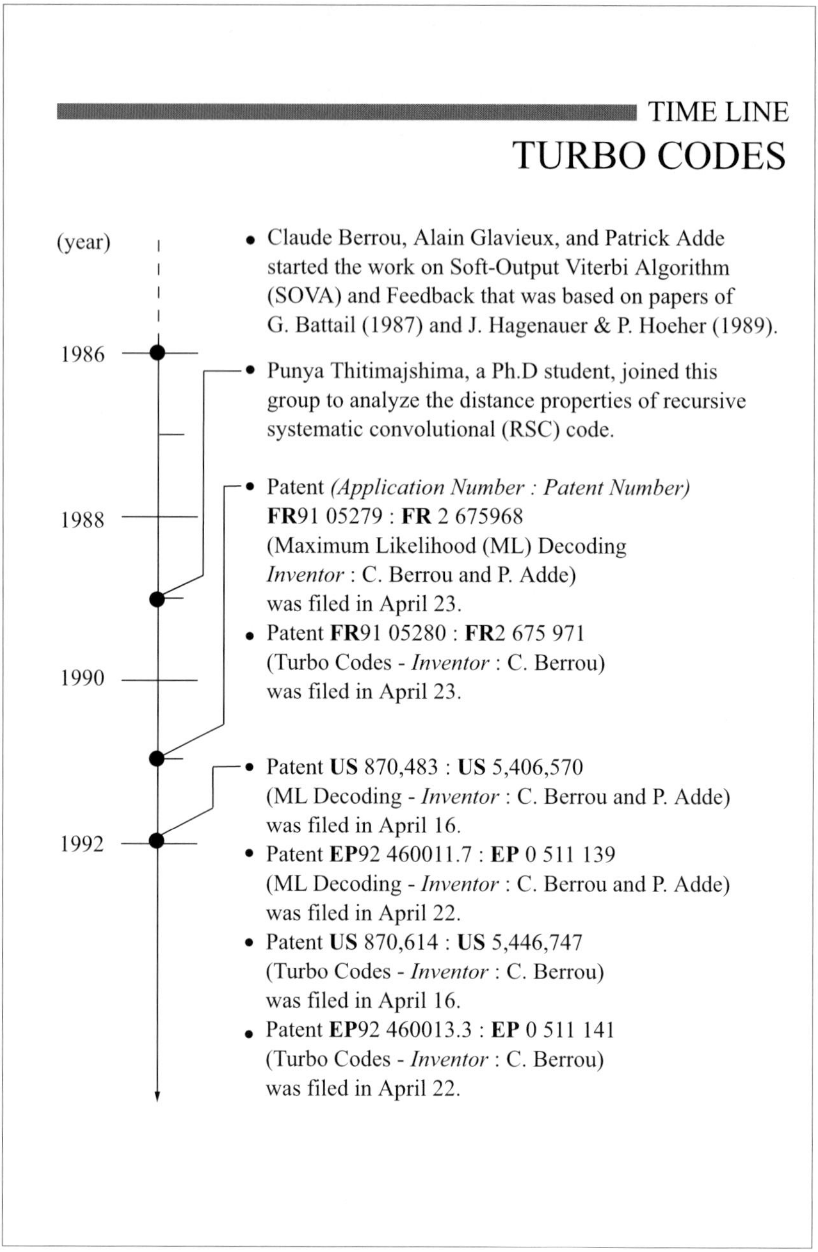

Fig. 1.1. Milestones of Turbo Codes.

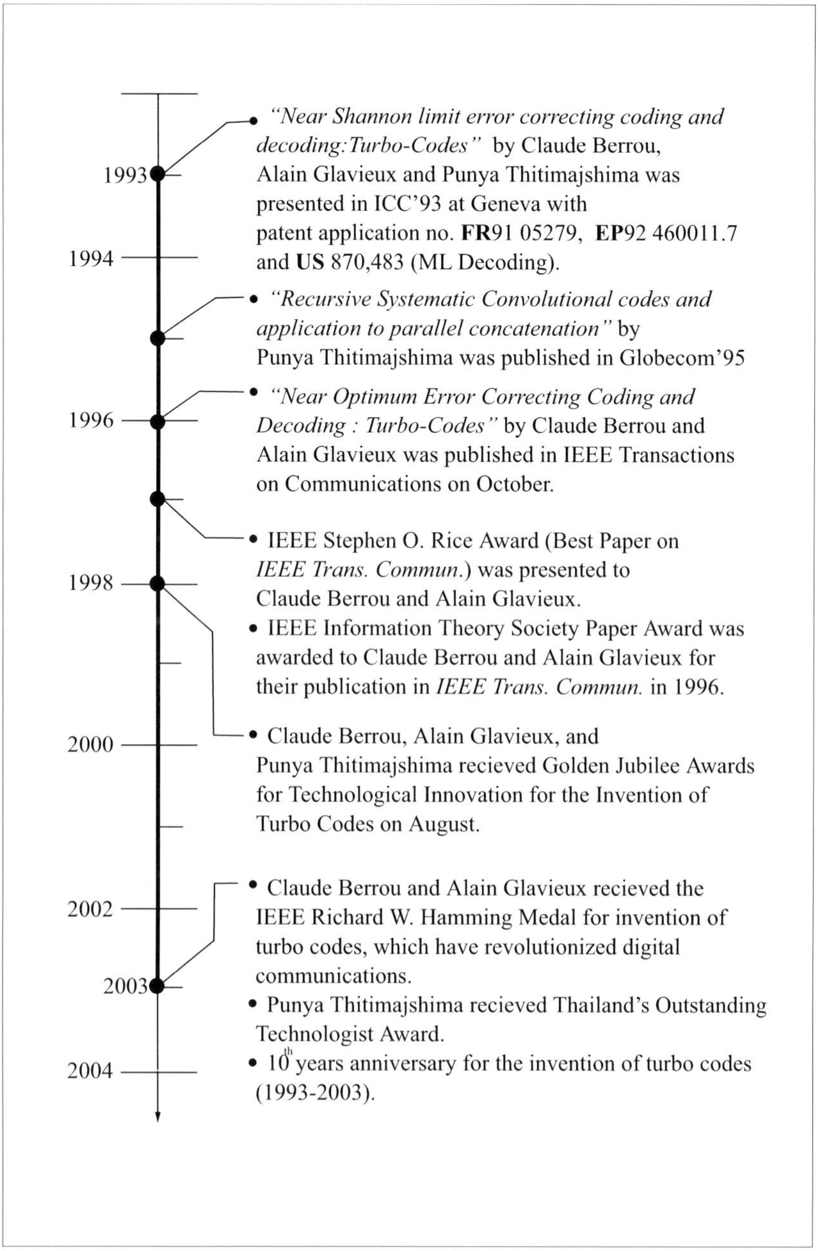

Fig. 1.1. Milestones of Turbo Codes (continued).

During this time of turbo codes foundation, a Ph.D student, *Punya Thitimajshima,* started joining this group to work on the distance properties analysis in the year 1989. His dissertation devotes to studying distance properties and of error probability of the recursive punctured systematic convolutional (RPSC) codes and their concatenation in serial and parallel styles. Certainly, it is combined with iterative decoding [13]. This work entitled *"Les codes Covolutifs Rcursifs Systmatiques et leur application la concatenation parallel",* as a dissertation at l'Universit de Bretagne Occidentale (UBO).

Gradually, the construction of original turbo codes was formed with related technical bricks. In order to solve obstruction in those initial works which reported on weighting problems, the beginning of SOVA was then replaced by Bahl-Cocke-Jelinek-Raviv (BCJR) algorithm [14] at the end of the discovery. It was mentioned that the first experiment with this novel coding construction was run in 1991 [15]. With the founding of following well known technical terms of extrinsic information, iterative decoding, recursive systematic convolutional codes, parallel concatenation, and non-regular interleaving, turbo codes was born finally.

There are two other main publications regarding turbo codes which appeared to the public after its introduction. First, a part of above dissertation was published in "Recursive Systematic Convolutional codes and application to parallel concatenation", which was presented at IEEE Globecom 1995 conference by *Thitimajshima* [16]. Moreover, at a year later another well known article was published as "Near optimum error correcting coding and decoding: turbo-codes" on the IEEE transactions on communications. That was issued on October 1996 and written by *Claude Berrou* and *Alain Glavieux* [17].

Since 1993, the legacy of turbo codes has opened new technical research areas continuously. It sparks new numerous ideas to improve its own performance. Moreover, its concept is combined with other communication techniques in order to improve overall system performance. Those examples of "turbo codes effect" are;

- *Turbo product codes / Turbo block codes* - a new iterative decoding algorithm for product (block) codes based on soft decoding and soft decision output of the component codes. It was invented as a new generation coding scheme with a high code rate.
- *Turbo equalization* - an iterative equalization technique that achieves highly impressive performance for communication through intersymbol interference (ISI) channels. That is for the multi-path propagation environment of wireless communications, or for other bandlimited-channel systems.
- *Turbo codes for multilevel or turbo trellis coded modulation (TTCM)* - the combined technique of turbo coding with high spectral efficiency modulation or non-binary (high) order signaling.

- *Space-time turbo codes* - the application of turbo codes with multiple transmit antennas for improving the data rate and/or the reliability of communications over fading channels for wireless communications.
- *Low-density parity-check codes (LDPC)* - a long time forgotten code that was invented much earlier in 1962. Turbo codes recall researchers to this date-back invention of LDPC codes. Then, to develop this complex coding scheme of the past to be a today competitive method for obtaining the better coding gain. This LDPC has returned to the society of communication engineering and has obtained the closer performance to that Shannon's limit. Obviously, LDPC was re-stimulated from the invention of turbo codes.

Moreover, turbo codes / turbo principle and their successors of above mentioned, have been applied successfully with other popular communication techniques. For examples, those are *multiuser detection, multiple-input multiple-output (MIMO)* - a technique that results to high spectral efficiency and capacity-approaching performance, and *orthogonal-frequency division multiplexing (OFDM)* - an efficient method capable of establishing high speed digital transmission through frequency selective fading channels. Details are presented in the upcoming chapters.

Finally, an obvious milestone of a young turbo code, has settled permanently along the road of digital communication development. Its successors, then, have been continuously following on the next miles and ahead.

1.1.2 Golden Patents and Awards

After the successful revolution in the year 1993, turbo code has been praised and crowned widely. Its impacts are not only in its technical communities but also found on economic, educational, and academic aspects. It affects to sparking of other technical ideas as mentioned before. Following its emergence, enhanced researchers worldwide generate a number of new related works. More than 400 patents involving its theory and applications have been filed afterward [18]. Successfully, it became one of the core technology for today's cutting edge communication products.

Prior to mentioning to the high valued patent of turbo codes, principle of the invention should be redefined with the construction concept comprising of a). Recursive Systematic Convolutional (RSC) coding and its parallel concatenation, b). iterative decoding, and c). extrinsic information.

Initially, it recalled us to the first glance of turbo code appearance in ICC'93, that was on the context with patent filing numbers of 91 05279 (France), 92 460011.7 (Europe), and 07/870,483 (USA) [4]. In fact, these numbers are entitled in French of *"Procédé de décodage d'un code convolutif à maximum de vraisemblance et pondération des décision, et décodeur correspondants"* for filing in France and Europe. *"Method for a maximum likelihood decoding of a convolutional code with decision weighting, and corresponding decoder"* is the coincided title that was filed in USA. They are all invented by *Claude Berrou* and *Patrick Adde* [19–21].

However, to follow above mentioned turbo coding concept, there are other numbers of concerned patents. The main or the golden patent should most match with that in the title of *"Procédé de codage correcteur d'erreurs à au moins deux codages convolutifs systématiques en parallèle, procédé de décodage itératif, module de décodage et décodeur correspondants"* or *"Error-correction coding method with at least two systematic convolutional coding in parallel, corresponding iterative decoding method, decoding module and decoder"*. The first one was first filed in France (number 91 05280) on April 23, 1991. Later, this number was used as a priority data for filing other two main patents for expanding the right on turbo codes covering over Europe and USA. *Claude Berrou* is solely the inventor of them. Details are collected in Table 1.1.

Legally, the exclusive right on a patent exists for twenty years from the filing date. The patent owner may give permission to, or license, other parties to use the invention on mutually agreed terms. However, the patented invention may be available for commercial exploration by others in the countries which the patent is not filed. Thus, above mentioned turbo code patents which filed over three places (France, Europe, and USA), are then free to use at other places as in Asian countries.

The exclusive right on turbo codes and other turbo code related patents have been licensed and used for various application platforms. Many industrial standards have been incorporated. Consequently, a lot of product models from a number of chip making manufacturers have been placed in the market. In early of 2000s, the commercialization of this innovation focuses mainly for the new generation mobile and satellite communication systems. Licensing of those patents has been reported with impressive stories on its values [15, 22, 23].

Apparently, turbo code has revolutionized the communication engineering. Its successful stories and impacts have been highlighted. To guarantee those accomplishments, below awards and honors to its invention are the witness.

- In 1997, information theory society paper award was announced for *Claude Berrou* and *Alain Glavieux.* That was based on their work of *"Near optimum error-correcting coding and decoding: Turbo codes,"* published in IEEE transaction on communications–October 1996 [24]. In the same event, an honorable mention was given to *Punya Thitimajshima* for his contribution to the first turbo code paper in ICC'93.
- Based on the same work, *Claude Berrou* and *Alain Glavieux* were recipients of 1997 Stephen O. Rice award for the best paper in IEEE transactions on communications.
- Again, turbo code was honored in the year 1998 as one of the seventeen of great innovations. It was presented in the fifty year anniversary of information theory that *Claude Berrou, Alain Glavieux* and *Punya Thitimajshima* captured the IEEE information theory society's golden jubilee award for

Table 1.1. Basic Information of Golden Turbo Code Patent

Institut National de la Propriété Industrielle (INPI), France	
Title of invention : Proédé de codage correcteur d'erreurs à au moins deux codages convolutifs systématiques en parallèle, procédé de décodage itératif, module de décodage et décodeur correspondants	
Inventor	Claude Berrou
Assignee	France Telecom and Telediffusion de France S.A.
Application number	91 05280
Patent number	2675971
Filing date	April 23, 1991
European Patent Office(EPO), Europe	
Title of invention : Proédé de codage correcteur d'erreurs à au moins deux codages convolutifs systématiques en parallèle, procédé de décodage itératif, module de décodage et décodeur correspondants	
Inventor	Claude Berrou
Assignee	France Telecom and Telediffusion de France S.A.
Application number	92 460013.3
Patent number	0 511141
Filing date	April 22, 1992
United States Patent Office (USPTO), USA	
Title of invention : Error-Correction coding method with at least two systematic convolutional coding in parallel, corresponding iterative decoding method, decoding module and decoder	
Inventor	Claude Berrou
Assignee	France Telecom and Telediffusion de France S.A.
Application number	870614
Patent number	5446747
Filing date	April 16, 1992

technological innovation. This award was among other great inventions which were invented earlier during the past fifty years. For examples, those are algebraic decoding algorithm, convolutional codes, concatenated codes, Reed-Solomon (RS) codes, trellis coded modulation (TCM), and the Viterbi algorithm [3].

- In 2003, *Clude Berrou* and *Alain Glavieux* received IEEE Richard W. Hamming medal, for the invention of turbo codes, which have revolutionized digital communications. *Punya Thitimajshima* was honored with the 2003 Thailand's outstanding technologist award for turbo code invention.

All above impressive turbo code stories, from its invention through the related technological development as well as the achievements, motivates us to organize for this editorial book. Also, it is in order to celebrate another successful milestone for the first fifty years of the information theory that was founded by *Claude E. Shannon.*

1.2 Outline of Book: *a journey from a paper to realization*

> *"It's not often in the rarefied world of technological research that an esoteric paper is greeted with scoffing. Its even rarer that paper proves in the end to be truly revolutionary." [15]*
>
> *Erico Guizzo (2004)*

Starting with the application in data storage systems, first two chapters present with the application of turbo and turbo-like codes in the magnetic and optical storage media respectively. Typically, the demand for higher capacity, transfer rate, and storage density, is the main target of research in the field. In the hard-drive system, although the increasing of storage capacity is leading by the advances in head and media technologies, however coding and signal processing are those the cost-efficient methods to improve this storage capacity as well. In Chapter 2, the traditional media of magnetic recording channels where the application of recent developed error-control codes including turbo codes and low density parity check (LDPC) codes, is reviewed under the turbo equalization structure. It is remarked in this chapter that the iterative detection and decoding technique is the most potential candidate for the next generation read channels.

Another storage media follows in the Chapter 3. It presents the environment of read/write system of binary and multilevel (ML) for optical recording systems. In this chapter, it provides interesting principle of optical recording system through the mechanism of multilevel. Turbo product codes are applied potentially in this high-density storage system comparing with other conventional schemes. In addition, the concatenated coding for future optical recording systems is also discussed. It is noted with the necessity of using Reed-Solomon (RS) code as the outer part, and with iterative decoding nature codes as the inner one.

For wire or land line communication systems, applications of turbo codes are provided in two chapters. Those are for classical metal line and in fiber optic systems respectively. Chapter 4 presents turbo and turbo-like codes that is designed for Asymmetric Digital Subscriber Line (ADSL) which allows household consumers to access high speed broadband internet. In ADSL channel, by employing turbo and turbo-like coding it is possible to operate DSL links with greater robustness and reliability under the imperfect channel conditions. This chapter illustrates the approach to improve transmission performance by incorporating turbo and LDPC coding into ADSL technologies. The results of those applications are provided and compared with that of the concatenated coding scheme in ANSI standard (T1.413 or Wei code).

To reduce the negative effect from various types of noise and dispersion in fiber optic communications, error control coding by using turbo codes is one of the solutions. Chapter 5 reviews the application of turbo product codes (TPC) in optical fiber networks for both long-haul applications using single-mode fibers, and for short-haul multimode fiber links. Well organized sections and a thorough review would give readers with a complete guide to understand the basic of using contemporary error control codes in this type of channel.

For the present popular wireless communication systems, the application of turbo code principle is reviewed in five chapters. In order to improve overall performance in the wireless environment, they combine the turbo or iterative principle with other techniques. Those are, for examples, the multiple-input multiple-output (MIMO), space time coding, and orthogonal-frequency division multiplexing (OFDM).

In Chapter 6, the fascination of iterative demodulation and decoding for large constellation channel is presented. Code-Division Multiple Access (CDMA) and the multiple antenna technique are illustrated as that type of channel. Incorporating with turbo decoding principle, it is shown as the extremely useful scheme. Moreover, iterative demodulation and decoding is considered as a very powerful methodology to work with large numbers of interfering signals, and as the undergoing significant research.

Chapter 7 discusses importance of the application of the iterative decoding principle to the demodulation and error control decoding operations within a coded MIMO OFDM system. The principle of turbo decoding, that of iterative exchange of extrinsic information, is extended to this system and its receiver architecture. These techniques result high spectral efficiency and capacity-approaching performance in wireless channels.

Chapter 8 introduces a new paradigm for MIMO signal transmission by summarizing single and multiple antenna turbo coded modulation for using in the future wireless communication systems. This chapter presents the combining application techniques of space-time coding, turbo coding, and high order modulation scheme as the space-time turbo coded modulation (ST-TTCM). In the same hand, it is an application of turbo codes to design space-time trellis codes.

Currently, wireless communication industries have shown considerable interest in the progress of development of MIMO products which is used to support for high speed wireless communication systems. MIMO is advocated to be used in the future wireless data networks such as wireless local area network of IEEE 802.11 standards. It also will likely to be included in the next phase of the third generation (3G) mobile communication standardization. Chapter 9 reviews the turbo or iterative techniques with the above mentioned MIMO (turbo-MIMO systems). Its concentration is on the trade-off between performance and complexity for different detection schemes. Specifically, it includes low complexity MIMO transceiver design and low-complexity detector designs that challenges for practical solutions.

Chapter 10 gives an interesting application among the broadband wireless access (BWA) communications. That is the metropolitan area networks (MANs) based on IEEE 802.16a standard. This wireless system is an alternative broadband service for urban or suburban areas when ADSL and cable modem communications are disrupted or almost saturated. Turbo code and its derivative will be a crucial component in this emerging wireless system, as they have been proposed as an option error coding scheme. In addition, related system architectures and the survey on concerned turbo code implementation are presented.

For the satellite communications, digital video broadcasting (DVB), and deep space communication systems, high potential coding scheme is quite necessary for improving the system performance in these low signal margin channels. Certainly, these communications focus to the application of turbo code concept.

Firstly, turbo codes have already revitalized the satellite industry with a very substantial increase in bandwidth efficiency. It is likely that all future systems will be based on turbo codes, or the successors. New applications of turbo codes are actively pursued, which will provide on-going and substantial improvements in communications over satellite. Chapter 11 gives an interesting and comprehensive review of the use of turbo codes in satellite communication applications. In addition to providing a review of systems where these codes are currently being used, it also reviews new applications of turbo codes being developed which could be used in future systems.

Chapter 12 presents the application of turbo codes in DVB system that was founded by the European Telecommunications Standards Institute (ETSI). This DVB system takes benefit of the previous satellite communication links for delivering the digital television services. In this type of communication channel, turbo codes have been adopted in the return channel via satellite (DVB-RCS) which is set for the additional services of internet and data delivering. Because the little margin signal on the uplink to the satellite, then the strong forward error correction as turbo codes, is desired. Consequently, with a circular duobinary turbo code that selected for this DVB-RCS, internet service via satellite became a serious competitor to cable modem and DSL services. Moreover, up to the latest standard of DVB-S2, LDPC has been adopted achieving a significant improvement in the satellite downlink.

Various aspects of turbo codes on a standard of Consultative Committee for Space Data Systems (CCSDS) are presented in Chapter 13. In this CCSDS, turbo codes were chosen as an option since the year 1999. Indeed, performance of turbo codes is shown as the excellent method to improve power efficiency over that of existing codes. Impressively, the principle of turbo code has been involving in many space exploring projects. Those are missions to the Moon, Mercury, Comet, and Mars.

Chapter 14 reviews main achievements of hardware implementation including the critical aspects of its design. The attempt to obtain turbo decoder with the cheaper, faster, or more energy efficient, has presented as one of the most fascinating cases of study or a formidable example for VLSI designers. In this final chapter, the related standards and products of turbo codes are surveyed as well. That would give readers through the understanding of turbo code implementation.

References

1. Claude Berrou, "Turbo Codes: Some Simple Ideas for Efficient Communications", *The Seventh International Workshop on Digital Signal Processing Techniques for Space Communications*, 1-3 October 2001, Sesimbra, Portugal
2. Man Young Rhee, *Error Correcting Coding Theory*, McGraw-Hill, 1989.
3. "IEEE Information Theory Society Golden Jubilee Awards for Technological Innovation," *IEEE Information Theory Society Newsletter,* special issue, September 1998.
4. C. Berrou, A. Glavieux, P. Thitimajshima: "Near Shannon limit error-correcting coding and decoding : turbo-codes" Proceedings of ICC'93, Geneva, pp. 1064-1070, May 1993.
5. Christian Schlegel and Lance Perez, *Trellis and turbo coding,* Piscataway, NJ Wiley-IEEE Press; 2003.
6. L. Hanzo, T.H. Liew, B.L. Yeap, *Turbo coding, turbo equalisation and space-time coding : for transmission over fading channels,*John Wiley, 2002.
7. Bernard Sklar, *Digital Communications: Fundamentals and Applications, (2nd Edition),* Prentice Hall (2001)
8. Claude Berrou and Alain Glavieux, "Reflections on the Prize Paper: Near optimum error-correcting coding and decoding: turbo codes," *IEEE Information Theory Society Newsletter*, Vol. 48, No.2, June 1998.
9. G. Battail, "Pondération des symboles décodés par l'algorithme de Viterbi" (in French), *Ann. Télécommun.*, Fr., 42, N 1-2, pp. 31-38, Jan. 1987.
10. J. Hagenauer and P. Hoeher, "A Viterbi algorithm with soft-decision outputs and its applications", *Proc. of Globecom '89,* Dallas, Texas, pp. 47.11-47.17, Nov. 1989.
11. A. J. Viterbi, "Convolutional codes and their performance in communication systems," *IEEE Trans. Com. Technology,* vol. COM-19, No. 15, pp. 751-772, Oct. 1971.
12. G. D. Forney, "The Viterbi algorithm," *Proc. IEEE,* vol. 61, N 3, pp. 268-278, Mar. 1973.
13. P. Thitimajshima, *"Les codes convolutifs récursifs systématiques et leur application à la concaténation parallèle"* (in French), thèse N 284, Université de Bretagne Occidentale, Brest, France, Déc. 1993.
14. L.R. Bahl, J. Cocke, F. Jelinek and J. Raviv, "Optimal decoding of linear codes for minimizing symbol error rate," *IEEE Trans. Inform. Theory,* IT-20, pp. 248-287, Mar. 1974.
15. Erico Guizzo, "Closing in on the perfect code," *IEEE Spectrum,* March 2004, pp.28-34.

16. Punya Thitimajshima, "Recursive Systematic Convolutional codes and application to parallel concatenation," *IEEE Globecom'95,* Singapore, pp.2267-2272.
17. Claude Berrou and Alain Glavieux, "Near optimum error correcting coding and decoding: turbo-codes," *IEEE Transactions on Communications,* vol.11, no.10, October 1996.
18. http://ep.espacenet.com and http://www.uspto.gov
19. Claude Berrou and Patrick Adde, *Procédé de décodage d'un code convolutif à maximum de vraisemblance et pondération des décision, et décodeur correspondants,* France patent: 2675968, application number: 91 05279, filing date: April 23, 1991.
20. Claude Berrou and Patrick Adde, *Procédé de décodage d'un code convolutif à maximum de vraisemblance et pondération des décision, et décodeur correspondants,* European Patent: 511139, application number: 92 460011.7, filing date: April 22, 1992.
21. Claude Berrou and Patrick Adde, *Method for a maximum likelihood decoding of a convolutional code with decision weighting, and corresponding decoder,* US patent: 5406570, application number: 92 0870483, filing date: April 16, 1992.
22. http://www.3gnewsroom.com/3g_news/dec_01/news_1606.shtml
23. Chris Hergard and Stephen B. Wicker, *Turbo Coding,* KAP, 1999, pp. 4-5.
24. IT Society Paper Award, *IEEE Information Theory Society Newsletter,* Vol. 48, No.1, March 1998

Part I

Data Storage Systems

Chapter 2

Iterative Codes in Magnetic Storage Systems

Hongwei Song[1] and B. V. K. Vijaya Kumar[2]

[1] Agere Systems, USA
[2] Carnegie Mellon University, USA

The development of information technology has spurred enormous demand for vast and reliable data storage. In the past decade, the areal density of commercial hard disk drives has increased at an unprecedented rate of 60% compound annual growth, basically doubling the data storage density every 18 months. The goal of future hard disk drives is to realize storage densities of 1 Terabits/in^2 and higher. The storage capacity increase up to now has been primarily driven by the advances in head and media technologies; however, coding and signal processing are being increasingly recognized as cost-efficient means of improving storage capacity as advances in VLSI circuit technology are enabling increasingly sophisticated signal processing methods at negligible cost increases.

2.1 Introduction

Fig. 2.1 shows a block diagram of a read/write channel in a hard disk drive. Encoder and decoder are denoted as "Enc" and "Dec" respectively. In the write process, user data is organized into sectors of 4096 bits, appended with several bytes of cyclic redundancy check bits, and fed to a Reed-Solomon (RS) [1] encoder that operates on 8 or 10-bit symbols. The modulation code [2] is a high-rate code that imposes a run-length constraint to facilitate timing recovery and for perpendicular recording, also imposes a running-digital-sum constraint to reduce baseline wander. The modulation encoder is followed by a parity encoder that appends one or more parity bits to the modulation codeword. The parity bits allow detection and correction of certain type of error events [3, 4]. In order to alleviate the effect of the nonlinearities in the writing process, a write precompensation is employed. The major causes of these nonlinearities are bandwidth limitations in the write path and the demagnetizing fields in the magnetic medium. These nonlinearities will cause data pattern-dependent displacements of recorded transitions relative to their nominal positions. The write precompensation compensates for these displacements by

K. Sripmanwat (ed.), Turbo Code Applications: a journey from a paper to realization, 17–44.

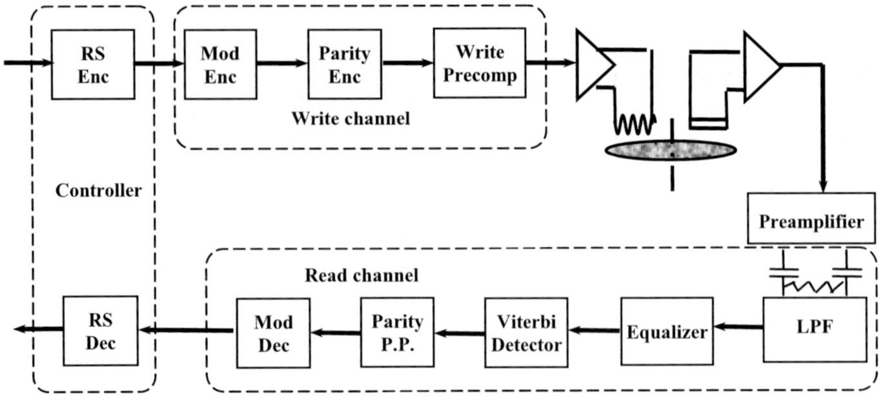

Fig. 2.1. Major Components in a Typical Read/Write Channel.

introducing data pattern-dependent compensating shifts into the signals. Digital information is stored on magnetic media by saturating the media in one of two magnetic directions. The recording is accomplished by alternating the direction of the current in the write head coil, which alternates the direction of the magnetic field emanating from the head while the magnetic media is in motion. The alternating head field magnetizes the medium accordingly, producing regions along the medium that are magnetized in either of two directions.

To retrieve data from the disk drive, the read head senses the transitions (i.e., changes) in the magnetization and converts the stored information back to an electronic waveform. The analog playback signal is amplified before feeding it to the read channel shown schematically in Fig. 2.1. A front-end low pass filter (LPF) is employed to suppress out-of-band noise and perform some preliminary equalization (e.g., pulse shaping). An analog-to-digital converter samples the analog signal from the LPF at the desired sampling phase that is adjusted by a timing recovery loop. The obtained symbol-rate-sampled signal is further shaped into a partial response (PR) signal [5] by the equalizer and fed to the channel detector. An optimal sequence detector for a parity-coded system is a Viterbi detector [6] that combines channel states and code states [4]. The decoder ensures that the states at the parity block boundary satisfy the parity constraint. A suboptimal decoding of a parity coded system begins with the detection of recorded bits using the Viterbi algorithm matched to the PR channel only. Then, a parity-based post processor (denoted as "Parity P.P.") is employed to correct a specified number of the most likely error events at the output of the Viterbi detector by exploiting the parity information in the incoming sequence. The post processor produces the final estimate of the PR channel input sequence. This sequence is passed to the modulation decoder that delivers estimates of the modulation encoder input to the RS decoder. Finally, the RS decoder corrects up to certain number of byte errors, and sends the recovered data to the host computer. The RS encoding and decoding are performed in the disk controller.

The role of write/read electronics in a digital data storage system is similar to that of the transmitter/receiver in a digital communication system. A recording channel can be modeled as a linear, intersymbol interference (ISI) channel with noise, subject to a binary input constraint. The fundamental problem of a recording channel lies in being able to encode and then read back bits of data at increasing data rates and areal densities while keeping error rates below an acceptable level, typically 10^{-15}. At high linear recording densities and high data rates, coping with noise and ISI becomes increasingly difficult. As a result, advanced equalization, estimation and detection, and coding techniques are needed to ensure reliable recovery of the stored information.

In his seminal paper [7], Forney proved that the optimal detection of data in an ISI channel can be achieved by a maximum likelihood sequence detector (MLSD). However, the complexity of MLSD grows exponentially with the length of the channel ISI. To reduce the ISI length, an equalizer is commonly used to shape the channel pulse response to some short well-controlled ISI target (called the partial response target), which combined with sequence detection is usually referred to as partial response maximum likelihood (PRML) technique. It was not until early 1990s that PRML became an industry standard for advanced magnetic recording, and made it possible for advanced coding and signal processing to be applied to the recording channel to achieve higher data capacity. A family of partial response (PR) targets of the form $(1 - D)^n(1 + D)^m$ has traditionally been used due to their good match with the real channel pulse response and integer coefficients. However, these partial response shapes do not match the recording channel well at high densities, leading to noise enhancement at the equalizer output. Generalized partial response (GPR) polynomials with real coefficients provide a better match to the magnetic recording channel than the conventional PR targets with integer coefficients do. In particular, the polynomials of the form $G(D) = (1-D)(1+p_1D+p_2D^2+...+p_nD^n)$ with a spectral null at DC can be designed to whiten the noise at the input of the channel detector, provided the equalizer and the GPR target are sufficiently long. This class of generalized partial response (GPR) targets is significant in practice and, when combined with sequence detection, gives rise to noise-predictive maximum-likelihood (NPML) systems [8].

No matter what equalization and detection are used, the performance of a linear recording channel cannot exceed the matched filter bound and this raw bit error rate (BER) is simply too large for data storage systems that require system bit error rates of 10^{-15} and better. To meet such very low BER requirement for magnetic recording systems, error-correcting codes (ECC) are commonly employed to improve the reliability of information storage on noisy recording channels.

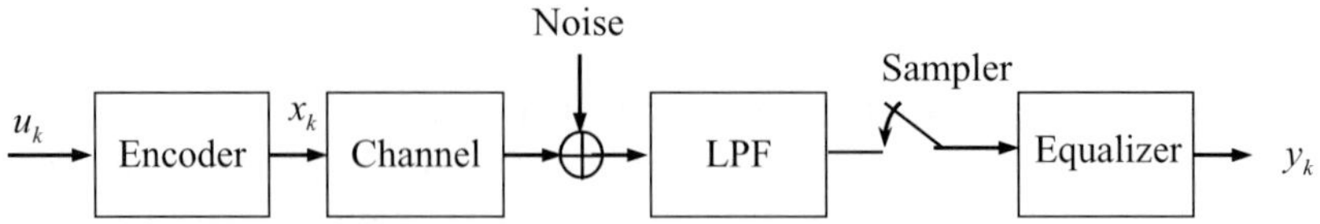

Fig. 2.2. Schematic of a magnetic recording channel.

Up until 1993, coding research in magnetic recording was limited to high rate codes in order to minimize the code rate loss. In 1993, with the invention of turbo codes by Berrou, Glavieux and Thitimajshima [9], which achieved error correction performance very close to the theoretical limits, turbo coding started receiving more attention than ever in the data storage industry. In the wake of the excellent performance of turbo codes, another class of codes called Low-Density Parity-Check (LDPC) or Gallager codes [10, 11], offering an alternative approach of increasing signal-to-noise ratio (SNR) gains, was rediscovered. Soon after, various forms of concatenated codes were proposed, which under iterative decoding, forms a class of iterative codes.

Another significant development following the introduction of turbo codes is the iterative equalization and decoding of ISI channels, also referred to as turbo equalization pioneered in [12]. The basic idea in turbo equalization is to treat the ISI channel as an inner code in a serial concatenated system and to apply an iterative soft decoding procedure to equalize the channel and decode the outer code. The emergence of iterative codes and turbo equalization has opened the door to new and intriguing possibilities of achieving significant coding gains in the magnetic recording channels. Current research efforts in coding for recording channel thus are focusing on the potential applicability of this emerging iterative coding technology and its implementation.

2.2 Turbo Equalization

A digital recording channel can be approximated by an equivalent baseband model, as shown in Fig. 2.2, where the channel and the receiver are represented by a linear filter. The readback waveform is passed through a front-end LPF, sampled at symbol rate, and further shaped by a linear equalizer which forces the transfer function of the overall channel response to a partial response polynomial $h(D) = \sum_{i=0}^{\nu} h_i D^i$. The equalizer output at time k may be written as

$$y_k = \sum_{i=0}^{\nu} h_i x_{k-1} + n_k \tag{2.1}$$

where $x_k \in \{-1, 1\}$ is the channel input bit (i.e., the output of the encoder), and n_k is a sample of a noise process at time k.

Ideally, the goal of an optimal receiver is to minimize the probability of error $P(\hat{u}_k \mid \mathbf{y})$ for each user bit, or equivalently, maximize the *a posteriori probability* (APP) given the received samples $\mathbf{y}$

$$\hat{u}_k = \arg \max_{u \in \{0,1\}} P(u_k = u \mid \mathbf{y}) = \sum_{\forall \mathbf{u} u_k = u} P(\mathbf{u} \mid \mathbf{y}) = \sum_{\forall \mathbf{u} u_k = u} \frac{p(\mathbf{y} \mid \mathbf{u}) p(\mathbf{u})}{p(\mathbf{y})} \tag{2.2}$$

which is computationally impractical since it depends on the entire user sequence $\mathbf{u}$.

In magnetic recording channel, this sequence estimation problem is traditionally performed in two steps. First, the equalized samples are fed to the channel detector (typically a Viterbi detector) tuned to the PR polynomial. The detector produces hard bit decisions, which are further processed by an ECC decoder to protect the integrity of the user data. Substantial performance degradation is induced through the separation of these inherently dependent tasks. This performance gap can be significantly narrowed using the idea of turbo equalization, which facilitates information exchange between the channel detector and the decoder by working on the same set of received samples.

The block diagram in Fig. 2.3 illustrates the receiver structure of a turbo equalization system, which consists of one soft-input soft-output (SISO) channel detector, one interleaver π, one deinterleaver π^{-1}, and one SISO decoder. The channel detector employs SISO algorithms such as the BCJR algorithm [13] or the soft output Viterbi algorithm (SOVA) [14] to compute the log-likelihood ratios $LLR(x_k) = \log \frac{P(x_k=+1|\mathbf{y})}{P(x_k=-1|\mathbf{y})}$ for the channel input sequence $\mathbf{x}$, and delivers the extrinsic information, $L_{ext}(x_k)$, defined as

$$L_{ext} = LLR(x_k) - L(x_k) = \log \frac{P(x_k = +1 \mid \mathbf{y})}{P(x_k = -1 \mid \mathbf{y})} - \log \frac{P(x_k = +1)}{P(x_k = -1)} \tag{2.3}$$

where the a priori LLR, $L(x_k)$, is zero for the first iteration assuming equal probability for +1 and -1, and is provided by the decoder in the 2nd iteration and so on. In a similar fashion, the SISO decoder accepts the interleaved detector output $L_{ext}(x_k)$ as a priori $L(u_k)$, computes the extrinsic information

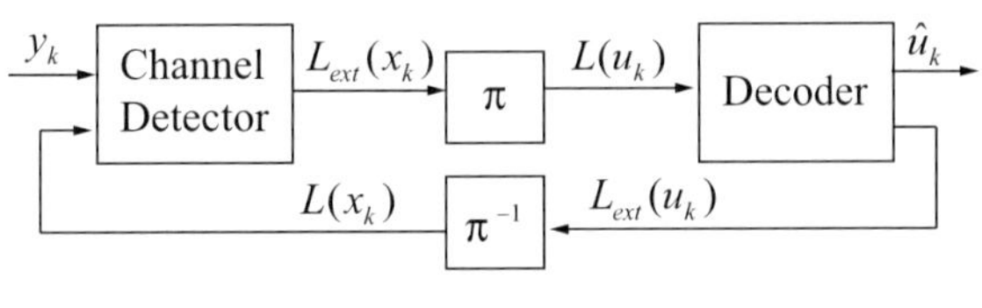

Fig. 2.3. Turbo Equalizer.

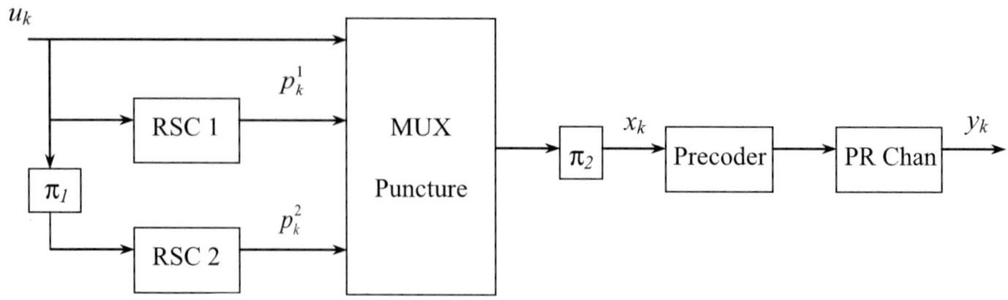

Fig. 2.4. Turbo encoder concatenated with PR channel.

$L_{ext}(u_k)$ and feeds it back to the channel detector for further iterations.

$$L_{ext}(u_k) = LLR(u_k) - L(u_k) = \log \frac{P(u_k = +1 \mid \mathbf{L}(u))}{P(u_k = -1 \mid \mathbf{L}(u))} - \log \frac{P(u_k = +1)}{P(u_k = -1)} \tag{2.4}$$

where $L(u)$ is the decoder input sequence. This extrinsic information exchange process is repeated several times over a block of received samples. In the final iteration, the decoder estimates the user data based on the *a posteriori* $LLR(u_k)$.

2.2.1 Turbo Codes Concatenated with Partial Response Channels

Fig. 2.4 shows a serial concatenation of a turbo encoder and a PR channel sandwiched with another interleaver π_2. Note that π_1 and π_2 have different sizes since the turbo code rate is not 1. Precoder is essentially a rate-1 recursive systematic code (RSC) encoder (e.g., $1/(1 \oplus D)$ where the symbol $\oplus$ denotes modulo-2 addition). Precoder can significantly affect the convergence and BER performance of the system, including the error floor.

Fig. 2.5 shows the decoding system corresponding to the encoding structure shown in Fig. 2.4. Three SISO decoders are employed: one channel MAP detector matched to the precoded PR channel, and two MAP decoders corresponding to the two RSC encoders. In general, use of precoder will not increase decoding complexity since it only modifies the input/output mapping of the detector trellis, and does not change the number of states in the detector. The channel MAP detector provides the soft information $LLR(x_k)$ about the channel bits x_k (x_k includes the information bits u_k and the parity bits p_k) that is subtracted by the output of the parallel decoder to get the extrinsic information $L_{ext}(x_k)$ (to be deinterleaved by π_2^{-1} and used for the parallel decoder). The depuncturer pads zeros to $\mathbf{y}_{p1}$ and $\mathbf{y}_{p2}$ where parity bits were thrown away by the puncturer in the encoder, and demultiplexes the extrinsic information into three streams: y_{u_k}, $y_{p_k^1}$ and $y_{p_k^2}$. MAP decoder1 uses extrinsic information L_{ext}^u (deinterleaved by π_1^{-1}) from the MAP decoder2 as well as $y_{p_k^1}$ to estimate the soft information $L(\hat{u}_k)$ about the information bits u_k

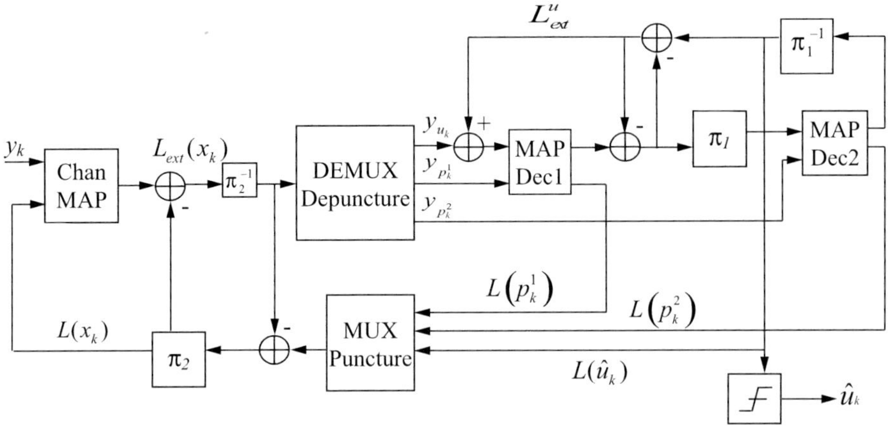

Fig. 2.5. Decoder architecture for turbo code concatenated with PR channel.

and the soft information $L(p_k^1)$ about the parity bits p_k^1. Then MAP decoder2 uses the π_1 interleaved extrinsic information from MAP decoder1 and $y_{p_k^2}$ to estimate the soft information $L(\hat{u}_k)$ about the information bits u_k and the soft information $L(p_k^2)$ about the parity bits p_k^2. The soft information $L(\hat{u}_k)$, $L(p_k^1)$ and $L(p_k^2)$ are then punctured and multiplexed to get the extrinsic information from the turbo decoder, which is fed back to channel detector as a priori information $L(x_k)$. In the end, the soft information $L(\hat{u}_k)$ is compared to a threshold (e.g., 0) to get the final hard decision $\hat{u}_k$.

The BER performance of this decoding structure is denoted as "full turbo"in the chapter. If the turbo decoder makes no attempt to take advantage of the channel MAP detector matched to the PR target, no iteration between the channel detector and turbo decoder is involved. In the plots, "partial turbo" is used to differentiate this decoding structure from "full turbo" architecture.

2.2.2 Single Convolutional Code Concatenated with PR Channel

Motivated by the excellent performance of serial concatenated convolutional codes [15] for additive white Gaussian noise (AWGN) channels, Souvignier et al [16] explored a reduced-complexity serial concatenated system for magnetic recording channels which uses a single convolutional code as the outer code and treats the PR channel as an inner code, as shown in Fig. 2.6. There is only one interleaver between the outer convolutional code and the inner precoded PR channel. The precoder is important to this serial concatenation system. PR channel itself does not possess recursive nature (e.g., EPR4 is $1+D-D^2-D^3$). Since precoder is RSC, precoding makes PR channel appear like an inner rate-1, non-binary RSC encoder. The corresponding receiver requires only two SISO decoders as shown in Fig. 2.7.

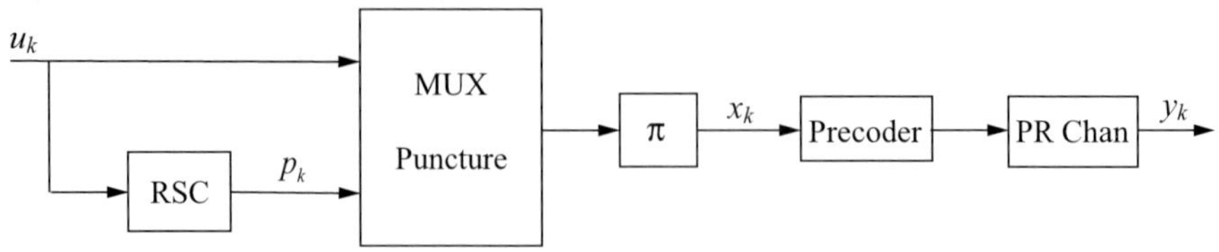

Fig. 2.6. Single convolutional code concatenated with PR channel.

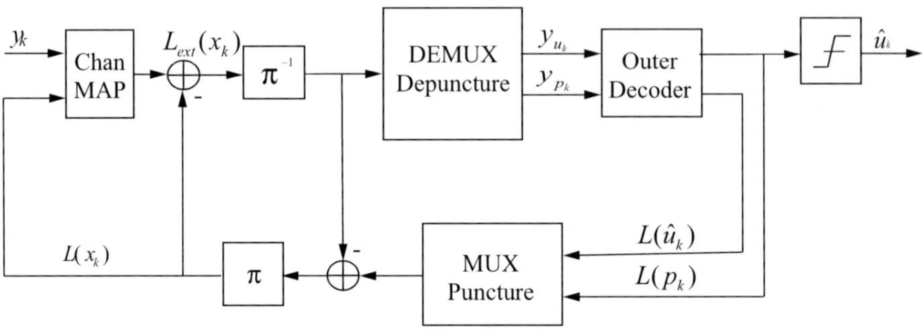

Fig. 2.7. Decoder structure for the concatenated system shown in Fig. 2.6.

Low-density parity check (LDPC) codes, first introduced by Gallager in his Ph.D. dissertation, were rediscovered as another category of codes approaching the Shannon capacity limit with practical decoding complexity. Gallager defined an (n, j, k) LDPC code with a matrix $\mathbf{H}$ of n columns that has j ones in each column, k ones in each row, and zeros elsewhere. The column weight is important to the performance of a regular LDPC code. The most common method for iterative soft decoding of LDPC coded data is known as the sum-product algorithm and is based on a bipartite graph representation of the LDPC code [17].

The application of LDPC codes for intersymbol interference (ISI) channel or PR channel is similar to applying turbo codes for PR channel–adding a SISO channel detector matched to the PR channel in front of the LDPC decoder so that the reliability information is exchanged between the LDPC decoder and the SISO channel detector as Fig. 2.8 shows.

In Fig. 2.8, the channel detector estimates the extrinsic information and delivers that to the LDPC decoder. The LDPC decoder accepts as channel values, propagates the beliefs within the associated Tanner graph iteratively to refine the estimates, and passes the refined extrinsic information back as *a priori* information for the channel detector. Different number of iterations in LDPC decoder is possible for every iteration between the channel detector and the LDPC decoder. Given that the computational complexity of the channel detector using the BCJR algorithm or the SOVA is much higher than that of

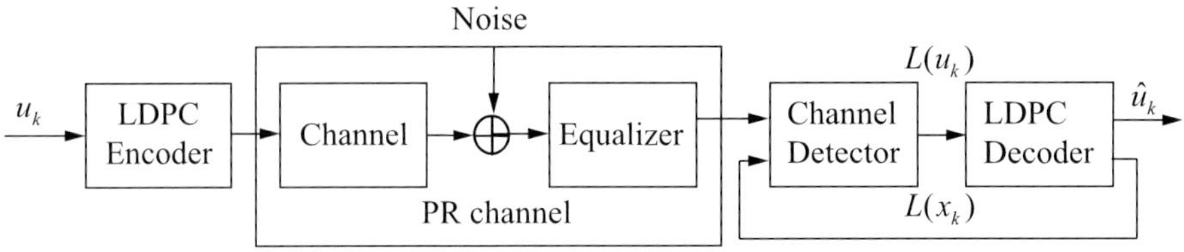

Fig. 2.8. LDPC codes concatenated with PR channel.

the LDPC decoder using the sum-product algorithm, more number of iterations in LDPC decoding and fewer number of iterations for channel detector is preferred. Iterations are terminated when a valid codeword is declared by the LDPC decoder. If the LDPC decoder does not find a valid codeword, the decoding process is terminated after a predetermined maximum number of iterations between the channel detector and LDPC decoder is reached.

2.2.3 EXIT Chart Analysis of Turbo Equalization

A large body of research has been undertaken to provide tools for iterative code design. One analysis approach proposed by ten Brink [18], known as the extrinsic information transfer (EXIT) chart, is to track the extrinsic information transformation between the channel detector and the decoder. Under i.i.d. and Gaussian assumptions, the channel detector and the decoder can be treated approximately as a signal-to-noise ratio transformers. Then, the turbo equalization receiver can be modeled as a dynamical system where each module functions like a SNR transformer [19].

Consider the input and output SNRs for the turbo equalization system at each iteration, as shown in Fig. 2.9. The channel detector delivers extrinsic information SNR^C_{out} to the decoder as input SNR^L_{in} information, while the decoder passes its extrinsic information SNR^L_{out} to the channel detector as a priori information SNR^C_{in}. The usual signal-to-noise ratio measuring the quality of the channel symbols is denoted as SNR. The transfer functions for the channel detector (denoted by f_C) and decoder (denoted by f_L) are defined as

$$SNR^C_{out} = f_c(SNR^C_{in}, SNR) \tag{2.5}$$

$$SNR^L_{out} = f_L(SNR^L_{in}) \tag{2.6}$$

The convergence of the turbo equalization receiver can be evaluated by measuring the evolution in the SNR of the extrinsic information from one iteration to the next.

For a detector/decoder using the BCJR algorithm, no simple analytical expression for the transfer function is available. The function f_c is assessed empirically by Monte Carlo simulation. Density evolution [20] or Gaussian approximation [21] technique is typically used to compute the transfer function of LDPC decoder.

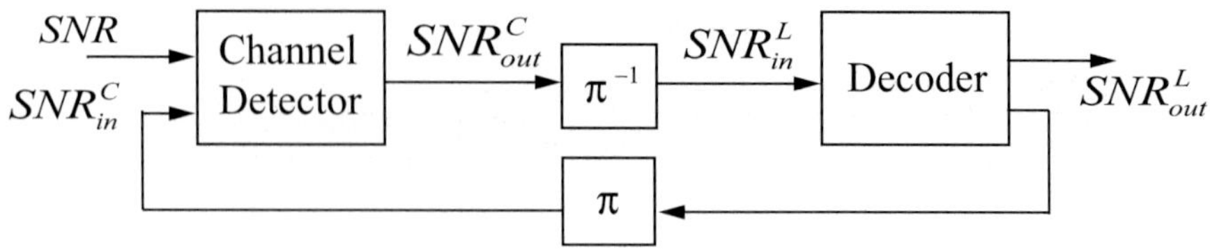

Fig. 2.9. Dynamical model for a turbo equalization system.

Assuming that the incoming messages to the bit nodes and check nodes involved in sum product algorithm have Gaussian densities and are independent, it is statistically sufficient to track their means at each iteration. The mean of the extrinsic information of bit nodes and check nodes at l-th LDPC decoding iteration can be shown as follows [21]:

$$m_b^{(l)} = m_{u_0} + (j-1)m_u^{(l-1)} \tag{2.7}$$

$$m_u^{(l)} = \phi^{-1}\left(1 - \left[1 - \phi(m_b^{(l)})\right]^{k-1}\right) \tag{2.8}$$

where j and k denote column weight and row weight of the LDPC code m_b, m_u, and m_{u_0} are the means of the density of message passed from bit to check, from check to bit and from channel detector to LDPC decoder, respectively. Note that $m_u^{(0)} = 0$ since the initial message from any check node is 0. The function $\phi(x)$ is defined as

$$\phi(x) = \begin{Bmatrix} 1 - \frac{1}{\sqrt{4\pi x}} \int_{-\infty}^{+\infty} \tanh\left(\frac{u}{2}\right) e^{-\frac{(u-x)^2}{4x}} du \; if x > 0 \\ 1 \qquad\qquad\qquad\qquad\qquad\qquad if x = 0 \end{Bmatrix} \tag{2.9}$$

Density evolution or Gaussian approximation was initially proposed to calculate the decoding thresholds of LDPC codes. For a fixed channel noise power, Eq. 2.7 and (2.8) are run iteratively until the density of bit node converges. The threshold is defined as the maximum noise level such that the probability of error goes to zero, i.e., $m_b \rightarrow \infty$, as the number of iterations tends to infinity. The value of threshold determines the boundary of the error-free region asymptotically, as the block length tends to infinity. For LDPC codes with small to moderate block lengths that are of practical interest in magnetic recording channels, such a threshold value is not very meaningful. However, the idea of tracking density of message can be used to analyze the convergence of iterative decoding and offers insights useful for code design.

2.3 Development

Although iterative codes and turbo equalization have been investigated extensively as potential candidates for magnetic recording channels [16, 22, 23], more investigations must be performed before they are ready to be implemented in a commercial read channel chip.

One major challenge in applying turbo/LDPC codes to hard disk drives is how to achieve 10^{-15} type corrected bit error rate (BER). Turbo codes suffer from a phenomenon known as "error floor". Due to a small number of codewords with low weight, the BER versus SNR performance curves often exhibit a region with small slope at BER about 10^{-6}. In this region, increasing SNRs bring only small improvements in BER. An outer ECC such as RS code may still be required to handle the residual errors after the turbo/LDPC decoder. Empirical simulations of turbo/LDPC codes concatenated with magnetic recording channels show that these codes exhibit long burst errors, requiring powerful RS codes (i.e., more redundancy) to correct the residual errors, thus eating up the potential coding gain. This is one of the major obstacles the chip manufacturers face in developing read channel with iterative codes. Low column weight LDPC codes [23] and single parity code based product codes [24] have been proposed to be used in conjunction with RS code, since they may work more in harmony with RS codes.

Iterative code concatenated with an outer RS ECC is widely investigated and expected as the first iterative coding solution in magnetic recording channel, since the RS code has been the standard ECC technology for many decades in data storage. For this concatenated coding scheme, reducing error propagation of the iterative code is crucial to achieve good post ECC performance.

Recently, Richardson [25] introduced a computational technique that predicts the performance of a given LDPC code in the error floor region with AWGN channel, and observed that the error floor does not typically arise from low-weight codewords. A concept of trapping set is introduced to analyze the error floor behavior. It remains an open question if LDPC codes exhibit similar error floor in turbo equalization system where interaction with channel detector is possible. Design of LDPC codes with large minimum distance is of great importance to magnetic recording application, due to the very low mis-correction probability requirement.

A problem with many magnetic read heads is the occurrence of thermal asperities and media defects. A thermal asperity occurs when, for example, a read head hits a dust particle on a disk, and may produce a long burst of errors. As the areal density increases, burst errors introduced by thermal asperities and media defects etc become difficult to deal with. It is challenging to design iterative codes which are robust again random errors as well as burst errors.

RS codes are the most widely used error correcting codes in digital communications and data storage. They have provable good minimum distance, and are robust against burst errors. Recently, several breakthroughs [26, 27] have been achieved in improving the error-correction capability of RS codes with soft decoding at the price of higher computation complexity. Therefore, complexity-reduced soft RS decoding has naturally become a hot research topic.

The stringent data rate and power dissipation specifications demanded of the read/write channel electronics present formidable hurdles for the practical implementation of iterative soft decoding. Simple decoding algorithms are desired without significant performance degradation. While LDPC codes can be iteratively decoded using the sum-product algorithm with less complexity than the BCJR algorithm used for turbo code decoding, the memory and bandwidth required to specify the nonzero elements in a randomly constructed low density parity check matrix can still be a major challenge for hardware implementation. Construction of good LDPC codes with hardware-friendly structure [28, 29] is particularly important for magnetic recording channels, where data rates of 2 Gbps is commercially achieved.

The trellis-based MAP/ML algorithms (BCJR and SOVA) provide excellent performance in the turbo equalization systems; however, these schemes often suffer from high computational burden for channels with long memory. Thus, one active research topic is to investigate low complexity turbo equalizers with small performance degradation. An important class of reduced-complexity alternatives to the MAP/ML algorithms consists of a combination of a linear filter and an interference canceller [30, 31]. In these structures, the received signal goes through a linear filter, and the residual ISI of the filter outputs is cancelled based on the decoder output.

In a typical magnetic recording system as shown in Fig. 2.1, the modulation code is designed to limit its error propagation such that the decoder does not generate burst errors which may fail the RS code. On the other hand, the modulation code has to be high rate (e.g., 30/31) to minimize the severe rate loss penalty associated with recording channel. The use of modulation coding poses a potential problem for the iterative soft decoder since the standard modulation decoders accept and deliver hard decisions. Therefore, methods for making the iterative soft decoders compatible with the modulation constraint are required. Reverse ECC architecture provides a solution to this problem, where the user data are first encoded by a modulation code before feeding to a systematic ECC encoder. Very high rate (e.g., 200/201) modulation codes can be used in reverse ECC because error propagation is no longer a concern, which results in 0.3∼0.5dB less rate penalty compared to the standard modulation code with low error propagation. Soft modulation decoders have also been proposed for iterative decoding system, where the demodulation step converts soft channel information into soft information about the demodulated output and passes it to the soft ECC decoder [32]. It is also an interesting research topic to design an iterative code with run length limited constraint.

2.4 Simulation Results

In this section, BER performance is investigated via computer simulations for a selected set of PR targets, precoder and decoding architecture choices. The SNR is defined as

$$SNR = 10 \cdot \log_{10}\left(\frac{E_b}{N_0}\right) = 10 \cdot \log_{10}\left(\frac{E_c}{2R\sigma^2}\right) \tag{2.10}$$

where E_b is the user bit energy, $E_c = R \cdot E_b$ is the code bit energy, R is the code rate, and $N_0/2$ is the two-sided power spectral density and σ^2 is the variance of the white Gaussian noise being added to the read-back samples after the equalizer. This SNR definition is used throughout the rest of the chapter unless it is explicitly stated differently. Random interleavers are used for all the simulation results in this chapter.

2.4.1 Bit Error Rate (BER) Performance

Fig. 2.10 shows the BER vs. SNR results for the EPR4 $(1+D-D^2-D^3)$ channel concatenated with (i) a turbo code, (ii) a single convolutional code and (iii) no outer code. A Viterbi channel detector is assumed for the uncoded EPR4 channel, while the BCJR algorithm is used for the channel detector for the other systems. Both the convolutional codes employed use the same 16-state RSC code with the encoder polynomials $(31,\ 23)_8$, where 31 is the feedback polynomial. No effort is applied to terminate the trellis. All the coded PR channels have the same overall code rate 8/9 and codeword length of $N = 4608$ to incorporate 4096 information bits. The iterative soft decoding employed 15 iterations although only the results of the last iteration are plotted. No precoder is used in conjunction with outer turbo codes. Also presented is the BER performance of the single convolutional code with the precoder $1/(1 \oplus D^2)$.

As seen in Fig. 2.10, over the range of SNRs investigated, the "full turbo" decoding appears to perform about 0.8 dB better than the "partial turbo" decoding by feeding the extrinsic information from the turbo decoder to the channel detector as a priori information, both showing more than 4.0 dB SNR gain over the uncoded EPR4 channel for a target BER of 10^{-5}. The EPR4 channel concatenated with the single convolutional code (denoted as "SCC") does not have as steep a BER vs SNR curve as the turbo codes. However, the single convolutional code combined with the precoder $1/(1 \oplus D^2)$ provides significant coding gain for the EPR4 channel, i.e., it outperforms the full turbo decoder over the BER range investigated. For larger SNR values, the simpler single code with a precoder solution exhibits an error floor, whereas the other two schemes exhibit a significantly faster decrease in BER with increasing SNR. A recently proposed approach [33] for computing the information rate of binary-input AWGN channels with memory allows us to compute that rate

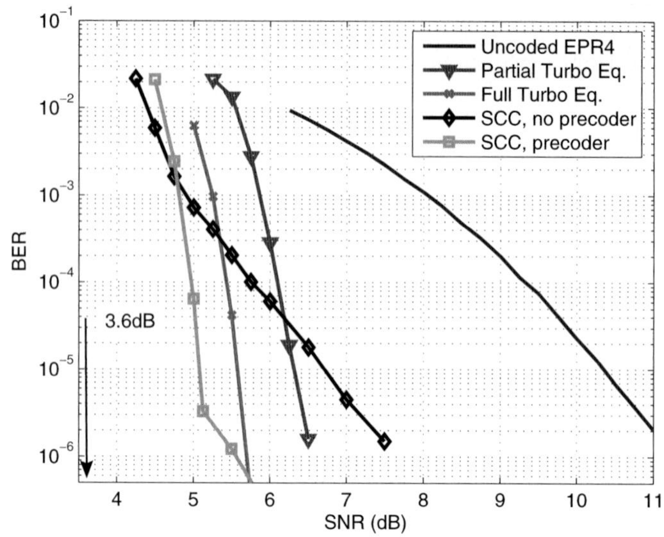

Fig. 2.10. BER vs. SNR for the EPR4 channel concatenated with a rate 8/9 turbo code, a rate 8/9 single convolutional code (SCC), and no outer code.

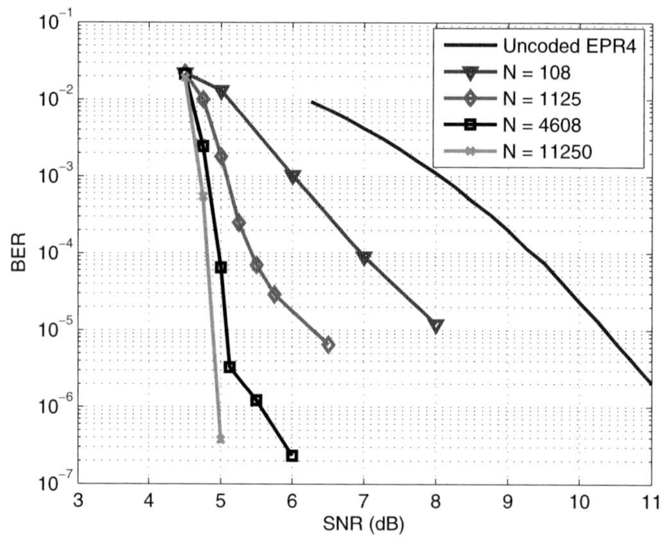

Fig. 2.11. BER versus SNR for the $\frac{1}{1 \oplus D^2}$ precoded EPR4 channel concatenated with a rate 8/9 single convolutional code of various block length N.

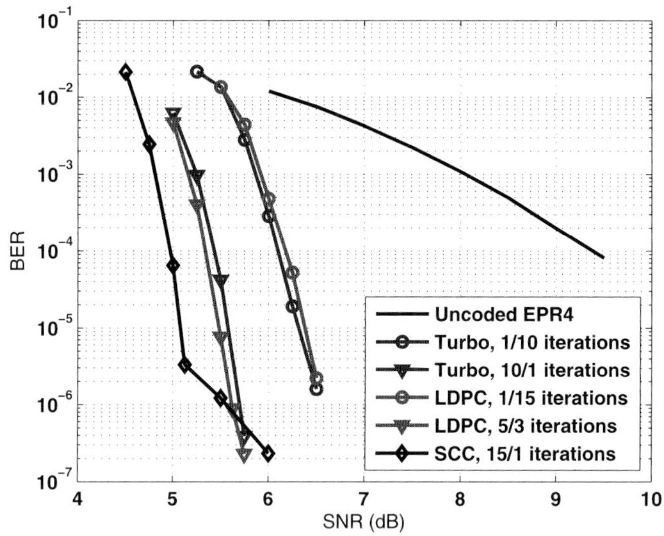

Fig. 2.12. BER performance of rate 8/9, j=3 LDPC code, turbo code and single convolutional code for EPR4 channel with different iteration schemes.

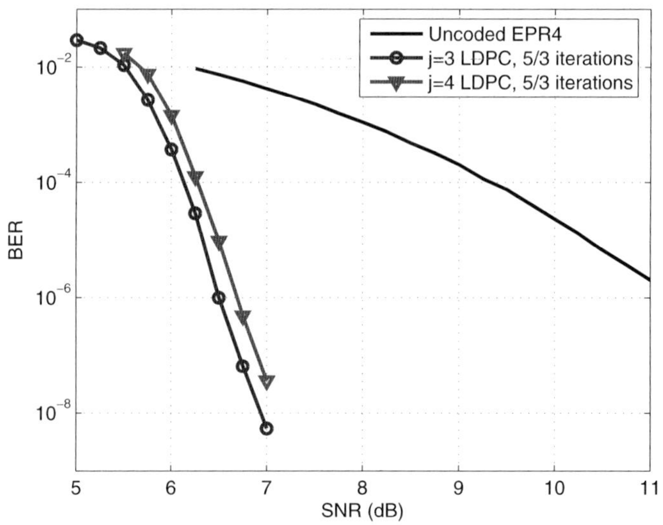

Fig. 2.13. BER performance of rate 0.94 LDPC codes for EPR4 channel with different column weights j=3 and 4, respectively.

8/9 coded EPR4 channel requires 3.6dB to achieve its capacity as indicated by the arrow in Fig. 2.10.

The effect of interleaver size on the BER performance for the EPR4 channel concatenated with the rate 8/9 single convolutional code is shown in Fig. 2.11. The precoder $1/(1 \oplus D^2)$ is used to achieve the recursive property. Simulated results are shown for four interleaver sizes of N=108, 1125, 4608 and 11250. Significant improvements are observed by increasing interleaver size from N=108 to N=4608. Small performance gain is achieved in the cliff region by further increasing interleaver size from N=4608 to N=11250; however, the error floor is lowered significantly. The interleaving gain is important to recording channels as it does not sacrifice code rate. Unfortunately, larger interleaver size does introduce larger memory and longer delay.

Fig. 2.12 compares the BER performance of column weight j=3 LDPC code with turbo codes for EPR4 channel. The performance of single convolutional code for EPR4 channel is also plotted as a reference. The concatenated systems have the same code rate 8/9 and codeword length N=4608. The iteration schemes are denoted by "x/y iterations" in the legend, meaning x number of global iterations between the channel detector and the decoder are performed with y number of iterations inside the decoder for each outer iteration. For example, "turbo, 1/10 iterations" denotes that the channel detector performs only once followed by 10 iterations of turbo decoding, which was denoted earlier by "Partial turbo". Fig. 2.12 indicates that LDPC codes have BER performance comparable to turbo codes for EPR4 channel, both losing less than 1 dB if the decoders do not exchange information with the channel detector. Single convolutional codes concatenated with precoded EPR4 channel perform best down to 10^{-6} BER among the concatenated systems investigated, but reach the error floor sooner than the others.

In an effort to check if LDPC codes exhibit error floor similar to turbo codes, long sequence of up to 10^{10} bits are simulated bit-by-bit for ideal EPR4 channel, and the BER performance are shown in Fig. 2.13. Two randomly constructed LDPC codes are used for this study with same iteration schemes (i.e., 5/3 iterations): a column weight j=3 LDPC code with code length N=4352, and a column weight j=4 LDPC code with code length N=4360, both of code rate R=0.94. No error floor is observed in the BER range simulated. At SNR=7.0dB, 10 blocks in error were collected among the 2.1 million trials for j=3 LDPC code. Six out of the 10 blocks are undetected errors with 4 error bits in each case (note that the tested j=3 LDPC code has minimum distance of 4), while the remaining blocks contain detected errors. At the same SNR=7.0dB, 1.1 million blocks are simulated for j=4 LDPC code to collect 4 detected errors with each containing around 50 error bits, no undetected errors are observed. If further iterations are allowed, the numbers of blocks with detected errors remain the same, though numbers of error bits in each block are reduced slightly. However, undetected errors are not reduced by increasing number of iterations. Fig. 2.13 suggests that both codes can achieve 10^{-7} BER for EPR4 channels with significant coding gains, but high rate

LDPC codes with column weight j=3, code length N=4352 are not sufficient to provide 10^{-15}BER.

2.4.2 Sector Failure Rate (SFR) Performance

In hard disk drive, a sector failure is declared if any recovered bit in the sector is in error. Therefore, sector failure rate is an important metric to measure the performance of a recording system. Fig. 2.14 compares the sector failure rate (SFR) performance of EPR4 channel with LDPC codes of block length N=4608, rate 8/9, and column weight j=3, 4 and 5, respectively. The SFR of a RS code with the same code rate is also presented as a reference. The column weight j=3 LDPC code performs best in the SFR region simulated. At SFR=2e-5, it outperforms the RS code by 1.8dB with 5 full iterations and 3 local iterations. Without global iteration (denoted as "1/50 iter"), the j=3 and 4 codes performs similar, both slightly outperforming the j=5 code at low SNR, but appear to have crossover with j=5 code at SFR=1e-4. Though not shown here, simulations across a family of PR targets indicate that LDPC codes have 1.5∼2.0dB gain over RS codes with the same redundancy at SFR=1e-5. Unfortunately, it is still not clear if similar gain holds at SFR=1e-12 which is typically required in commercial hard disk drives. Hardware platforms such as a field programmable gate array (FPGA) are required to verify the performance of LDPC codes at low SFR region.

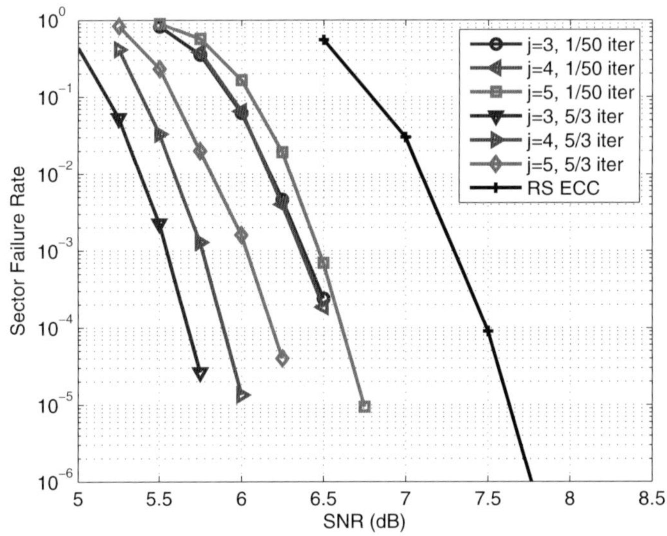

Fig. 2.14. SFR performance of rate 8/9 RS code and LDPC codes for EPR4 channel with different column weights j=3 and 4, respectively.

2.4.3 Block Error Statistics

The BER comparisons around 10^{-5} do not reveal the overall system performance in magnetic recording systems since a RS ECC is usually required to provide 10^{-15} type BER. If an outer RS ECC is employed in addition to the turbo/LDPC code, then, it is important to make sure that the number of errors after the decoder is well within the error correction capability of the RS code. Therefore, the block error statistics of serial concatenated partial response systems are investigated via simulations.

As one illustration of the importance of observing the block failure statistics, Fig. 2.15 shows the number of blocks as a function of the number of errors per block for a single convolutional coded EPR4 without precoding. A large number of blocks have no errors, and the total number of blocks with no errors is not being shown in this figure for the sake of clarity. Also listed in the subplot are the SNR and BER, the number of blocks simulated. For the three cases simulated, the number of errors per block (given that the block is in error) is typically constrained within a small number of bits. As SNR increases, the maximum number of errors per block observed decreases steadily. Fig. 2.16 presents similar block error statistics as in Fig. 2.15 but for EPR4 channel with $1/(1 \oplus D^2)$ precoder. Although coded EPR4 channel with precoding outperforms EPR4 channel without precoder in terms of BER performance at SNR=5 dB, inclusion of precoder results in two blocks with more than 100 errors out of 5000 trials, which may be beyond the error correction capability of an outer RS code. Therefore, caution should be exerted in using precoder in magnetic recording channels if an outer RS code is used to correct the residual errors.

Fig. 2.17 shows the block error statistics of j=2 LDPC coded EPR4 without precoding. Similar to the error statistics of single RSC code shown in Fig. 2.15, the number of errors per block is typically constrained within a small number of bits. This is due to the fact that j=2 LDPC code and RSC code have similar extrinsic information transfer characteristic, thus similar BER performance as well as block error statistics. At SNR=6.5dB, a significant number of undetected errors with Hamming weight 4 were observed, which contributes to the high peak in the bottom subplot.

Fig. 2.18 shows block error statistics plots but for LDPC codes with column weight j=3, 4 and 6 respectively. Although these codes outperform j=2 code in terms of BER performance, they have some blocks with more than 100 errors which will fail the conventional outer RS ECC. As the column weight increases, the probability of getting blocks with burst errors increases for the same SNR. Therefore, LDPC codes with large column weights are not suggested to use in conjunction with an outer RS code.

It should be noted that although simulation results suggest that single RSC code and j=2 LDPC code may work in more harmony with the outer RS ECC, the statistics are nonetheless insufficient. Due to the suboptimal iterative soft decoding, it is difficult to predict the BER performance and

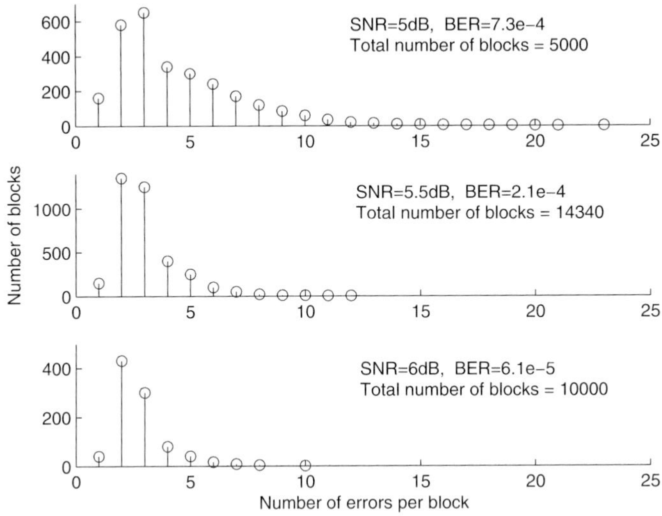

Fig. 2.15. Number of blocks as a function of the number of bit errors for single convolutional code concatenated with EPR4 channel without precoding.

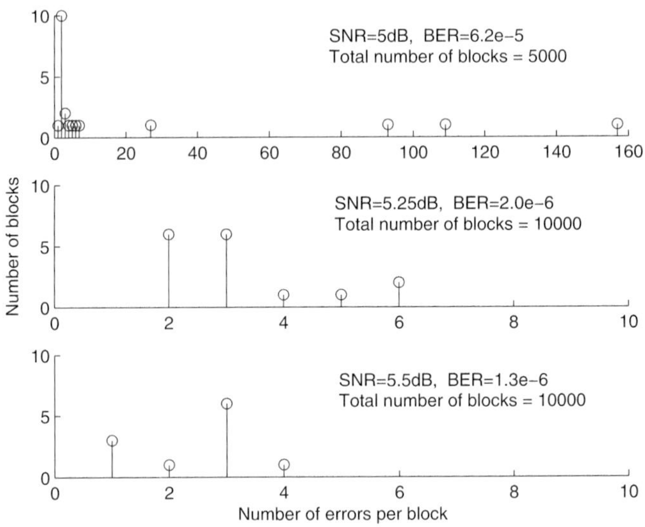

Fig. 2.16. Number of blocks as a function of the number of bit errors for single convolutional code concatenated with EPR4 channel $1/(1 \oplus D^2)$ precoder.

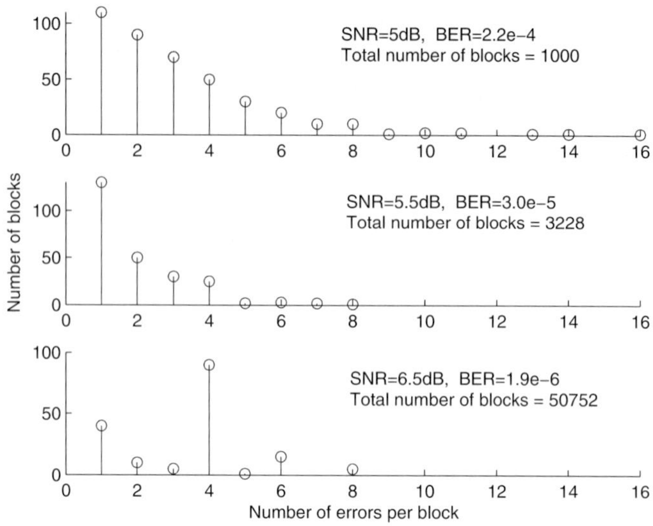

Fig. 2.17. Number of blocks as a function of the number of bit errors for j=2 LDPC decoded EPR4 channel without precoder, 5/3 iterations.

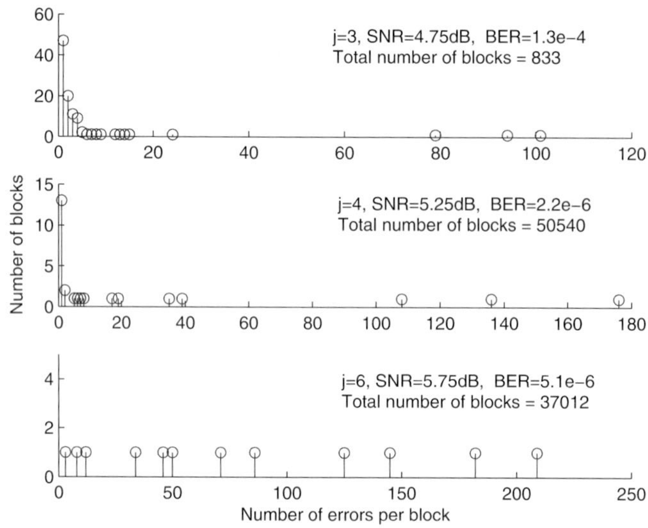

Fig. 2.18. Number of blocks as a function of the number of bit errors for j=3, 4 and 6 LDPC decoded EPR4 channel without precoder, 5/3 iterations.

block error statistics behavior accurately. Therefore, hardware tools such as an FPGA are needed before a convincing argument can be made.

2.4.4 Transfer Functions of the Channel Detector and Decoder

Fig. 2.19 shows the transfer functions $f_c(SNR^C_{in}, SNR)$ of the channel detector at $SNR = 4.75dB$ for $1 - D^2$ (PR4), $1 + D - D^2 - D^3$ (EPR4), $1 + 2D - 2D^3 - D^4$ (EEPR4), $1 + D$ (PR1), $(1 + D)^2$ (PR2) and $(1 + D)^3$ (PR3) PR channels without precoder and an EPR4 target with a precoder $1/(1 \oplus D^2)$. PR4 and PR1 target have identical transfer functions, both exhibit the largest output SNR^C_{out} if $SNR^C_{in} = 0$, i.e., when *a priori* information is all zero. Although the transfer function curves for the six PR channels start from different output SNR^C_{out} when $SNR^C_{in} = 0$, they all flatten out to zero slope and saturate at the same output SNR^C_{out} if SNR^C_{in} is sufficiently large. This indicates that PR4 and PR1 targets outperform the other PR targets for the first iteration ($SNR^C_{in} = 0$) at $SNR = 4.75dB$, but the six PR channels eventually have similar performance if SNR^C_{in} is improved upon LDPC decoding. Similar behavior was observed for various PR targets and at large range values of SNR. Fig. 2.19 also shows SNR^C_{out} versus SNR^C_{in} for EPR4 channel with a precoder. In contrast to the transfer function for the PR channel without precoding, the transfer function of the precoded EPR4 channel approaches a straight-line asymptote with slope 1. The distinct transfer functions of PR channel with and without precoder stem from the recursive nature of the precoder. A precoder is essentially rate-1 recursive concolutonal code which shapes the overall distance spectrum by mapping an input sequence with small information weights to sequence with large weights. It is shown in [34] that inclusion of precoder always results in a loss in information at the first iteration compared with no precoding. On the other hand, recursive nature is crucial to generate additional extrinsic information for the precoded PR channel when *a priori* information is highly reliable.

Fig. 2.20 shows the measured extrinsic information transfer functions for a rate $R = 8/9$, block length N=4608 RSC codes with generator polynomial $(31, 23)_8$. For comparison, Fig. 2.20 also shows the transfer functions after three iterations for j=2 and j=6 LDPC codes calculated using Eq. 2.7 and 2.8. The transfer functions of both LDPC codes cross the functions of RSC code, i.e., they gain smaller SNR improvement at low SNR input, but more at high SNR input than the convolutional codes. Note that the behavior of j=6 LDPC code is significantly different from the rest, as it possesses very sharp transition at the input SNR around 5.5. The j=2 code has transfer function more similar to that of $(31, 23)_8$ code compared to j=6 code, though it is constructed in a markedly different way than the $(31, 23)_8$ code.

Fig. 2.21 plots the transfer function f_L of the decoder for j=2 LDPC code after 3 decoding iterations, as well as the channel detector inverse transfer functions $f_c^{-1}(SNR^C_{out}, SNR)$ for channel SNR increasing from 4.5dB to 7.5dB in steps of 0.5dB from the left to the right, and the evolution of the SNR

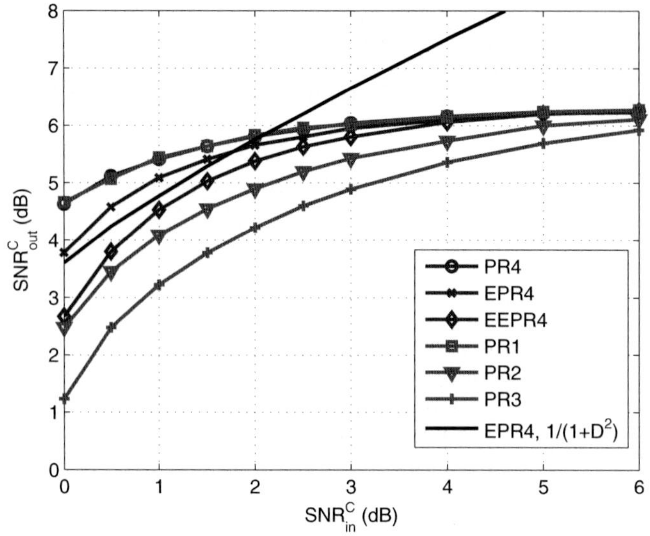

Fig. 2.19. Extrinsic information transfer function of channel detector for various PR targets at $SNR = 4.75dB$.

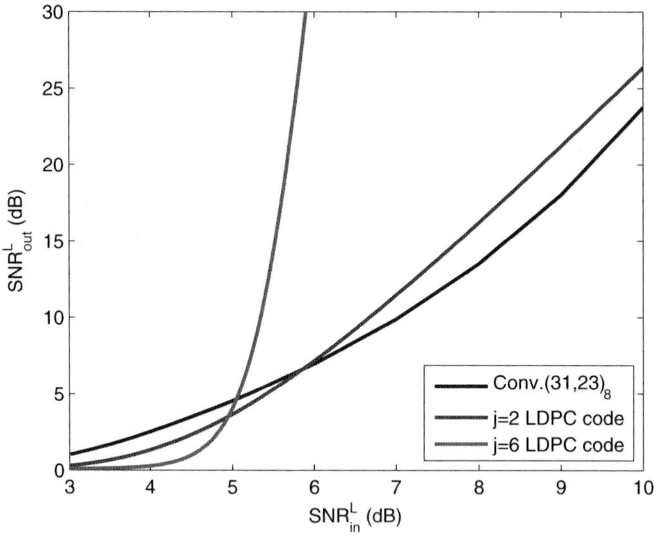

Fig. 2.20. Extrinsic information transfer functions of three outer decoders.

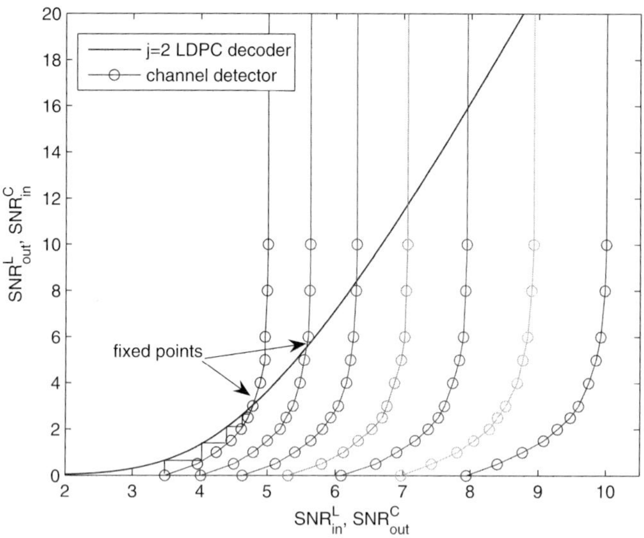

Fig. 2.21. Extrinsic information transfer chart for LDPC coded PR4 channel over a set of channel SNR values (4.5 dB to 7.5 dB in steps of 0.5 dB from the left to the right).

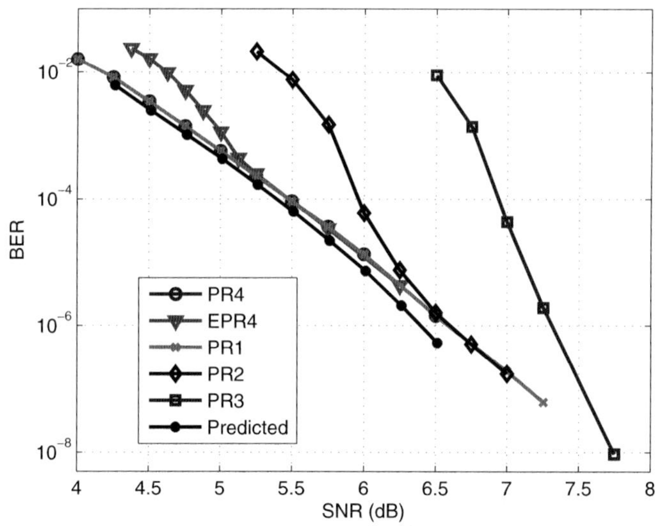

Fig. 2.22. BER versus SNR plots for j=2, rate 8/9 LDPC code, concatenated with PR4, EPR4, PR1, PR2 and PR3 channels. Predicted: PR4 channel.

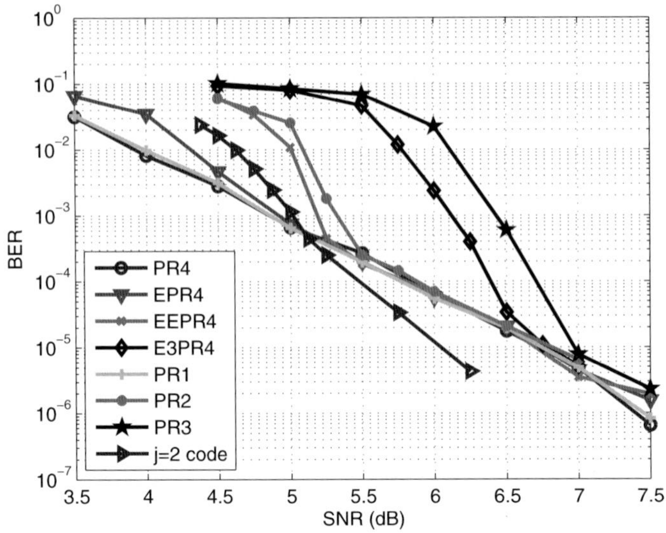

Fig. 2.23. BER versus SNR plots for a rate R=8/9 convolutional code with polynomial $(31, 23)_8$ concatenated with different PR channels without precoder.

between the two modules as well. Iteration starts with the channel detector delivering a nonzeroto the LDPC decoder. The decoding trajectory denoted by the staircase path graphically depicts the progress of the iterations between the channel detector and the LDPC decoder. For the example shown, SNR improvements at successive iterations are eventually trapped at fixed points where the two curves cross. Then, the overall SNR of the system is calculated as

$$SNR_{all} = SNR^L_{out}(fixed) + SNR^L_{in}(fixed) \tag{2.11}$$

where $SNR^L_{in}(fixed)$ and $SNR^L_{out}(fixed)$ are the input and output SNR values for the LDPC decoder at the fixed point. Under the i.i.d. Gaussian assumption, the BER of the system can be approximated as $BER = Q(\sqrt{SNR_{all}})$.

Fig. 2.22 shows the predicted BER versus SNR results for PR4 channel. Also plotted are the simulation results for j=2 LDPC coded PR4, EPR4, PR1, PR2 and PR3 channels without a precoder. The predicted BER for PR4 matches well with the simulated results in the studied BER region. Notice that PR4 and PR1 channels have identical BER performance, both outperforming the other PR targets at low SNR. While PR2 channel outperforms PR3 channel by about 1dB and experiences less than 1dB loss compared to EPR4 channel in high BER region, all the investigated PR channels appear to exhibit the same BER performance once they catch up with the BER curve of the PR4 channel at medium-to-high SNRs. This matches well with the transfer function analysis of Fig. 2.19.

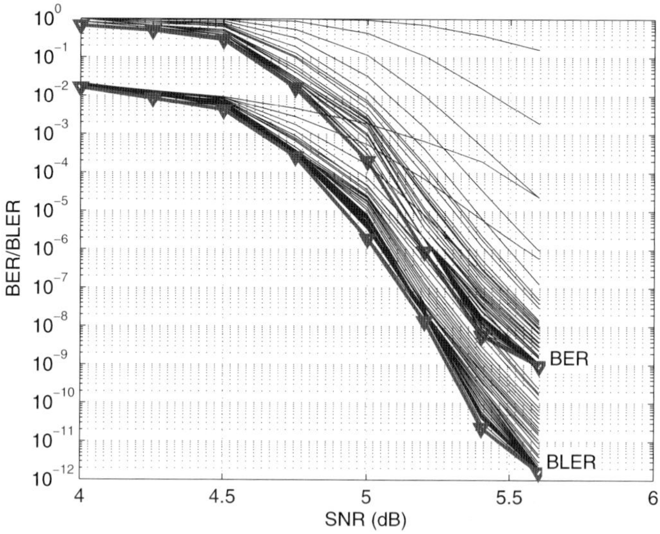

Fig. 2.24. BER/BLER performance of a column weight j=5, code rate R=8/9, block length N=4637 array code with 50 iterations for AWGN channel.

Fig. 2.23 presents simulated BERs of RSC $(31, 23)_8$ code concatenated with different PR channels without a precoder. The BER performance of various PR targets is different only in the waterfall region. They have almost identical BER performances in the high SNR region. Fig. 2.23 also presents the BERs of j=2 LDPC code for EPR4 channel without a precoder for comparison. Both codes suffer from "error floor" since their transfer functions intersect with that of the channel detector and form fixed points which prevent further accumulation of extrinsic information, but j=2 LDPC code has steeper BER curve due to its steeper transfer function shown in Fig. 2.20.

2.4.5 FPGA Based Reconfigurable Platform for LDPC Code Evaluation

In an effort to check if LDPC codes exhibit error floor similar to turbo codes, a high-throughput, fully-reconfigurable field programmable gate array (FPGA) based platform has been developed by Prof. Kumar's group at Carnegie Mellon University [35] to evaluate the BERs of LDPC codes down to 10^{-12} which is essentially impractical in software simulations. Fig. 2.24 shows the FPGA simulated bit error rate (BER) as well as block error rate (BLER) for a column weight j=5, code rate R=8/9, and block length N=4637 array code [36] for AWGN channel. The plotted BER and BLER performance correspond to 50 LDPC decoder iterations with the 50th iteration results highlighted. For this array code, the hardware LDPC decoder achieves throughput up to 1.4Gbps per iteration with 100 MHz system clock.

2.5 Conclusions

Coding and signal processing are increasingly being recognized as a cost-efficient approach to achieving substantial areal density growth while preserving the high integrity of magnetic storage systems. While it is well understood that the data dependent noise prediction technique [37] employed in current read channel achieves near maximum likelihood detection, the recent advances in iterative codes hold the promise to push the areal density to the ultimate coding limit for recording channels. In this chapter, the application of turbo codes and low density parity check (LDPC) codes in magnetic recording channel is reviewed under the turbo equalization structure. Iterative codes concatenated with partial response (PR) channels achieve performance gains of 1.5~2.0 dB at sector failure rate of 10^{-4} over the state-of-the-art Reed-Solomon (RS) error correcting codes (ECC) that is currently the standard in hard disk drives. However, commercial disk drives require an overall sector failure rate of 10^{-12}. The challenge of predicting the onset of error floors and the coding gain at the required sector failure rate (SFR) remain an open problem, which is the main impediment in applying iterative codes to magnetic recording as ECC. It is also critical to design iterative codes which are robust again both random errors and burst errors for magnetic recording systems. The stringent data rate and power dissipation specifications demanded of the read/write channel electronics present hurdles for the practical implementation of iterative soft decoding. Design and analysis of iterative codes with good tradeoff between performance, implementation complexity and latency for recording channels is an area of active research. Despite the aforementioned challenges, iterative detection and decoding technique is widely accepted as the most potential candidate for the next generation read channels.

References

1. W.W. Peterson and E. J. Weldon, Jr., *Error-Correcting Codes.* MIT Press: Cambridge, Mass., 1972.
2. K. Immink, P. Siegel, and J. Wolf, "Codes for digital recorders," *IEEE Trans. Inform. Theory*, vol. 44, no. 6, pp. 2260–2299, 1998.
3. T. Conway, "A new target with parity coding for high density magnetic recording channels," *IEEE Trans. on Magnetics*, vol. 34, no. 4, pp. 2382–2386, July 1998.
4. Z. Wu, P. A. McEwen, K. K. Fitzpatrick, and J. M. Cioffi, "Interleaved parity check codes and reduced complexity detection," in *Proc. ICC'1999*, pp. 1648-1652, June 1999.

5. H. K. Thapar and A. M. Patel, "A class of partial response system for increasing storage density in magnetic recording," *IEEE Trans. Magn.*, vol. Mag-25, pp 3666–3668, Sep. 1987.
6. A. J. Viterbi and J. K. Omura, *Principles of Digital Communication and Coding.* New York: McGraw-Hill, 1979.
7. G. D. Forney, "Maximum-likelihood sequence estimation of digital sequences in the presence of intersymbol interference," *IEEE Trans. Inform. Theory*, vol. IT-18, no. 3, pp. 363–378, May 1972.
8. P. Chevillat, E. Eleftheriou, and D. Maiwald, "Noise predictive partial-response equalizers and applications," in *Proc. ICC1992*, pp. 942–947. June 1992.
9. C. Berrou, A. Glavieux, and P. Thitimajshima, "Near Shannon limit error-correcting coding and decoding: Turbo codes," in *Proc. ICC'1993*, pp. 1064–1070, 1993.
10. R. G. Gallager, *Low-Density Parity Check Codes.* MIT Press, Cambridge, MA, 1963.
11. D. J. C. Mackay and R. M. Neal, "Good codes based on very sparse matrices", *Cryptography and Coding. 5th IMA Conference*, 1995. Available from http://wol.ra.phy.cam.ac.uk.
12. C. Douillard, M. Jezequel, C. Berrou, A. Picart, P. Didier and A. Glavieux, "Iterative correction of intersymbol interference: Turbo-equalization," *European Transaction on Telecommunications*, vol. 6, pp. 507–511, Sep. 1995.
13. L. Bahl, J. Cocke, F. Jelinek, and J. Raviv, "Optimal decoding of linear codes for minimizing symbol error rate," *IEEE Trans. Inform. Theory*, vol. IT-20, pp. 284–287, Mar. 1974.
14. J. Hagenauer and P. Hoeher, "A Viterbi algorithm with soft-decision outputs and its applications," in *Proc. IEEE Globecom. Conf.*, pp. 1680–1686, Nov. 1989.
15. S. Benedetto, D. Divsalar, G. Montorsi, and F. Pollara, "Serial concatenation of interleaved codes: Performance analysis, design, and iterative decoding," *IEEE Trans. on Inform. Theory*, vol. IT-44, no. 3, pp. 909–926, May 1998.
16. T. Souvignier, A. Friedmann, M. Oberg, P. Siegel, R. E. Swanson, and J. K. Wolf, "Turbo codes for PR4: parallel Versus serial concatenation," in *Proc. ICC'1999*, pp. 1638–1642, June 1999.
17. F. Kschischang, B. Frey, and H. Loeliger, "Factor graphs and the sum-product algorithm", *IEEE Trans. Inform. Theory*, vol. 47, no. 2, pp. 498–519, Feb. 2001.
18. S. Ten Brink, "Convergence of iterative decoding," *Electronics Letters*, vol. 35, no. 10, pp. 806–808, May 1999.
19. H. El Gamal and A. R. Hammons Jr., "Analyzing the turbo decoder using the Gaussian approximation," *IEEE Trans. Inform. Theory*, vol. 47, pp. 671-686, Feb., 2001.
20. T. J. Richardson, and R. Urbanke, "The capacity of low-density parity-check codes under message-passing decoding," *IEEE Trans. Inform. Theory*, vol. 47, No.2, pp. 599–618, Feb. 2001.
21. S. Y. Chung, "On the construction of some capacity-approaching coding schemes," *Ph.D. dissertation*, Massachusetts Institute Technology, Sep. 2000.
22. J. Fan, E. Kurtas, A. Friedmann, S.W. McLaughlin, "LDPC codes for magnetic recording," *Proc. of 36th Allerton Conf.*, Monticello, Illinois, pp. 1314–1323, Sept. 1999.

23. H. Song, J. Liu and B. V. K. Vijaya Kumar, "Low complexity LDPC codes for partial response channels," in *Proc. IEEE Globecom. Conf.*, pp. 1294–1299, Taipei, Nov. 2002.
24. J. Li, K. Narayanan, E. Kurtas, and C. N. Georghiades, "On the performance of high-rate TPC/SPC Codes and LDPC codes over partial response channels," *IEEE Trans. Commun.*, vol. 50, no. 5, May 2002, pp. 723–734.
25. T. J. Richardson, "Error floors of LDPC codes," *Proc. of 41th Allerton Conf.*, 2003.
26. V. Guruswami and M. Sudan, "Improved decoding of Reed-Solomon and algebraic-geometric codes," *IEEE Trans. Inform. Theory*, vol. 45, no. 6, pp. 1757–1767, September 1999.
27. R. Koetter and A. Vardy, "Algebraic soft-decision decoding of Reed-Solomon codes," *IEEE Trans. Inform. Theory*, vol. 49, no.11, pp. 2809–2825, Nov. 2003.
28. Y. Kou, S. Lin and M. Fossorier, "Low density parity check codes based on finite geometries: a rediscovery," *Proc. IEEE International Symposium on Information Theory*, Italy, June 2000.
29. H. Song, "Iterative soft detection and decoding for data storage channels," *Ph.D. dissertation*, Dept. of ECE, Carnegie Mellon University, Pittsburgh, PA, 2002.
30. TC. Laot, A. Glavieux, and J. Labat, "Turbo equalization: adaptive equalization and channel decoding jointly optimized," *IEEE Journal on Sel. Areas in Comm.*, vol 19, pp. 1744–1752, September 2001.
31. M. Tuchler, R. Koetter and A. C. Singer, "Turbo equalization: principles and new results," *IEEE Trans. Comm.*, vol. 50, no. 5, pp. 754–767, May 2002.
32. J. Fan, *Constrained coding and soft iterative decoding*, Kluwer Academic Publishers, July 2001.
33. D. Arnold and H. Loeliger, "On the information rate of binary-input channels with memory," in *Proc. ICC2001*, Helsinki, Finland, pp. 2692–2695, June 2001.
34. K. R. Narayanan, "Effect of precoding on the convergence of turbo equlization for partial response channels," *IEEE Journal on Sel. Area. Commun.*, vol 19, pp.686–698, April 2002.
35. S. Lin, H. Song and B.V.K. Vijaya Kumar, "Performance evaluation of near-shannon limit performance code using reconfigurable computing," submitted to *2005 ACM/SIGDA 13th international Symposium on FPGA*, Aug. 2004.
36. J.L. Fan, "Array codes as Low-Density Parity Check Codes" 2nd *International Symposium on Turbo Codes and Related Topics (Brest, France)*, September 2000.
37. A. Kavcic and J. M. Moura, "The Viterbi algorithm and Markov noise memory," *IEEE Trans. Inform. Theory*, vol. 46, pp. 291–301, Jan. 2000.

Chapter 3

Turbo Product Codes for Optical Recording Systems

Pornchai Supnithi

King Mongkut's Institute of Technology Ladkrabang, Thailand

Since the introduction of read-only compact disc (CD-ROM) in the early 80s, optical recording has found its place in mainstream data storage. They are quite a success as removable media particularly in the entertainment and data storage industry. As explosive amount of information in the 90s fueled an ever increasing demand for data storage, further development of optical recording systems in terms of higher capacity, transfer rate and areal density have become major goals in research and development and commercial communities. An evolution of "Blu-ray" and its rival " High-density DVD (HD-DVD)" have recently been envisioned as the next generation optical disc to achieve the recording of a two-hour program in digital HiVision. The standards agree on the capacity of 25 GB with improvement of numerical aperture (NA) and the wavelength λ alone. As of September 2004, HD-DVD aims to provide 15-GB single-layer disc, and 30-GB dual-layer disc, while Blu-ray aims for upto 27-GB and 54 GB for single-layer and dual-layer discs, respectively [1]. Current formats such as Read-only/Rewritable compact discs (CD-R/RW) and read-only/rewritable digital video discs (DVD-R/RW) are based on storing binary symbols on the optical media. The technology path from CD to DVD employs combined signal processing and coding techniques such as Reed-Solomon codes or RS codes and Runlength-limited codes or RLL codes. To achieve the goal of even higher storage capacity, more sophisticated signal processing and coding techniques would be required. In the early 1990s, there was an evolution of the field of error-correction codes with the introduction of "turbo codes" [2]. For additive white Gaussian noise (AWGN) channel, the novel code provides coding gain that almost reaches the theoretical Shannon capacity. This planted the seed for active research and development in the area of iterative decoding, variants of turbo codes, and turbo equalization in many communication systems. The advanced signal processing and coding such as turbo codes and turbo-like codes thus have a potential of providing additional coding gain as required in the future generations of high-density optical storage systems. In this chapter, the focus is on the use of a variant of

K. Sripmanwat (ed.), Turbo Code Applications: a journey from a paper to realization, 45–63.

turbo codes so called "Turbo Product Codes" [3] in multileveled (ML) optical recording systems.

3.1 Optical Recording Systems

An overview of the WRITE/READ system of binary and ML optical discs at a system level is illustrated in Fig. 3.1. The systems consists of the WRITE side where the information is coded and then written on the physical disc, and the READ side where the stored data is read, usually in altered form, and decoded as the original information.

3.1.1 Writing

On the write side, a binary input sequence is first encoded by the error-correction codes (ECC), which acts as an outer code to handle bursty errors.

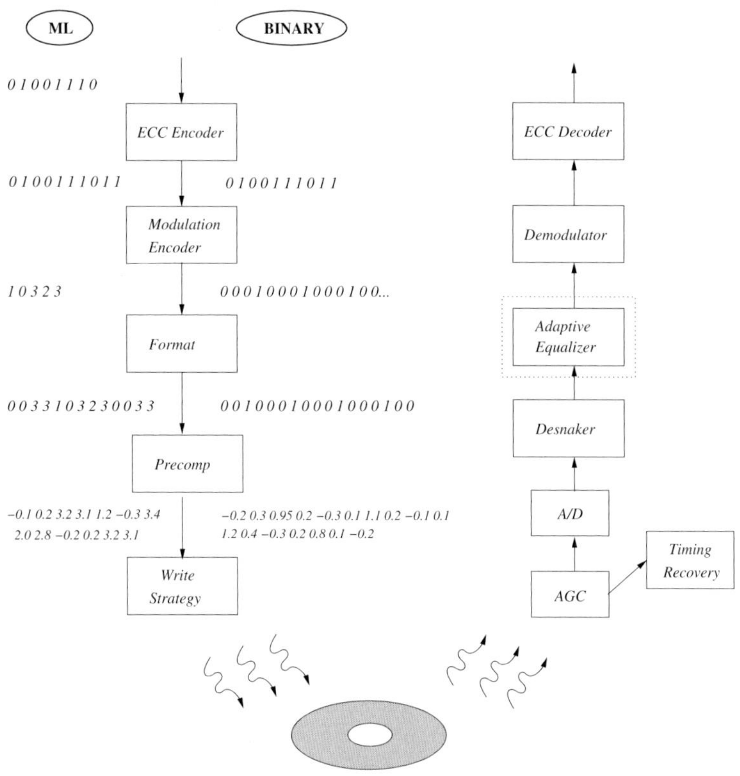

Fig. 3.1. Block diagram of typical binary and ML optical recording system.

Bursty errors are errors that occur next together after the decoding of modulation decoder either due to the nature of modulation decoder or bursty errors on the disc itself. Scratches and dust particles on the surface layer are some of the examples. The ECC is usually a form of concatenated Reed-Solomon (RS) code [4]. The modulation code is designed to add redundancy so that the intersymbol interference (ISI) is minimized as well as for timing recovery purpose. The binary CD/DVD uses runlength-limited (RLL) code [5], while the ML version of CD/DVD employs trellis-coded modulation (TCM), and turbo and turbo-like codes. Details of advanced modulation code like turbo and turbo-like codes will later on be explained in Sec. 3.4. After the modulation block, binary or multilevel data are then processed according to the specified formats, normally extra information is added for headers, addressing, dc control, synchronization and etc. To reduce the nonlinearity of the channel, the precompensation and write strategy blocks determine the power level to write the marks corresponding to the written levels and their neighbors as previously explained in Sec. 3.2.3.

Binary and ML optical recording systems differ on how marks are written on the media as shown in Fig. 3.2. The binary system writes only mark '0' and '1' to the disc using non-return-to-zero-inverted (NRZI) format or pulse-width modulation (PWM). The ML system, on the other hand, writes more levels using pulse-amplitude modulation (PAM). Information in a binary CD is thus encoded in edges while that in an ML CD-RW is encoded in levels. The smallest cell size of the binary CD is 0.83μm, while on an ML CD-RW, each mark is positioned inside the cell of a fixed size of 0.6 μm. The mark size as a function of reflectivity is determined before the writing process using the Write strategy and Write compensation. Comparison of optical parameters of

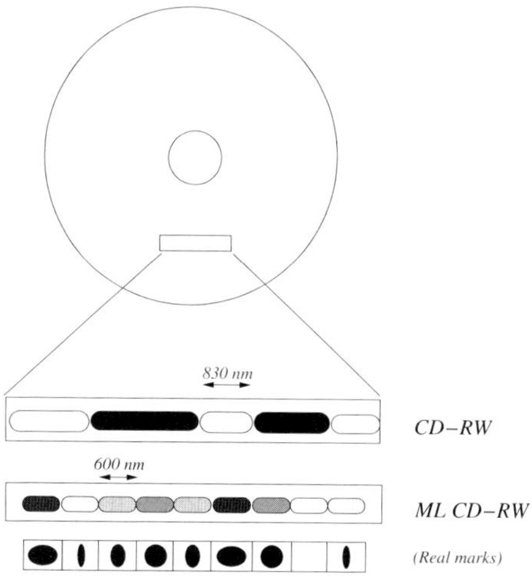

Fig. 3.2. Marks on a CD-R/RW and ML CD-R/RW.

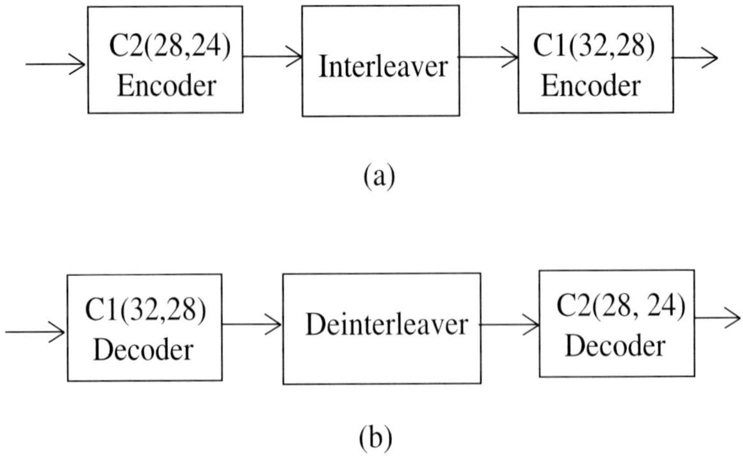

Fig. 3.3. Cross-Interleaved Reed-Solomon codes (a) Encoder (b) Decoder.

182×208 output bytes, the size of a standard ECC block. The total minimum Hamming distance of the DVD code is $\delta = 17 \times 11 = 187$, thus the error correcting capability, t, is 93 bytes. This powerful code is purposefully used as an outer code to exploit erasure correction ability of RS codes. It is designed to correct errors and erasures from the EFM/EFMPlus decoder in binary optical recording systems. In later section, we will consider the use of turbo product codes in ML optical recording system, any random and burst errors from the output of turbo decoder will be handled by the RSPC decoder.

As mentioned earlier, at present some form of RS codes are used in combination with RLL codes. RLL codes are, however, not designed for error correction. Future high-density optical discs require powerful error correction control. Thus, the use of turbo codes or turbo-like codes proved to acheive exceptional coding gain in other communication systems promise similar coding gains in the optical recording systems. Later we will discuss a variant of turbo codes so called turbo product codes (TPC) for use in ML optical recording systems. It is, however, practical to use the TPC in combination with RS product codes because the required bit error rate in the system is extremely low like 10^{-12} and TPC alone is not enough.

3.2 The Physics of Optical Recording

The concept of optical recording involves the use of optical stylus, a tightly focused laser beam, to heat the optical medium for writing and reading purposes. The minimum size of focused spot is controlled by diffraction effects. The smallest possible mark size, often called "full width at half maximum density (FWHM)," depends on the laser wavelength (λ) and the numerical

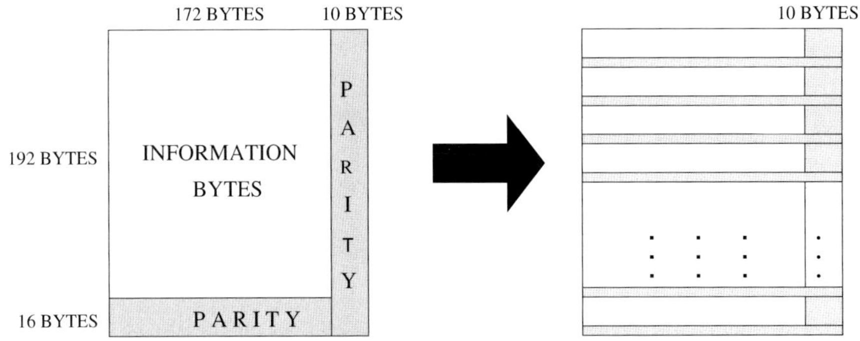

Fig. 3.4. Construction of the standard DVD code.

aperture(NA) of the lens, i.e.,

$$FWHM \cong 0.6\lambda/NA \tag{3.1}$$

The smallest resolvable mark size, on the other hand, l_{MF}, is given by $\lambda/4NA$ [7]. Numerical aperture is a measure of solid angle of a focused beam; typical values are 0.45-0.6 (CD uses $NA = 0.45$, DVD uses $NA = 0.6$). Common diode laser wavelengths are 635-780 nm. Smaller mark size requires the laser with smaller wavelength and/or the lens with larger numerical aperture. Recently, blue laser with wavelength of 460 nm, and $NA = 0.85$ have been considered and proposed for use in "High definition/High density DVD (HD-DVD)" and "Blu-ray" disc. Current optical recording technology include magneto-optical (MO) and phase change (PC) recording technology, both are suitable for write-once and rewritable discs. Below are short reviews of these two technologies, readers are encouraged to refer to [7] [8] for detailed information.

3.2.1 Recording on Phase-Change Media

Phase change media exist in two stable states, amorphous and crystalline. The amorphous state is "dull" and has low reflectivity, while the crystalline state is "shiny" with high reflectivity. Thus when illuminated with a small enough laser, the phase change material reflects light in two ways depending on the state of the medium, making it suitable for binary recording. Since the transition between these two states is reversible, information can be erased and/or overwritten many times. A GeTe thin film medium, for example, has a contrast of 20% in reflectivity level at the temperature of 150 and 200 Celsius. Widely used phase change materials belong to one of the two families, GeSbTe and AgInSbTe alloy [9]. The phase change material is a "stack" as illustrated in Fig. 3.5.

The recording layer is sandwiched between dielectrics layers which help protect the phase change media from the environment. The protective layers

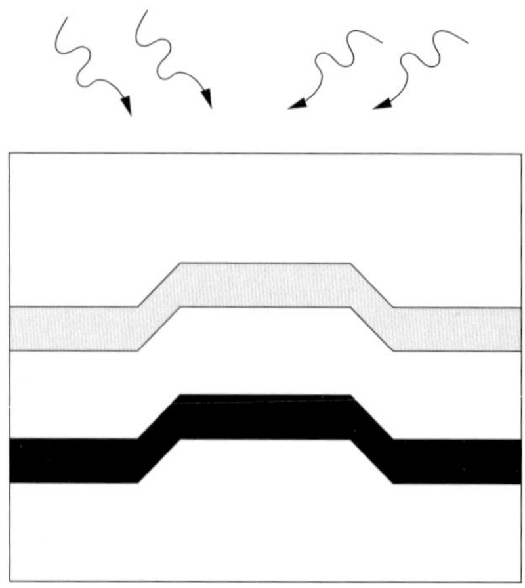

Fig. 3.5. A multilayer structure of a phase-change disc.

can also optimize the optical contrast between the amorphous and crystalline state of the phase change layer. Together with other layers in the stack, sufficient write sensitivity and cooling rate must be well controlled.

A phase change media can be used for both write-once and erasable ML discs. For a write-once disc, an annealing process turns the disc of an amorphous state into a stable crystalline one. However, for a recordable/rewritable disc, recording is achieved by heating the media with a focused laser, then locally cooling it, quenching the medium into an amorphous state. [10]Recorded amorphous marks are erased by operating the laser at an intermediate continuous erase power level, thus heating the recording layer between the glass transition temperature, T_g, and the melting point, T_m, where $T_m < T_g$. Between these temperatures, the media can turn into a crystalline state.

For overwrite applications, fast switching times between states are essential. The recrystallization of the amorphous state should occur within the dwell time of the laser spot, and needs to be stable at room temperature; T_m must be at least 100^oC. In order to support direct overwrite (DOW) cyclability, increased noise level due to imperfect recrystallization and track crosstalk needs to be minimal [10]. Furthermore, the materials must be tested at various linear velocities and data rates.

3.2.2 Magneto-Optical Media

At room temperature, it is difficult to change the magnetization of the recording layer. External bias magnetic field is required to coerce the magnetization, or so called "coercivity." Erasibility occurs above the Curie temperature, T_{curie}, where the magnetic structure disappears and the change of magnetization is dependent on the applied external field. Two common types of magnetization domains are logitudinal magnetization and vertical magnetization. The readout process relies on the Kerr magnetio-optical effect which senses the magnetic field from the surface layer. Another method that is used is the Faraday effect. The Kerr effect has an advantage that the readout can occur on the same side as the recording material.

The magneto-optical (MO) media are also suitable for both write-once and rewritable applications. Quite a few overwrite methodds exist but here two important approaches are detailed: Modulated field recording and Demagnetization field recording. In modulated field recording, the laser heats the surface above the T_{curie} so that the memory is erased. The modulated bias magnetic field is then applied to change the magnetization and , in effect, rewrite the data.

The demagnetization field method, on the other hand, does not require external bias field. The high laser power heats the recording surface until the coercivity disappears, the resulting demagnetization field then automatically switches the magnetization and causes *stable* domains. This functions as *recording* mode. Low laser power, however, causes *unstable* domains and serve as *erasing* mode.

3.2.3 Multilevel Recording (ML) on Optical Media

In Sec. 3.1 the details on writing/reading multilevel (ML) signals on a conventional phase change media will be presented. In short, an ML signal is achieved by writing smaller binary amorphous and crystalline marks, and carefully controlling their size and relative position.

Write strategy initialization enables ML writing on binary medium, while write precompensation reduces nonlinearity in ML reading. The write strategy initialization is essential to how each mark is written on the disc. The process begins at the plot of contours of equal reflectivity where we select the reflectivity section with high linearity. Corresponding power levels are then recorded. The write precompensation is the process which determines the power levels necessary to ensure a written mark have a linear response on the receiver side. For example, if a mark '1' needs less power than a mark '8,' during the write process, the mark '1' in the sequence '8 1 8' requires less power than '3 1 3' because the reading lens obtain additional reflectivity from neighboring mark '8's. For example, an ML CD-RW disc, with $M = 8$, produced by Calimetrics, Inc. utilizes the write precompensation technique based on two neighboring marks. The precompensation table contains 256 entries since for each mark, there are 64 possible combinations of neighboring marks. Fig. 3.6 shows a sequence of steps required to achieve an ML response from a binary medium.

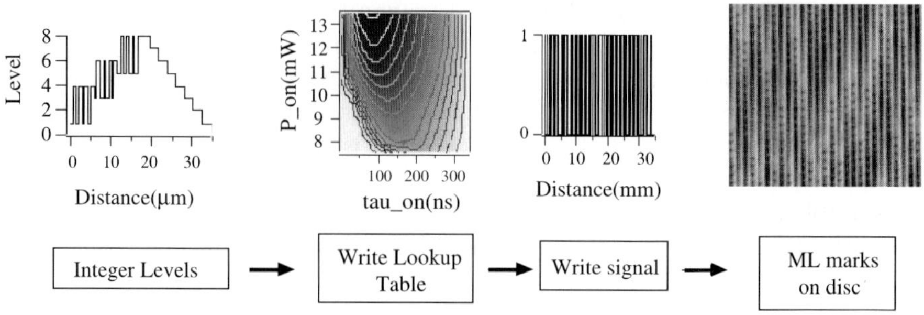

Fig. 3.6. Sequence of writing steps on a CD-R/RW.

3.3 Channel and Noise Modeling

Given the input is the sequence $\{a_i\}$ ($a_i \in \{0, 1, 2, \ldots, M-1\}$), which gets multiplied by pulses $p_T(t)$ with a duration of T seconds. The output is given by

$$r(t) = \left(\sum_i a_i p_T(t - iT) \right) * h_c(t) + \tilde{n}(t), \tag{3.2}$$

where $h_c(t)$ is the impulse response of the optical recording channel and T is the duration of one channel symbol.

The noise $\tilde{n}(t)$ is a combination of electronic and media noise. In optical recording systems, noise sources are modeled as white Gaussian noise (WGN) and media noise. Shot noise, electronic noise, laser noise and others are modeled as additive white Gaussian noise (AWGN). The media noise includes write noise, and variations in reflectivity and absorption, typically caused by physical properties which are intrinsic to the discs [8]. Media noise is a major contributor to the total noise in the system [11][12]. It is common to see 80%-90% of noise to be media noise in an optical recording channel. In Fig. 3.8, a detail of the noise signal $\tilde{n}(t)$ is given. The electronic noise is a white Gaussian random process $n_1(t)$ that is scaled by $\sqrt{\beta}$ then added to the output of the recording channel. The media noise is generated by another Gaussian noise $n_2(t)$ scaled by $\sqrt{1-\beta}$ before its spectrum shaped by $|H_c(f)|^2$. The factor β gives the proportion of the electronic noise in the noise signal. Then the total noise $\tilde{n}(t)$ is zero-mean, band-limited noise with power spectral density

$$S_n(f) = \frac{1-\beta}{K} \frac{N_o}{2} |H_c(f)|^2 + \beta \frac{N_o}{2}, \tag{3.3}$$

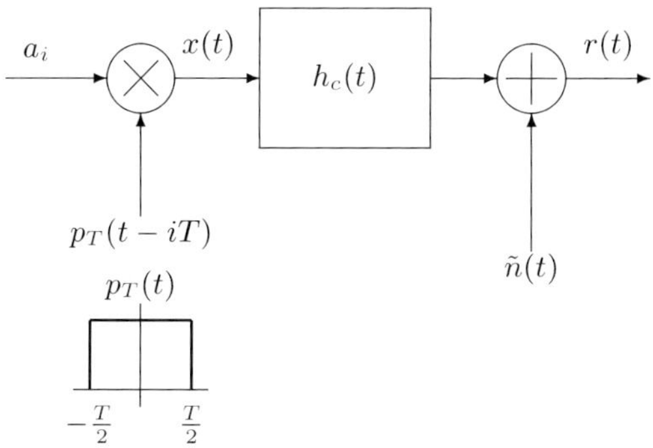

Fig. 3.7. Model for the Optical Recording Channel.

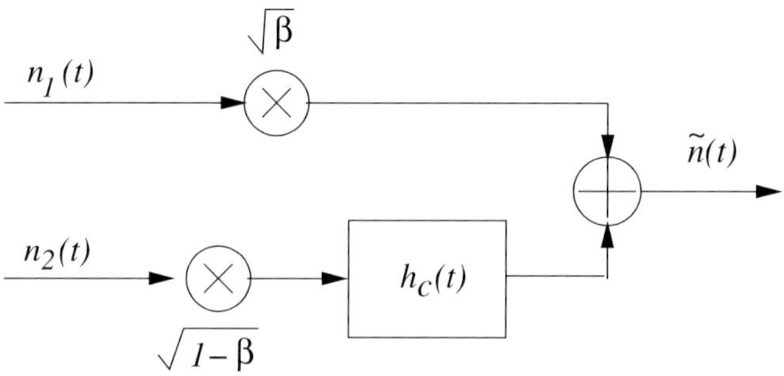

Fig. 3.8. Model for Noise Signal $\tilde{n}(t)$.

where $K = (\int_{-W}^{W} |H_c(f)|^2 \, df)$ and W is optical recording channel bandwidth. The assumption that the noise is band-limited can be justified by the fact that the noise is filtered

3.3.1 Optical Recording Channel Modeling

The analysis of readout signal of an optical recording channel requires knowledge of diffraction theory as described in [8]. In essence, an impulse response of an optical recording channel is determined by the type of irradiance distribution or the distribution of the optical field incident on the objective aperture of a read spot dependent on the spatial frequency, the disc reflectance, and the collection aperture. For phase-change media and magneto-optical media two common irradiance profiles are Airy profile and truncated Gaussian profile. In real applications, it is desirable to balance between the two. Assuming a circular aperture under uniform illumination with the Airy profile, the frequency response of a binary optical recording channel is the modulation transfer function(MTF) [13] defined as

$$H_c(f_n) = \frac{2}{\pi}\left(\arccos|f_n| - |f_n|\sqrt{1-f_n^2}\right) \tag{3.4}$$

where $f_n = f/f_c$ is a normalized frequency variable, such that $f_n \in [-1, 1]$. This means that $H_c(f)$ is non-zero only for $f \in [-f_c, f_c]$. The frequency f_c is the critical frequency of the optical recording channel given by

$$f_c = \frac{2NA}{\lambda} v \quad [Hz] \tag{3.5}$$

where NA is the numerical aperture of the focusing lens, v is the linear velocity and λ is the wavelength of the laser.

As for ML disc on the phase-change media [13] the channel response, $h_c(t)$ of which the frequency response, $H_c(f_n)$, is given by

$$H_c(f_n) = \left(\frac{2}{\pi}\left(\arccos|f_n| - |f_n|\sqrt{1-f_n^2}\right)\right)^2 \tag{3.6}$$

Two earlier channel models are in the form of frequency responses; their corresponding time responses are, however, difficult to derive. Another channel model is [14][15]

$$h_c(t) = \frac{1}{2}[u(t) - u(t - T_b)] \tag{3.7}$$

where

$$u(t) = erf\left(\frac{2\sqrt{\ln(2)}vt}{FWHM}\right) \tag{3.8}$$

and $\text{erf}(x) = \frac{2}{\sqrt{\pi}}\int_0^x e^{-t^2}dt$ and FWHM is the full width at half-maximum (FWHM) as defined in Eq. 3.1. In terms of the critical frequency, f_c, Eq. 3.8, $u(t)$ is then

$$u(t) = erf\left(\frac{\sqrt{\ln(2)}f_c t}{0.55}\right) \tag{3.9}$$

The normalized density S is defined as

$$S = FWHM/T_b \tag{3.10}$$

where T_b is the bit interval.

Similar to magnetic recording, a technique on the receive side to equalize to partial response (PR) targets has received research and practical interests. The PR targets for optical recording [11] in the D-transform are

$$h(D) = (1 + D)^n \tag{3.11}$$

where D is a unit time delay and n is an integer. Another form of PR target investigated is PR(a,b,b,a) [16], where the variables a and b are integers in

$$h(D) = a + bD + bD^2 + aD^3 \tag{3.12}$$

3.4 Turbo Product Codes in Optical Recording Systems

Like any other communication systems, optical recording systems suffer signal degradation due to nonlinearities and noises. Other impediments include scratches and dusts on the disc surface. Besides, readibility and reliability on drives produced by various manufacturers are design challenges not found in magnetic recording systems. Powerful error correction design is thus needed. Current products include concatenated RS codes or 2-D RS codes as well as RLL code as the means to meet stringent specifications. The bit error rate required is in the range of 10^{-12} or lower [7]. Next generation optical discs can exploit advanced coding algorithms, upto now, turbo codes and turbo-like codes have been studied for optical recording systems [15, 17–19]. Some were investigated via modeling and simulations, others have been tested on real discs. Recently, turbo product code (TPC) [3] was demonstrated to be a powerful codes. In [3], upto a 7-dB coding gain was obtained over uncoded system for a code rate 0.65 with QPSK modulation with AWGN channel at BER of 10^{-5} In this section, TPC is investigated and applied to ML optical recording systems.

3.4.1 Turbo Product Codes (TPC)

Turbo product codes are *N-dimensional* codes of which constituent codes are linear block codes. In the original paper [3], Bose-Chaudhuri-Hocqenghem (BCH) codes were demonstrated. The product code under consideration is a serially concatenated code. It consists of two systematic linear codes $\mathbf{C^1}(n_1, k_1, \delta_1)$ and $\mathbf{C^2}(n_2, k_2, \delta_2)$ where n_i, k_i, and δ_i are code length, message length, and minimum Hamming distance, respectively. The product code $\mathbf{P} = \mathbf{C^1} \otimes \mathbf{C^2}$ is conveniently obtained by coding k_2 rows using code $\mathbf{C^1}$ and coding k_1 columns using $\mathbf{C^2}$, as shown in Fig. 3.9. In addition, the $(n_2 - k_2)$ parity rows and $(n_1 - k_1)$ columns are encoded similarly. The final size of the product code is then $n_1 n_2$, thus the code rate is $k_1 k_2/(n_1 n_2)$. To reduce complexity, one can resort to use the same constituent codes, i.e., $\mathbf{C^1} = \mathbf{C^2}$.

The decoding involves soft decoding of block codes based on Chase's algorithm [20] and exchanges of soft information of each bit from rows to columns and vice versa. Using the Maximum-likelihood decoding (MLD), for an AWGN channel, the optimum decision $\mathbf{D} = (d_1, d_2, ..., d_n)$ where n is the codeword length is given by

$$\mathbf{D} = \mathbf{C^i} \text{ if } |\mathbf{R} - \mathbf{C^i}|^2 \leq |\mathbf{R} - \mathbf{C^l}|^2 \quad \forall l \in [1, 2^k], \tag{3.13}$$

where $\mathbf{C^i} = (c_1^i, c_2^i, ..., c_n^i)$ is the ith codeword of the code $\mathbf{C}$, $\mathbf{R} = (r_1, r_2, ..., r_n)$ is the received sequence and $|\mathbf{R} - \mathbf{C^i}|^2 = \sum_{j=1}^{n} (r_j - c_j^i)^2$ is the squared Euclidean distance. Note that l is an index of possible codewords $\mathbf{C^l}$, the total is 2^k, where k is the binary message length. The complexity of exhaustive search of $\mathbf{D}$ is exponential. Pyndiah proposed a complexity reduced search algorithm as follow [3]:

Step 1: Determine the $p = \lfloor \delta/2 \rfloor$ least reliable positions and values of $\mathbf{R}$, where δ is the minimum Hamming distance of the code. The reliability of a position j is defined as $(2/\sigma^2)r_j$ where σ^2 is the noise variance.

Step 2: Decode 2^p possible codewords from step 1, using an algebraic (hard) decoder, and determine $\mathbf{D}$.

At high SNRs, the computation of soft output, r'_j requires two codewords, one is $\mathbf{D}$ and the other is the codeword $\mathbf{C}'$, of which c_j has an opposite sign to d_j in $\mathbf{D}$ as shown below:

$$r'_j = \left(\frac{|\mathbf{R} - \mathbf{C}'|^2 - |\mathbf{R} - \mathbf{D}|^2}{4} \right) d_j, \tag{3.14}$$

When a codeword $\mathbf{C}$ cannot be found, use

$$r'_j = \beta \times d_j, \tag{3.15}$$

with $\beta \geq 0$. The values of β may be iteration dependent and obtained from experiments. At the end of an iteration, the hard decision of r'_j gives hard decoded bit of 0 or 1. To start the next iteration we pass the extrinsic information of bit j, w_j, defined as

$$w_j = r'_j - r_j \tag{3.16}$$

For example, at iteration 2,

$$\mathbf{R}(2) = \mathbf{R} + \alpha(2)\mathbf{W}(2), \tag{3.17}$$

where $\alpha(2)$ is a scaling factor and $\mathbf{W}(2)$ is the extrinsic information after iteration 1.

Other decoding algorithms are combined Chase-GMD algorithms (CGA) [21] and an adaptive versions [22].

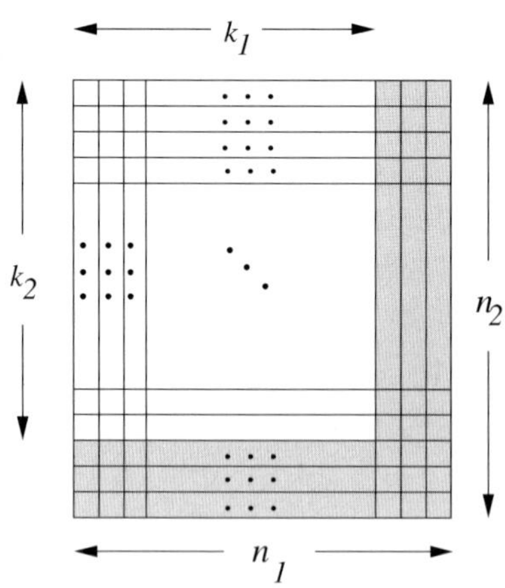

Fig. 3.9. A turbo product code $\mathbf{P} = \mathbf{C^1} \otimes \mathbf{C^2}$.

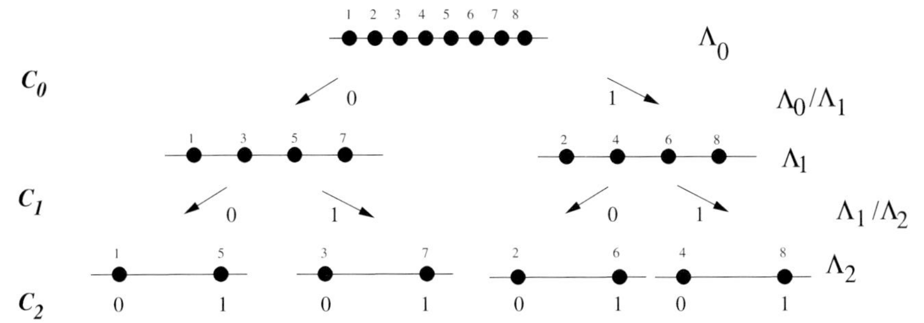

Fig. 3.10. Signal partitioning of PAM symbols with $M = 8$ and separate codes for each level.

3.4.2 Multilevel Coding

Multilevel coding systematically assigns component codes $\mathbf{C_0}, \mathbf{C_1}, ..., \mathbf{C_M}$ to each of the partition levels, $\mathbf{\Lambda_0}/\mathbf{\Lambda_1}, \mathbf{\Lambda_1}/\mathbf{\Lambda_2}, ..., \mathbf{\Lambda_{L-1}}/\mathbf{\Lambda_L}$, in the partition chain $\mathbf{\Lambda_0}/\mathbf{\Lambda_1}/...\mathbf{\Lambda_L}$. For classical trellis-coded modulation (TCM), the code protects higher-level bits while lower-level bits in the signal partitioning tree are unprotected. To achieve the same equivelent channel capacity, $\mathcal{C}$, one can resort to coding each stage of the partition separately [23]. For a regular parition chain $\mathbf{\Lambda_0}/\mathbf{\Lambda_1}, \mathbf{\Lambda_1}/\mathbf{\Lambda_2}, ..., \mathbf{\Lambda_{L-1}}/\mathbf{\Lambda_L}$,

$$\mathcal{C}(\mathbf{\Lambda_0}) = \sum_{i=0}^{L-1} \mathcal{C}(\mathbf{\Lambda_i}/\mathbf{\Lambda_{i+1}}) + \mathcal{C}(\mathbf{\Lambda_L}), \tag{3.18}$$

where $\mathcal{C}(\mathbf{\Lambda_i}/\mathbf{\Lambda_{i+1}}) = \mathcal{C}(\mathbf{\Lambda_i}) - \mathcal{C}(\mathbf{\Lambda_{i+1}})$.

As an example, an 8-*ary* optical recording system is shown in Fig. 3.10, where 8 symbols are partitioned into 3 levels or bits; each level or bit is coded by component codes $\mathbf{C_0}, \mathbf{C_1}$, and $\mathbf{C_2}$ which can be turbo-like codes. The use of separate code rates for each level allows the code design with a flexible total code rate.

3.5 System Performance

In this section the performance results of uncoded and coded optical recording systems are compared. For coded systems, the use of trellis-coded modulation (TCM) and turbo product codes on real ML discs are implemented and the performance points are shown.

3.5.1 Performance Parameters

Two of the commonly used parameters in an optical recording system before the outer ECC block are sigma-to-dynamic ratio (SDR) and signal-to-noise ratio (SNR). SDR for an M-ary signal is defined as

$$SDR = \frac{\sigma}{A} = \frac{\sigma}{2 * (M-1)} \tag{3.19}$$

where σ is the noise standard deviation and A is a dynamic range value SNR which is common in communication systems; for a channel with unit magnitude, SNR at the channel output is defined as

$$SNR = \frac{E(X^2)}{\sigma^2} = \frac{M^2-1}{3\sigma^2} \tag{3.20}$$

where $X \in \{-(M-1), ..., -1, 1, ..., M-1\}$ and $E(X^2)$ is the average energy of transmitted symbols. Other forms of SNR also appear in the literature, for example, $SNR = E_b/N_0$ where E_b is the average bit energy, and $N_0 = 2\sigma_n^2$, where σ_n^2 is the noise variance. When the code rate, R, is considered, one can use $SNR = E_b/RN_0$ These two parameters typically are measured after the equalization, i.e., at the input of the Viterbi detector. For example, an ML disc with $M = 8$ has $A = 14$ and a typical value of SDR in an ML CD-R/RW system is $1\% - 2\%$. The corresponding SNR range is thus 24-30 dB. Although the SNR parameter is useful, one should note that unlike most communication systems, we cannot modify SNRs by adjusting the transmitted power.

3.5.2 Performance Results

Current research of turbo and tubo-like codes exist for binary and multi-leveled optical recording systems. As for binary optical recording systems, the use of turbo code were recently investigated [15] [17][18][19]. In [15], the performance of the serially concatenated turbo codes(SCTC) and low-density parity-check codes (LDPC) is investigated on PR equalized targets. The PR channels are typcially considered inner codes with the precoder for each PR target is included. turbo codes offer coding gain of 4-5 dB over uncoded PRML systems. The SCTC provides better performance than the LDPC codes at $S = 2.0$, but comparable to the later at lower density $S = 1.0$. Furthermore, turbo codes provide resilience to transition noise and lose only 1 dB at 15% transition noise. In [17], work on a combination of SCTC and RLL codes are investigated for DVD systems. The PR targets of $(1+D)$ and $(1+D+D^2+D^3)$ are used together with the PRML technique. The turbo-coded system provides better formance with media noise and offers higher density gains.

As far as the author is concerned, up to date, the only product that is tested turbo-like codes on real optical discs are the MultiLevel disc or ML optical discs [6] by Calimetrics, Inc. (now part of Hewlett Packard) The ML™ discs were first demonstrated using red laser on phase-change media. In [13] the use of various equalizers and error correction control shows the performance gain relative to the capacity at $M = 8$ as a function of signal-to-noise ratio as shown in Fig. 3.11. For uncoded systems, there are three sets of 3-point plots for the case of, from left to right, $M = 2$, $M = 4$, and $M = 8$, respectively. It is evident that the MLSE performs the best for all Ms considered.

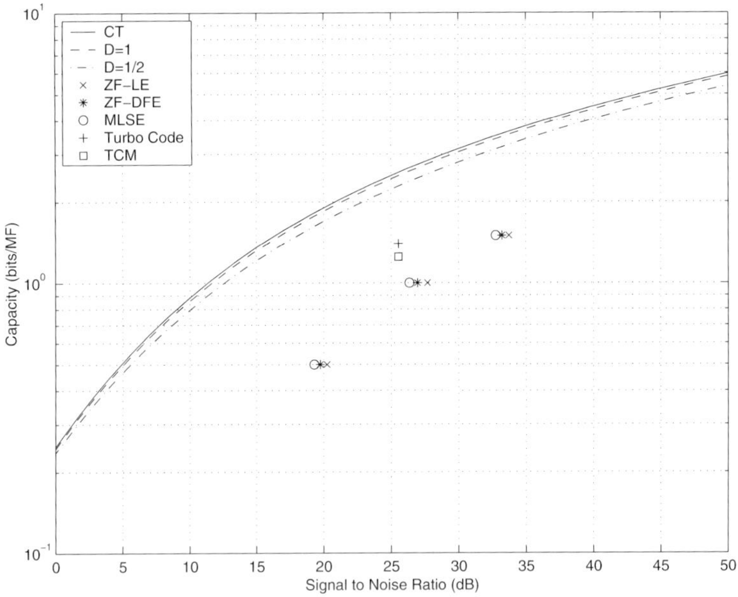

Fig. 3.11. The performance of ML optical recording system with various detectors and codes.

For coded system, an actual 8-*ary* ML CD-RW uses the TCM and turbo code as channel code (inner code in the complete optical recording system). The turbo code which is represented by '+' on the plot is actually the turbo product code designed for 8-level ML discs. The TCM code of a code rate of 5/6 with 256 states offers 2.5 bits/mark, but the turbo product codes designed for each separate level of the partitioning of PAM signals with M = 8 yields 2.75 bits/mark. Coded systems shows the gain of at least 6.5 dB when compared with the simulation result of uncoded systems at the BER of $P_b(E) = 10^{-5}$. Original turbo codes [2] are not considered due to the decoding complexity requirement. Later on, improvement from 8 to 12 levels are verified in combination with the use of MultiLevel Coded Modulation (MCM) [24]. This results in a 10-GB DVD-RW system with coding efficiency of 3.08 bits/data cell using red laser.

Fig. 3.11 compares the system performance versus the theoretical capacity. Three capacity curves are plotted. CT is for the continous-time channel model of MTF^2, while for $D = 1$ and $D = 1/2$ are for discrete-time version of the channel model. $D = 1$ is when the mark size is the same as minimum feature, and $D = 1/2$ is when the mark size is twice the size of the minimum feature(MF). The minimum feature is theoretically the smallest mark size one can write and the read lens can distinguish. Thus, with discrete-time channel, we lose a bit of capacity.

It is, however, important to note that in the near future, an RS code is still needed as the outer code and the codes with iterative decodig nature are used as the inner code.

3.6 Approved DVD Standards

As of the end of 2004, two rival formats are being developed. The first of which, HD-DVD, is supported by DVD Forum with over 200 members like Hitachi, NEC, and etc. The single-layer type holds 15 GB, while the dual layer one holds 30 GB. The other camp, the Blu-Ray Disc Founders Association, with members like Dell, Sony, and etc., offer DVD with upto 27GB for a single-layer disc and 54 GB for the dual layer disc [1]. Many companies are members of both groups.

Acknowledgment

The author would like to acknowledge Professor Steven W. McLaughlin for his extensive work in this field of optical recording systems and be grateful to his advisement during the author's academic work at Georgia Institute of Technology, Atlanta, GA, USA during 1997-2002 and also to Calimetrics, Inc., Alameda, California, which was the first company to develop the testbed for ML CD/R-RW and ML DVD/R-RW. The company offered the author an internship in during April-November of 2001.

References

1. Website:, *http://www.dvdforum.org.*
2. C. Berrou, A. Glavieux, and P. Thitimajshima, "Near shannon limit error-correcting coding and decoding: Turbo codes," in *Proc. 1993 Int. Conf. Commun.*, 1993, pp. 1064–1070.
3. R. Pyndiah, "Near-optimum decoding of product codes : Block Codes," *IEEE Trans. on Communications*, vol. 46, no. 8, pp. 1003–1010, Aug. 1998.
4. K. Immink, *Coding Techniques for Digital Recorders.* Englewood Cliffs, New Jersey: Prentice-Hall Internation(UK) Ltd., 1991.
5. ——, "Runlength-limited sequences," *Proceedings of IEEE*, vol. 78, no. 11, pp. 1745–1759, 1990.
6. S. McLaughlin, D. Lee, T. Lo, C. Pepin, P. Supnithi, and D. Warland, "Coding and signal processing for multilevel CD-RW, CD-R and DVD-RW," in *Optical Data Storage Conference*, 2001.
7. A. Marchant, *Optical Recording: A Technical Overview.* New York: Addison-Wesley, 1990.
8. T. McDaniel and R. Victora, *Handbook of Magneto-optical data recording.* New Jersey: Noyes Publications, 1997.

9. K. Daly-Flynn and D. Strand, "InSbTe phase change materials for high performance multi-level recording," in *Optical Data Storage Conference*, Hawaii, 2002.
10. H. Borg and R. van Woudenberg, "Trend in optical recording," *Journal of Magnetism and Magnetic Materials*, vol. 193, pp. 519–525, 1999.
11. R. Lynch, "Channels and codes for magnetooptical systems," *IEEE Journal of Selected Areas in Communications*, vol. 10, no. 1, pp. 57–72, Jan. 1992.
12. D. Treves and D.A.Bloomberg, "Signal, noise, and codes in optical memories," *Opt. Eng.*, vol. 25, no. 7, pp. 881–891, 1986.
13. E. Licona, P. Supnithi, and S. McLaughlin, "$M - ary$ signaling on rewritable/write-once optical discs: fundamental limits and actual performance," in *Globecom'01*, San Antonito, TX, 2001, pp. 102–107.
14. M. Levenson, R. Lynch, and S. Tan, "Edge detection for magneto-optical data storage," in *Appl. Opt.*, 1988, pp. 723–727.
15. H. S. et. al., "Iterative decoding for partial response (PR, equalized, magneto-optical (MO) data storage channels," *JSAC*, vol. 19, no. 4, pp. 774–782, Apr. 2001.
16. M. G. K. et al., "Performance comparisons of detection methods in magneto-optical disk system with (1,7) RLL code," in *Globecom'95*, 1995, pp. 1389–1393.
17. L. McPheters and S. McLaughlin, "Concatenated codes and iterative (turbo) decoding for PRML optical recording channels," in *Optical Data Storage Conference*, 1999, pp. 342–343.
18. H. Song, B. V. Kumar, E. Kurtas, and Y. Yuan, "Turbo decoding for optical storage," in *ICC 2000*, vol. 1, New Orleans, LA, 2000, pp. 104–108.
19. L. McPheters, "Turbo-coded optical recording for channels with dvd minimum mark size," *IEEE Trans. on Magnetics*, vol. 38, no. 1, pp. 298–302, Jan. 2002.
20. D. Chase, "A class of algorithms of decoding block codes with channel measurement information," *IEEE Trans. on Information Theory*, vol. IT-18, pp. 170–182, Jan. 1972.
21. M. Fossier and S. Lin, "Chase-type and gdm coset decodings," *IEEE Trans. on Communications*, vol. 48, pp. 345–350, Mar. 2000.
22. A. Mahran and M. Benaissa, "Iterative decoding of product codes via adaptive application of the combined Chase-2 and GMD decoding algorithm," in *ICC'01*, Helsinki, Finland, 2001, pp. 2256–2260.
23. J. Huber and U. Wachsmann, "Capacities of the equivalent channels in multilevel coding schemes," *IEE Electronics Letters*, vol. 30, pp. 557–558, Mar. 1994.
24. S. McLaughlin, Y. Lo, C. Pepin, and D. Warland, "MultiLevel DVD: coding beyond 3 bits/data cell," in *Optical Data Storage Conference*, 2002.

Part II

Wireline Communications

Chapter 4

Turbo and Turbo-like Code Design in ADSL Modems

Hamid R. Sadjadpour[1] and Sedat Ölçer[2]

[1] University of California, Santa Cruz, USA
[2] IBM Research Division, Switzerland

The invention of the Turbo codes in 1993 by Berrou et al. [1] represents perhaps one of the most important advances in coding theory in the past decade. The application of the turbo coding techniques for error correction in practical data transmission and recording systems has since then attracted considerable interest. In this chapter, a new application of turbo coding to transmission systems that utilize multi-carrier modulation is presented. In particular, the proposed method is used in Asymmetric Digital Subscriber Line (ADSL) modems that employ the Discrete Multi-tone (DMT) technology. DMT modulation can be regarded as a variation of the Orthogonal Frequency Division Multiplexing (OFDM) technique used mostly in the context of wireless communications [2].

The ANSI standard T1.413 [3], which is the first ADSL standard published in 1995, specified the use of Reed-Solomon (RS) coding for forward error control (FEC). Later, the ANSI standard introduced the concatenation of RS coding with a 16-state 4-dimensional trellis code, known as Wei code, as an inner coding technique. In this chapter, the performance of turbo codes is compared with that of the Wei code and is shown to lead to a higher coding gain than the recommended Wei code. The described turbo coding scheme is also bandwidth efficient, which is an important feature given that signal bandwidth for transmission over twisted-pair wires is limited. This scheme employs a parallel concatenated convolutional code in combination with multi-tone Quadrature Amplitude Modulation (QAM) suitable for ADSL modems.

One challenge in the design of such codes is the length of the interleaver. Turbo codes have a near-Shannon limit error correcting performance when the length of the interleaver is in the order of several thousand bits. However, many practical applications, including ADSL, have limitations in terms of delay or latency. Such delay constraints force the designer to limit the length of the interleaver in turbo codes. For this reason, special attention must be given to the design of interleavers. In this chapter, several pseudo-random interleaver designs with short block length suitable for turbo codes are presented that result in high coding gains. The proposed pseudo-random interleavers avoid

K. Sripmanwat (ed.), Turbo Code Applications: a journey from a paper to realization, 67–93.

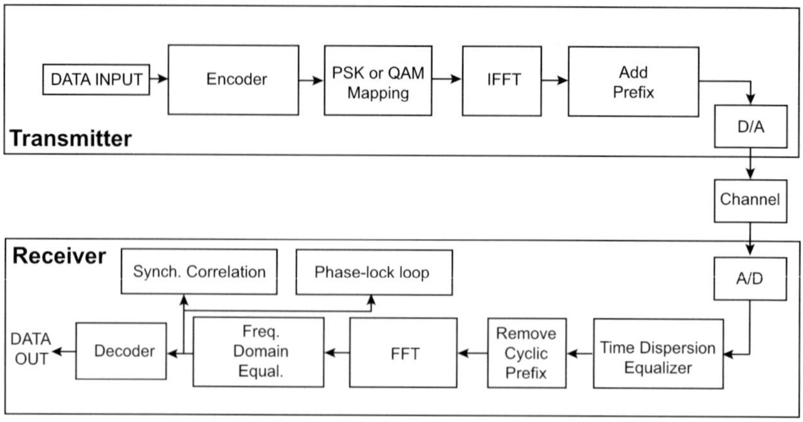

Fig. 4.1. Block diagram of an ADSL modem.

error floor effects in the practical range of bit error rates (BERs) for ADSL applications.

Since the symbol constellation size in different sub-carriers of the ADSL modem can be different, the code rate for each QAM constellation varies within one DMT symbol (frame). For this reason, special attention should be paid to decode this system having possibly a different code rate in each sub-carrier. A simple encoding and decoding procedure is here introduced that allows different constellation sizes and code rates to be used within one turbo block.

This chapter also describes the application of low-density parity-check (LDPC) codes to ADSL transmission as this class of codes represents an important alternative to turbo codes for near-Shannon limit communications. Binary LDPC codes have been known since the early 1960s [4, 5] but their capacity-approaching performance has only been demonstrated in the past decade [6, 7]. There is currently considerable research activity to explore efficient constructions of such codes as well as to understand their theoretical limits.

Turbo codes can be viewed as LDPC codes and the decoding algorithms used for both types of codes can be derived from the formalism of factor graphs [8]. Other turbo-like codes, such as serial turbo codes, turbo product codes, or repeat accumulate codes will not be considered in this chapter. However, the techniques explained can be extended to encompass these classes of codes as well.

Fig. 4.1 shows a block diagram of an ADSL modem. The encoder (decoder) in this system consists of an outer RS encoder (decoder) and an inner encoder (decoder) for a turbo or an LDPC code.

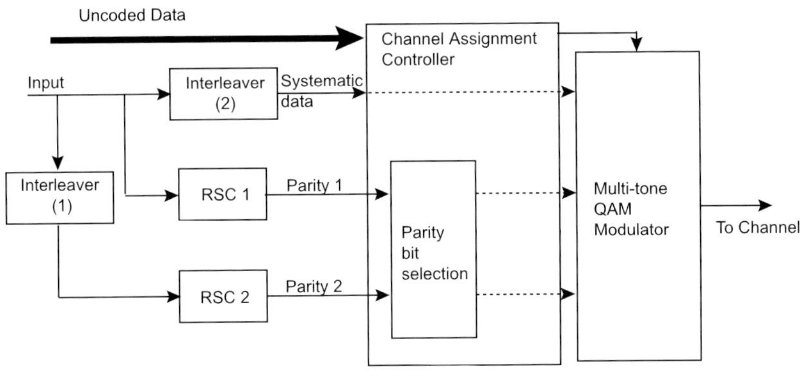

Fig. 4.2. Block diagram of the turbo encoder.

4.1 Turbo Encoder Design for ADSL Modems

Fig. 4.2 describes a block diagram of a turbo encoder that can be used for inner coding in the system shown in Fig. 4.1. The structure of the turbo encoder is similar to that of the original scheme in [1] with an additional interleaver in the path of systematic data. The turbo encoder receives the information bits as input and it generates one interleaved systematic data bit and two parity bits for each information bit. The additional interleaver provided in the path of systematic data is designed to combat impulse noise that sometimes affects signals transmitted over twisted pair wires. The parity bits are punctured based on the information provided by the channel assignment controller (CAC) . During the initialization of the modem, the CAC collects the information about the noise level and channel condition of each sub-carrier. Based on these measurements for a particular loop, the power mask, and the power budget constraint, CAC allocates a combination of uncoded data, interleaved systematic data, and punctured parity bits to each sub-carrier. In particular, a bit loading algorithm [9, 10] is used to assign the number of information and parity bits for each sub-carrier, and hence determines the constellation size for each sub-carrier. The CAC determines the appropriate combination of uncoded bits, interleaved systematic bits, and punctured parity bits for mapping onto a M-QAM symbol. Note that some information bits are not passed through the turbo encoder. These bits are usually assigned to the most protected bits in a M-QAM symbol to ensure higher protection. The most protected bit is the one that has the least probability of incorrect detection compared to the other bits in one symbol. This can be defined for a symbol or for each dimension (I or Q) of a complex symbol. Gray code mapping is the most common mapping for assignment of the bits to an M-QAM constellation. In this case, each I or Q component of a M-QAM symbol is mapped using Gray coding separately. In Gray code mapping, the most

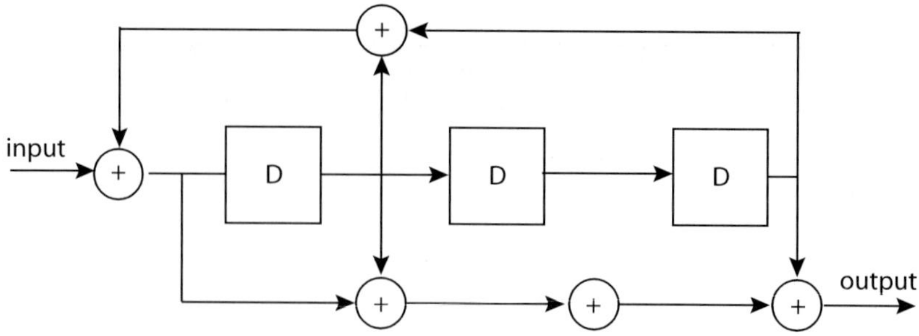

Fig. 4.3. RSC encoder for turbo codes with generator matrix $[1, \frac{17_{oct}}{15_{oct}}]$.

protected bit is the most significant bit. The main advantage of Gray code mapping is its simplicity in terms of implementation complexity. From the data that is generated at the output of the turbo encoder, parity bits are assigned to the remaining more significant bits and systematic bits to the least significant bits. There should be an equal number of puncturing of parity bits from each component of the turbo encoder.

Each turbo block contains N information bits. Also the total number of bits that can be transmitted during one DMT frame in ADSL depends on the characteristics of the particular twisted pair wire that is carrying the signal. In general, there is no relationship between the turbo block size, N, and the number of bits loaded in each ADSL frame, N_f. N_f depends on the channel characteristics, i.e., the channel attenuation function and the noise and interference level in a particular line. The CAC function will ensure that the boundaries of the turbo block are properly processed in the transmitter, so that the receiver can decode each turbo block accordingly. The CAC also controls the flow of data in the receiver. In the receiver, the conditional probability of each transmitted bit can be directly computed from the M-QAM constellation. The details of the decoder will be discussed later.

For the components of the turbo encoder, recursive systematic convolutional (RSC) codes can be used similar to [1]. Each RSC encoder has a generator matrix $[1, \frac{g_{ff}(D)}{g_{fb}(D)}]$, where $g_{ff}(D)$ and $g_{fb}(D)$ are the feedforward and feedback polynomials of the RSC encoder respectively. The feedback polynomial is usually selected as a primitive polynomial [11]. Fig. 4.3 demonstrates one example of the RSC encoder recommended for turbo codes with feedback and feedforward generator polynomials equal to 15_{oct} and 17_{oct} respectively. More details on the encoder and decoder structure of turbo codes can be found in [1].

For example, a 4096-QAM (12 bits) symbol with a spectral efficiency of 10 bits/s/Hz can be constructed with or without uncoded bits. Table 4.1 demonstrates the mapping for such a symbol if uncoded bits are not used.

Table 4.1. Mapping for 4096-QAM constellation with code rate $\frac{5}{6}$ using only turbo encoded data.

Systematic encoded data d_k	d_1	d_2	d_3	d_4	d_5	d_6	d_7	d_8	d_9	d_{10}
parity y_k^1	y_1^1	-	-	-	-	-	-	-	-	-
parity y_k^2	-	-	-	-	-	y_6^2	-	-	-	-
4096-QAM symbol	$(d_1, d_2, d_3, d_4, d_5, y_1^1, d_6, d_7, d_8, d_9, d_{10}, y_6^2)$									

Table 4.2. Mapping for 4096-QAM constellation with code rate $\frac{5}{6}$, with nine uncoded bits and three bits from the turbo encoder.

uncoded data u_k	u_1	u_2	u_3	u_4	u_5	u_6	u_7	u_8	u_9
Systematic encoded data d_k	d_1	-	-	-	-	-	-	-	-
parity y_k^1	y_1^1	-	-	-	-	-	-	-	-
parity y_k^2	y_1^2	-	-	-	-	-	-	-	-
4096-QAM symbol	$(d_1, y_1^1, y_1^2, u_1, u_2, u_3, u_4, u_5, u_6, u_7, u_8, u_9)$								

Based on this Table, for each symbol, a total of eighteen parity bits from the two RSC encoders are punctured to construct one symbol.

If uncoded bits are used, then many different combinations are possible. For example, if we want to minimize puncturing, we can have a total of nine uncoded bits together with a systematic bit and two parity bits from the output of the turbo encoder mapped to the 4096-QAM constellation. By changing the number of uncoded bits to construct a 4096-QAM symbol, we can have a different level of puncturing pattern in the turbo encoder. As it is clear from these examples, the use of only up to two parity bits per symbol is preferred in order to maintain high spectral efficiency.

4.2 Turbo Decoder Design for ADSL Modems

The original turbo decoder utilized a maximum *a posteriori* (MAP) algorithm also known as BCJR [12] algorithm for soft decoding. The MAP algorithm is used in a recursive manner in the turbo decoder [1] by computing the Log-Likelihood Ratio (LLR) of each bit based on the entire observed data block of length N, which is also the interleaver length. The LLR is computed as

$$\Lambda_1(d_k) = \log \frac{\Pr(d_k = 1|R_1^N)}{\Pr(d_k = 0|R_1^N)} = \log \frac{\sum_s \sum_{s'} \gamma_1(R_k, s', s)\alpha_{k-1}(s')\beta_k(s)}{\sum_s \sum_{s'} \gamma_0(R_k, s', s)\alpha_{k-1}(s')\beta_k(s)}, \tag{4.1}$$

where $R_k = \{d_k^r, y_k^{r_i}\}$ is the observed data at the receiver, d_k^r and $y_k^{r_i}$ are the received systematic data and parity information from the i-th RSC encoder at time k, $\gamma_j(R_k, s^{'}, s)$ consists of the transition probability of the discrete Gaussian memoryless channel and transition probabilities of the encoder trellis [12], $\alpha_k(s)$ and $\beta_k(s)$ are two parameters that can be computed by forward and backward recursions respectively based on $\gamma_j(R_k, s^{'}, s)$, and s is the state of the trellis at any given time. The details of this derivation is described in [1]. It can be shown that $\gamma_j(R_k, s^{'}, s)$ can be written as

$$\begin{aligned}\gamma_j(R_k, s^{'}, s) = \Pr(R_k|d_k = j, S_k = s, S_{k-1} = s^{'}) \times \Pr(d_k = j|S_k = s, S_{k-1}s^{'}) \\ \times \Pr(S_k = s|S_{k-1} = s^{'}) \end{aligned} \quad (4.2)$$

The second term in Eq. 4.2 is the transition probability of the discrete channel, the third term is equal to 1 or 0 depending on whether it is possible for $d_k = j$ when the system transition is from state $s^{'}$ to state s, and the fourth term is the transition state probabilities and for equiprobable binary data, it is equal to $\frac{1}{2}$. The second term in Eq. 4.2 can be divided into two terms:

$$\begin{aligned}\Pr(R_k|d_k = j, S_k = s, S_{k-1} = s^{'}) = \Pr(d_k^r|d_k = j, S_k = s, S_{k-1} = s^{'}) \\ \times \Pr(y_k^{r_i}|d_k = j, S_k = s, S_{k-1} = s^{'})\end{aligned} \quad (4.3)$$

Accurate computation of $\gamma_j(R_k, s^{'}, s)$ from the received signals is very important for obtaining the other parameters of the MAP decoding algorithm. Note that, in DSL application the received signal is an M-QAM ($M = 2^m$) symbol. The symbol consists of m bits from parity, systematic, and uncoded data. In a Gray code labeling, each bit d_k in a M-QAM symbol is represented by the real (I) or imaginary (Q) part of the M-QAM symbol. Therefore, if this bit is represented by the real part of the QAM symbol, the second term in Eq. 4.3 can be written as

$$\Pr(d_k^r|d_k = j, S_k = s, S_{k-1} = s^{'}) = \sum_{k_1=0}^{\frac{m}{2}} \Pr(I_{k_1}^r|I_{k_1} = i, S_k = s, S_{k-1} = s^{'}) \quad (4.4)$$

Here m is assumed to be an even number. If the bit (d_k) was represented by the imaginary (Q_{k_1}) part of the QAM signal, then I_{k_1} will be replaced by Q_{k_1} in the above equation. $\Pr(y_k^{r_i}|\ldots)$ can be computed using a similar approach. Once these conditional probability values are computed, then we can use the MAP decoding algorithm for the binary case as described in [1, 12].

The MAP algorithm is computationally complex and sensitive to SNR mismatch and inaccurate estimation of the noise variance [13]. This algorithm requires non-linear functions for computation of the probabilities and both multiplications and additions are needed. For example, the fixed-point representation of the MAP decoding variables usually require between 16 to 24 bits for a QPSK signal constellation. Based on the above hardware requirements, the MAP algorithm is not practical to implement in a chip. The logarithmic version of the MAP algorithm [14–16] and the Soft Output Viterbi

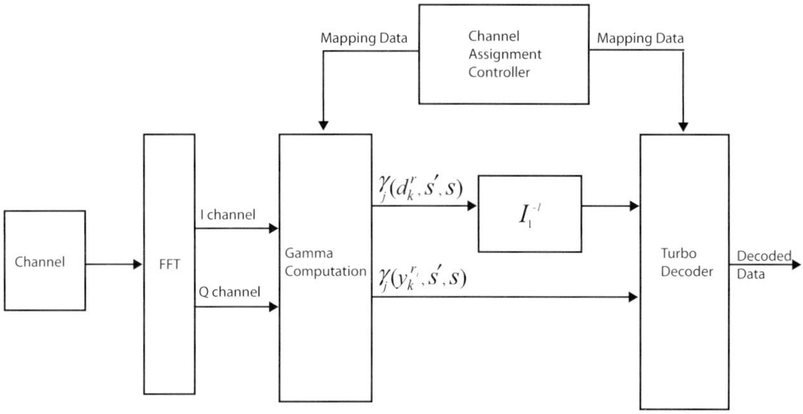

Fig. 4.4. Block diagram of turbo decoder.

Algorithm (SOVA) [17, 18] are the practical decoding algorithms for implementation. These algorithms are less sensitive to SNR mismatch and inaccurate estimation of the noise variance, and fixed-point representation of their variables require approximately 8 bits for a QPSK signal constellation. All different logarithmic versions of the MAP algorithm only involve additions and a max-operation that can be realized utilizing a simple look-up table [16] or a threshold detector [19]. Fig. 4.4 demonstrates the block diagram of the turbo decoder at the receiver of the ADSL modem. The sampled received signal is processed by an FFT block. At the output of this block, complex QAM symbols representing information in the real (I) and imaginary (Q) dimensions are obtained. This information is used to compute the Gamma variables and the results are fed into the turbo decoder after de-interleaving the Gamma variables related to the systematic data. The CAC controls the boundaries of the FFT frame and the turbo code.

Extensive simulations have been conducted to investigate the effectiveness of the turbo code for ADSL transmission. Fig. 4.5 compares the bit error rate (BER) performance of turbo code to Wei code when the turbo RSC component codes are eight-state codes with a block length of 1000 bits and with the same generator polynomial as in Fig. 4.3. The simulation results are based on only three iterations using the Log-MAP decoding algorithm. The additional coding gain due to the use of turbo code compared to Wei code varies from 4.1 dB for 4-QAM to 1.8 dB for 256-QAM constellation. For this example, there is one parity bit in each dimension and the remaining bits are systematic data from the output of the turbo encoder. No uncoded bit is used to construct these QAM symbols and consequently, the remaining parity bits from the output of the turbo encoder are punctured. For 1, 2, 3, 4, and 6

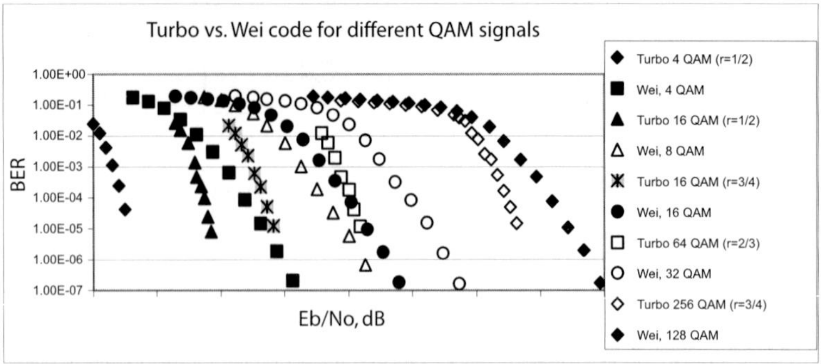

Fig. 4.5. BER performance of the turbo and Wei codes for various QAM modulation levels.

information bits per sub-carrier, constellation sizes of 4, 16, 16, 64, and 256 QAM with code rates of 0.5, 0.5, 0.75, 0.667, and 0.75 are used respectively.

Another approach to construct QAM symbols is to combine coded and uncoded data. Fig. 4.6 demonstrates the symbol error rate (SER) performance of turbo code using the normalized signal-to-noise ratio (SNR_{norm}) as defined in [20]. The normalized SNR for a modulation and coding scheme operating at a given rate of η (in bits per two-dimensional symbol) is defined as

$$SNR_{norm} = \frac{SNR}{2^\eta - 1} = \frac{\eta}{2^\eta - 1}\frac{E_b}{N_o} \tag{4.5}$$

Note that if the coding scheme achieves Shannon capacity, then $SNR_{norm} = 0$ dB. The turbo block length for this example is 2022 bits and the simulation results are for 20 iterations of the Log-MAP decoding algorithm. For this example, the QAM symbol mapping is done using one systematic bit and one parity bit in each dimension and the remaining bits are uncoded bits. The RSC component codes used here are similar to the previous example. The code rates for this example are 0.5, 0.5, 0.75, and 0.833 for 4, 16, 256, and 4096 QAM constellation sizes respectively.

4.3 Interleaver Design for Turbo Code in ADSL Modems

An interleaver π is a permutation $i \mapsto \pi(i)$ that maps a data sequence of N input symbols $d_1, d_2, \ldots, d_N$ into the same sequence in a new order. If the input data sequence is $\mathbf{d} = [d_1, d_2, \ldots, d_N]$, then the permuted data sequence is $\mathbf{d}P$, where P is an interleaving matrix with a single 1 in each row and column, all other entries being zero. Every interleaver has a corresponding de-interleaver (π^{-1}) that acts on the interleaved data sequence and restores it to its original order. The de-interleaving matrix is simply the transpose of the interleaving matrix (P^T).

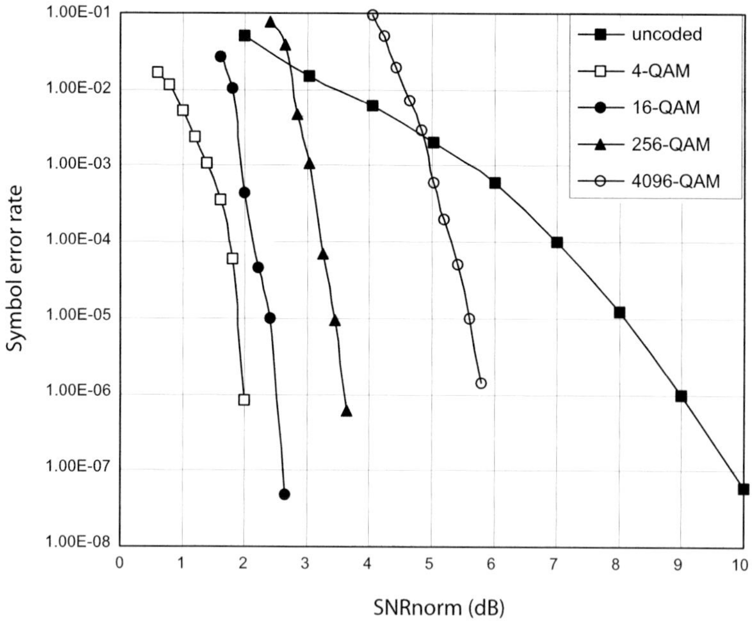

Fig. 4.6. SER Performance of turbo code for interleaver length of 2022 bits and a total of 20 iterations of the Log-MAP algorithm.

A random interleaver is simply a random permutation π. For large values of N, most random interleavers utilized in turbo codes perform well. However, as the interleaver block size decreases, the performance of the turbo code degrades substantially, up to a point when its BER performance is worse than that of a convolutional code with similar computational complexity. Thus the design of short interleavers for turbo codes is an important problem [21–24].

An S-random interleaver (where $S = 1, 2, 3, \ldots$) is a "semi-random" interleaver constructed as follows. Each randomly selected integer is compared with S previously selected random integers. If the difference between the current selection and S previous selections is smaller than S, the random integer is rejected. This process is repeated until N distinct integers have been selected. Computer simulations have shown that if $S \leq \sqrt{\frac{N}{2}}$, then this process converges [25] in a reasonable time. This interleaver design ensures that the short cycle events are avoided. A short cycle event occurs when two bits are close to each other both before and after interleaving.

The proposed turbo code for ADSL modems requires two interleavers. It is well-known that for a given block length, the design of the interleaver can have a significant effect on the decoder performance, due to the different values of the minimum effective free distance that each interleaver can have. Therefore, in this section, two types of interleaver design are described. The first type is the interleaver that is at the input of the RSC encoder (see Fig. 4.3). For short turbo block length, this interleaver is responsible to produce large minimum effective free distance, d_{min}, at the output of the turbo encoder. When the turbo block length is short, then a random interleaver can create small d_{min} which results in poor BER performance for the turbo code. There are various short interleaver designs for turbo codes described in the literature. Here, a few of these interleaver designs are briefly described.

In [22], the author considers three objectives: 1) the low weight input data sequences do not create low weight at the output of at least one of the RSC encoders; 2) the design provides a large span for the input sequences that result in a zero terminating phase for all the component codes; and finally 3)the design provides large distance for the terminating right edge of the sequences at the output of one or both encoders. The design includes a linear sum assignment optimization problem that can be solved using the Hungarian method [26].

An interleaver design was proposed based on the performance of iterative decoding in turbo codes [27]. Turbo codes utilize an iterative decoding process based on the MAP or other algorithms that can provide soft output. At each decoding step, some information related to the parity bits of one decoder is fed into the other decoder together with the systematic data sequence and the parity bits corresponding to that decoder. Fig. 4.7 shows this iterative decoding scheme. The inputs to each decoder are the input data sequence, d_k, the parity bits y_k^1 or y_k^2, and the logarithm of the likelihood ratio (LLR) associated with the parity bits from the other decoder (W_k^1 or W_k^2), which is used as *a priori* information. All these inputs are utilized by the decoder to create three outputs corresponding to the weighted version of these inputs. In Fig. 4.7, $\hat{d}_k$ represents the weighted version of the input data sequence, d_k. Also d_n in the same figure demonstrates the fact that the input data sequence is fed into the second decoder after interleaving. The input to each decoder from the other decoder is used as *a priori* information in the next decoding step and corresponds to the weighted version of the parity bits. This information will be more effective in the performance of iterative decoding if it is less correlated to the input data sequence (or interleaved input data sequence). Therefore it is reasonable to use this as a criterion for designing the interleaver. For large block size interleavers, most random interleavers provide a low correlation between W_k^i and input data sequence, d_k. The correlation coefficient, $r^1_{W^1_{k_1},d_{k_2}}$, is defined as the correlation between $W^1_{k_1}$ and d_{k_2}.

It has been shown [21] that $r^1_{W^1_{k_1},d_{k_2}}$ can be analytically approximated by

$$\hat{r}^1_{W^1_{k_1},d_{k_2}} = \begin{cases} a \exp^{-c|k_1-k_2|} & \text{if } k_1 \neq k_2 \\ 0 & \text{if } k_1 = k_2 \end{cases} \tag{4.6}$$

where a and c are constants that depend on the encoder feedback and feed-forward polynomials. The correlation coefficient at the output of the second

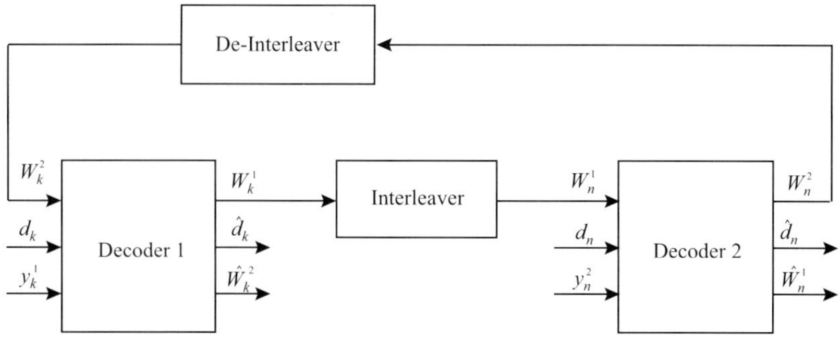

Fig. 4.7. Turbo decoder.

decoder, $\hat{r}^2_{\mathbf{W}^2,\mathbf{d}}$, is approximated by

$$\hat{\mathbf{r}}^2_{\mathbf{W}^2,\mathbf{d}} = \frac{1}{2}\hat{\mathbf{r}}^1_{\mathbf{W}^1,\mathbf{d}}P(I + \hat{\mathbf{r}}^1_{\mathbf{W}^1,\mathbf{d}}) \tag{4.7}$$

where the two terms on the right hand side of Eq. 4.7 correspond to the correlation coefficients between $\mathbf{W}^2$ and the input data, i.e., $\mathbf{W}^1$ and $\mathbf{d}$ [21]. In this section, $\hat{\mathbf{r}}^2_{\mathbf{W}^2,\mathbf{d}}$ represents the correlation coefficient matrix and $\hat{r}^2_{W^2_{k_1},d_{k_2}}$ represents one element of this matrix.

Similar correlation coefficients can be computed for the de-interleaver. The correlation matrix corresponding to de-interleaver, $\hat{\mathbf{r}'}^2_{\mathbf{W}^2,\mathbf{d}}$, is the same as in Eq. 4.7 except that P is replaced by P^T.

Then $\mathbf{V}_{k_1}$ is defined to be

$$\mathbf{V}_{k_1} = \frac{1}{N-1}\sum_{k_2=1}^{N}(\hat{r}^2_{W^2_{k_1},d_{k_2}} - \bar{\hat{r}}^2_{W^2_{k_1},\mathbf{d}}) \tag{4.8}$$

where

$$\bar{\hat{r}}^2_{W^2_{k_1},\mathbf{d}} = \frac{1}{N}\sum_{k_2=1}^{N}\hat{r}^2_{W^2_{k_1},d_{k_2}} \tag{4.9}$$

V'_{k_1} is defined in a similar way using $\hat{\mathbf{r}'}^2_{\mathbf{W}^2,\mathbf{d}}$. The iterative decoding suitability (IDS) measure is then defined as

$$IDS = \frac{1}{2N}\sum_{k_1=1}^{N}(V_{k_1} + V'_{k_1}) \tag{4.10}$$

A low value of IDS is an indication that the correlation properties between $\mathbf{W}^1$ and $\mathbf{d}$ are equally spread along the data sequence of length N. An interleaver design based on the IDS condition is proposed in [27].

Another interleaver design, a 2-step S-random interleaver, is presented in [28]. The goal is to increase the minimum effective free distance, d_{min}, of the turbo code while decreasing or at least not increasing the correlation properties between the information input data sequence and W_k^i. Hokfelt et al. [21, 27] introduced the IDS criterion to evaluate this correlation properties. The two vectors for the computation of IDS in Eq. 4.10 are very similar and for most interleavers it is sufficient to only use one of them, i.e., V_{k_1}. Instead, a new criterion based on decreasing the correlation coefficients for the third decoding step can be defined, i.e., the correlation coefficients between extrinsic information from the second decoder and information input data sequence. In this regard, the new correlation coefficient matrix, $\hat{\mathbf{r}'}^2_{\mathbf{W}^2,\mathbf{d}}$, is defined as

$$\begin{aligned}\hat{\mathbf{r}'}^2_{\mathbf{W}^2,\mathbf{d}} &= \frac{1}{2}\hat{\mathbf{r}}^2_{\mathbf{W}^2,\mathbf{d}}P^T(I+\hat{\mathbf{r}}^2_{\mathbf{W}^2,\mathbf{d}}) \\ &= \frac{1}{4}(\hat{\mathbf{r}}^1_{\mathbf{W}^1,\mathbf{d}}+\hat{\mathbf{r}}^1_{\mathbf{W}^1,\mathbf{d}}P\hat{\mathbf{r}}^1_{\mathbf{W}^1,\mathbf{d}}P^T) \\ &\times (I+\frac{1}{2}\hat{\mathbf{r}}^1_{\mathbf{W}^1,\mathbf{d}}P+\frac{1}{2}\hat{\mathbf{r}}^1_{\mathbf{W}^1,\mathbf{d}}P\hat{\mathbf{r}}^1_{\mathbf{W}^1,\mathbf{d}})\end{aligned} \tag{4.11}$$

$V_{k_1}^{'(new)}$ can now be computed in a similar way to Eq. 4.8 by using Eq. 4.11. The new iterative decoding suitability (IDS_1) is then defined as

$$IDS_1 = \frac{1}{2N}\sum_{k_1=1}^{N}(V_{k_1}+V_{k_1}^{'(new)}) \tag{4.12}$$

A small value for IDS_1 only guarantees that the correlation properties are spread equally throughout the data sequence. However, this criterion does not attempt to reduce the power of correlation coefficients, i.e., $(\hat{r}^2_{W^2_{k_1},d_{k_2}})^2$ and $(\hat{r'}^2_{W^2_{k_1},d_{k_2}})^2$. Therefore, the following additional condition is recommended as a second iterative decoding suitability criterion:

$$IDS_2 = \frac{1}{2N^2}\sum_{k_1=1}^{N}\sum_{k_2=1}^{N}((\hat{r}^2_{W^2_{k_1},d_{k_2}})^2+(\hat{r'}^2_{W^2_{k_1},d_{k_2}})^2) \tag{4.13}$$

Finally, the average of these two values is used as a new IDS criterion.

$$IDS_{(new)} = \frac{1}{2}(IDS_1+IDS_2) \tag{4.14}$$

Minimizing Eq. 4.14 is then one of the goals in optimizing the interleaver.

Unlike [27], where the interleaver design is just based on the IDS criterion, this interleaver is designed in two stages [28]. In the first stage, an interleaver that satisfies the S-random criterion together with an additional condition is designed [28]. In the second stage, the minimum effective free distance (d_{min}) of the turbo code is increased while considering the $IDS_{(new)}$ constraint. The

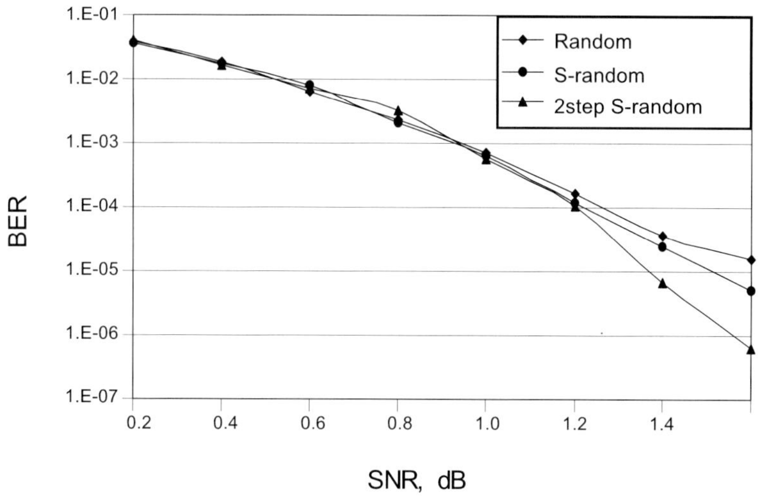

Fig. 4.8. Performance of turbo code for different interleavers of size 400 bits and BPSK signal.

design is as follows. First the values of S_1 and S_2 are selected.

Step 1: Each randomly selected integer $\pi(i)$ is compared with the previous selections $\pi(j)$ to check that if $i - j \leq S_1$ then $|\pi(i) - \pi(j)| > S_1$. The interleaver must also satisfy $|i - \pi(i)| > S_2$.

Besides the above conditions, the last m tail bits used for trellis termination in the first decoder are chosen to satisfy $\pi(1) = N$, and if $\pi(i) = N - k$ with $k < m$ then $i < N/2$. This condition will guarantee that trellis termination for the first decoder is sufficient and there will not be any low weight sequence at the output of the second decoder caused by failure of trellis termination.

Step 2: The procedure in the second step guarantees that all the input data sequences with a minimum weight less than or equal to w_{det} have a minimum effective free distance greater than $d_{min,w_{det}}$ where these parameters are defined in [28].

For the deterministic interleaver that is in the path of systematic data in Fig. 4.3, a row-column interleaver can be used to combat impulse noise.

Fig. 4.8 compares the BER performance of a turbo code with component codes of Fig. 4.3. Turbo code utilizes 18 iterations of Log-MAP algorithm and three types of interleaver of length 400 bits with BPSK modulation are implemented. The simulation result clearly shows that proper choice of interleaver has significant effect on the performance of the turbo code.

4.4 LDPC Codes and LDPC Encoder Design for ADSL Modems

A binary (N, K) linear code is a set of binary vectors of length N, called codewords, that satisfy a set of $M \geq (N - K)$ parity-check equations. The length of the information vector is denoted by K. A parity-check equation can be conveniently represented as a binary vector of length N in which a "1" in position n, $1 \leq n \leq N$, implies that the n-th codeword symbol participates in the parity check. The set of M parity checks can then be represented by an $M \times N$ parity-check matrix H. A "1" in the (m, n)-th position in H indicates that the m-th parity check involves the n-th symbol of the codeword, or equivalently, that codeword symbol n participates in the m-th parity check. This suggests a natural graphical representation of the parity-check matrix H as a "bipartite" graph with two kinds of nodes: N symbol or variable nodes that correspond to the codeword symbols, and M check nodes that correspond to the parity checks represented by the rows of matrix H [6, 30]. The connectivity of the bipartite graph, known as the Tanner graph, is such that H is its incidence matrix, i.e., for each "1" in the (m, n)-th position, the graph has an edge connecting check node m with symbol node n. No two nodes of the same kind are connected by an edge.

A binary (N, K) LDPC code [4, 5] is a linear block code described by a sparse (N, K) parity-check matrix H, i.e., H has a low density of 1s. Each symbol is checked by a small number of parity checks, and each parity check includes a small number of symbols. An LDPC code is called (d_s, d_c)-regular if in the corresponding bipartite graph every symbol node is connected to d_s check nodes and every check node is connected to d_c symbol nodes; otherwise it is called an irregular LDPC code. Fig. 4.9 shows an example of the Tanner graph of a $(d_s = 3, d_c = 4)$-regular LDPC code of block length $N = 12$ having $M = 9$ parity-check equations. In this figure, the left nodes represent symbol nodes and the right nodes represent parity-check nodes.

The parity-check matrix of a regular LDPC code thus contains d_s ones in each column and d_c ones in each row. By contrast, the number of ones in each column, respectively in each row, can vary widely for an irregular LDPC code. Excellent performance can be achieved with well-designed regular LDPC codes. The regularity of the code usually allows reducing hardware complexity and leads to efficient decoder implementations. However, irregularity improves performance further and makes this class of codes capacity approaching.

Irregular LDPC codes [31, 32] are conveniently specified by a pair of "degree distribution" polynomials: a degree distribution polynomial $\lambda(x)$ is associated with the symbol nodes and a degree distribution polynomial $\rho(x)$ with the check nodes. In general,

$$\lambda(x) = \sum_{i=1}^{d_s^*} \lambda_i x^{i-1} \quad \text{and} \quad \rho(x) = \sum_{i=1}^{d_c^*} \rho_i x^{i-1}$$

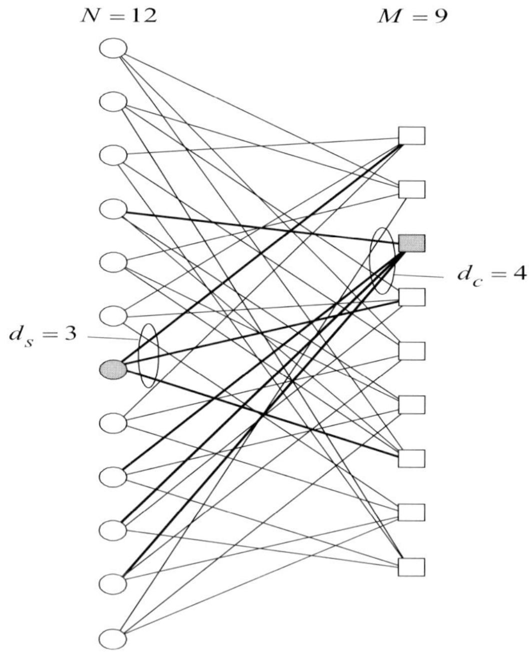

Fig. 4.9. An example of a $(d_s = 3, d_c = 4)$-regular Tanner graph.

where λ_i and ρ_i denote the fraction of edges connected to a symbol node, respectively to a check node, of degree i. Furthermore, $d_s^*(d_c^*)$ represents the maximum symbol (check) degree. For example, for the $(d_s = 3, d_c = 4)$-regular LDPC code in Fig. 4.9, $\lambda(x) = x^2$ and $\rho(x) = x^3$.

As mentioned earlier, in DMT modulation for ADSL, each subcarrier is modulated by symbols taken from constellations of possibly very different sizes, such as BPSK, QPSK, 16, 32, 64-QAM, etc., ultimately up to a 2^{15}-symbol constellation. One alternative for LDPC coding in a multilevel modulation system consists of using nonbinary LDPC codes [33]. Here, the focus is on the use of binary LDPC codes for multilevel modulation [34] as this approach makes the LDPC encoding and soft iterative decoding functions independent of the number of modulation levels employed. Therefore, at the receiver, the number of modulation levels only affects soft demapping, i.e., the extraction of soft information on individual code bits from the received noisy symbols.

Fig. 4.10 shows a block diagram of the encoder for an LDPC code that can be used to realize the inner coding function in Fig. 4.1. Information bits representing data or control messages are encoded into a binary codeword of length N. The symbol mapper collects groups of coded bits, possibly along

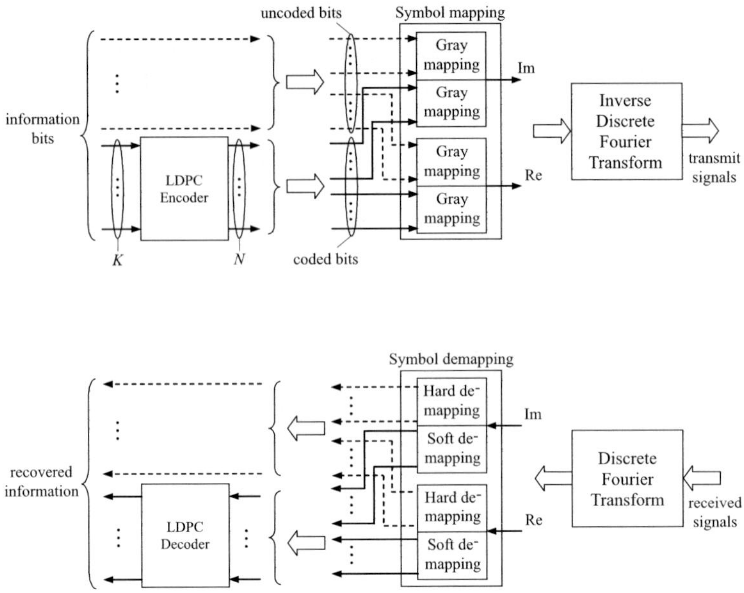

Fig. 4.10. LDPC encoding/symbol mapping and symbol demapping/decoding functions for DMT-based DSL transmission. Mapping and demapping functions are shown for the real (Re) and imaginary (Im) parts of a complex QAM symbol.

with uncoded information bits as shown in the figure, and builds QAM symbols for frequency-domain modulation by an inverse discrete Fourier transform operation. As discussed earlier in connection with turbo encoding, uncoded bits are mapped to the most protected bits, i.e., the more significant bits, of a QAM symbol. Furthermore, as explained below, a double Gray-code labeling is assumed for symbol mapping, wherein the less significant coded bits and the more significant uncoded bits are Gray-coded separately. As usual, the size of the QAM constellation used on each subchannel is determined through a "bit-loading" algorithm [35].

The symbol-mapping function is now explained in more detail with reference to the block diagram of Fig. 4.11. If the constellations employed have a square shape, soft demapping at the receiver is greatly simplified because the real and imaginary parts of the received noisy complex signals can be demapped independently. Thus the transmit symbols are chosen from a 2^b-QAM symbol constellation, where $b = 1$ or $b = 2m$, with m a positive integer.

When $b = 2m$, two binary m-tuples $v = (v_{m-1}, v_{m-2}, \ldots, v_1, v_o)$ and $w = (w_{m-1}, w_{m-2}, \ldots, w_1, w_o)$ independently select two L-ary real symbols, $L = 2^m$, representing the real and imaginary parts, respectively, of the complex

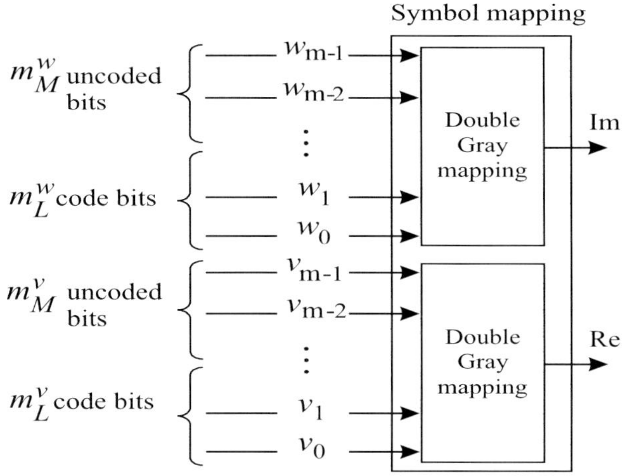

Fig. 4.11. A more detailed representation of the multilevel encoding and symbol-mapping function shown in Fig. 4.10.

QAM symbol to be transmitted. The L-ary symbols belong to the set

$$\mathbf{A} = \{A_l = 2l - (L-1), l = 0, 1, \ldots, L-1\}$$

The real part of a QAM symbol conveys m_L^v code bits and m_M^v uncoded bits, with $m_L^v + m_M^v = m$. Similarly the imaginary part conveys m_L^w code bits and m_M^w uncoded bits, with $m_L^w + m_M^w = m$. Symbol mapping relies on the partition of the set $\mathbf{A}$ into $2^{m_L^v}[2^{m_L^w}]$ subsets such that the minimum Euclidean distance between the symbols within each subset is maximized. The $m_L^v[m_L^w]$ less-significant bits (LSBs) of $v[w]$ label the subsets of A following a Gray-coding rule. The remaining more-significant bits (MSBs) $m_M^v[m_M^w]$ label symbols within a subset following a separate Gray-coding rule.

When $b = 1$, only the code bit v_o is employed. This case corresponds to BPSK modulation.

The number of uncoded and coded bits per transmit symbol is a design parameter that involves a trade-off in terms of performance and decoding complexity.

Double Gray-code labeling offers a good trade-off in terms of achievable performance and implementation complexity. The efficiency of symbol mapping based on this labeling technique can be assessed by computing the capacity of the binary-valued-input and continuous-valued-output AWGN channel that also includes the bit-to-symbol mapping function. More precisely, assuming ideal interleaving of the binary input sequence prior to the mapping of m consecutive bits into a multilevel symbol, the capacity of the equivalent "bit-interleaved" channel [36] can be determined. The results show that for 8-level

where the normalization constant μ_n is chosen such that $q_{n\to m}(0)+q_{n\to m}(1) = 1$.

Step (iii) (decision): Quantize $\hat{\mathbf{x}} = [\hat{x}_1, \hat{x}_2, \ldots, \hat{x}_N]$ such that $\hat{x}_n = 0$ if $q_n(0) \geq 0.5$ and $\hat{x}_n = 1$ if $q_n(0) < 0.5$. If $\hat{\mathbf{x}}\mathbf{H}^\mathbf{T} = 0$, then halt the algorithm with $\hat{\mathbf{x}}$ as the decoder output; otherwise go to Step (i). If the algorithm does not halt within some maximum number of iterations, then declare a decoder failure.

Because the check-node update in Step (i) requires the computation of sums of products of probabilities, the algorithm is called the "sum-product" algorithm.

A number of variations to the above sum-product algorithm have been proposed in order to simplify its implementation in practical systems. For example, using LLRs as messages offers implementation advantages over using probabilities or likelihood ratios because multiplications are replaced by additions and the normalization step is eliminated.

More precisely, let us define the LLRs $Z_{n\to m}(x_n) \log(q_{n\to m}(0)/q_{n\to m}(1))$ and $L_{m\to n}(x_n) = \log(r_{m\to n}(0)/r_{m\to n}(1))$. Then, the LLR-based sum-product algorithm is obtained, and can be summarized as follows.

Initialization: Each symbol node n is assigned an *a posteriori* LLR $L(x_n|y_n) = \log(\frac{P(x_n=0|y_n)}{P(x_n=1|y_n)})$. For every position (m,n) such that $H_{n,m} = 1$, set

$$Z_{n\to m}(x_n) = L(x_n|y_n),$$

$$L_{m\to n}(x_n) = 0$$

Step (i) (check-node update): For each m, and for each $n \in N(m)$, compute

$$L_{m\to n}(x_n) = \Big(\prod_{n'\in N(m)\backslash n} \mathrm{sign}(Z_{n'\to m}(x_{n'}))\Big) f\Big(\sum_{n'\in N(m)\backslash n} f(|Z_{n'\to m}(x_{n'})|)\Big)$$

Step (ii) (symbol-node update): For each n, and for each $m \in M(n)$, compute

$$Z_{n\to m}(x_n) = L(x_n|y_n) + \sum_{m'\in M(n)\backslash m} L_{m'\to n}(u_n)$$

For each n, compute

$$Z_n(x_n) = L(x_n|y_n) + \sum_{m\in M(n)} L_{m\to n}(x_n).$$

Step (iii) (decision): Quantize $\hat{\mathbf{x}} = [\hat{x}_1, \hat{x}_2, \ldots, \hat{x}_N]$ such that $\hat{x}_n = 0$ if $Z_n(x_n) \geq 0$ and $\hat{x}_n = 1$ if $Z_n(x_n) < 0$. If $\hat{\mathbf{x}}\mathbf{H}^\mathbf{T} = 0$, then halt the algorithm with $\hat{\mathbf{x}}$ as the decoder output; otherwise go to *Step (i)*. If the algorithm does not halt within some maximum number of iterations, then declare a decoder failure. In the check-node update of Step (i), the function $f(x)$ is defined as

$$f(x) = \log \frac{e^x + 1}{e^x - 1} = -\log[\tanh(\frac{x}{2})]$$

Clearly, the check-node updates represent the computationally most demanding part of the LLR sum-product algorithm. Therefore, a number of solutions have been proposed in the literature to simplify the check-node processing step of the algorithm and to achieve decoder architectures that are suited for high-speed digital applications. The reader is referred to [38] and the references therein for details on the topic.

Finally, the APPs needed to initialize the sum-product algorithm are obtained through the following soft demapping operation. Let y denote the real (equivalently, the imaginary) part of the noisy received signal:

$$y = A + v,$$

with v a sample of an AWGN process with variance σ_n^2. The APP of code bit i conveyed by symbol A being equal to $x = 0, 1$ is obtained as:

$$P(i = x|y) = \frac{\sum_{A_l \in \mathbf{A}_i, x} \exp^{-\frac{(y-A_l)^2}{2\sigma_n^2}}}{\sum_{A_l \in \mathbf{A}} \exp^{-\frac{(y-A_l)^2}{2\sigma_n^2}}} \qquad i = 0, 1, \ldots, m_L^v - 1,$$

where $\mathbf{A}_i, x$ denotes the subset of all symbols in $\mathbf{A}$ with label value $x = 0, 1$ in position i.

4.6 Performance

It appears that high-rate LDPC codes with medium block length, whose parity-check matrices are constructed similarly to those of array codes [39], exhibit as good a performance as random LDPC codes do. Array-code-based LDPC coding was shown in [40] to offer a number of advantages for DSL transmission. The LDPC parity-check matrix is specified, in that case, by a small set of parameters and constructed deterministically without requiring "preprocessing" operations. Furthermore, array LDPC codes are amenable to linear-time encoding.

The array-code parity-check matrix is specified by three parameters: a prime number p and two integers k and j such that $k, j \le p$. It has dimensions $jp \times kp$ and is given by

$$H_A = \begin{bmatrix} I & I & I & \ldots & I \\ I & \alpha & \alpha^2 & \ldots & \alpha^{k-1} \\ I & \alpha^2 & \alpha^4 & \ldots & \alpha^{2(k-1)} \\ \vdots & \vdots & \vdots & \ddots & \vdots \\ I & \alpha^{j-1} & \alpha^{2(j-1)} & \ldots & \alpha^{(j-1)(k-1)} \end{bmatrix}$$

where I is the $p \times p$ identity matrix and α is a $p \times p$ permutation matrix representing a single left or right cyclic shift of I. Powers of α in H_A indicate multiple cyclic shifts, with the number of shifts given by the value of the exponent.

To achieve efficient linear-time encoding for array-like LDPC codes, it was proposed in [40] to transform the above parity-check matrix into the following triangular form

$$H = \begin{bmatrix} I\ I \ldots I & I & \ldots & I \\ 0\ I\ \alpha & \ldots \alpha^{j-2} & \alpha^{j-1} & \alpha^{k-2} \\ 0\ 0\ I & \ldots \alpha^{2(j-3)} & \alpha^{2(j-2)} & \alpha^{2(k-3)} \\ \vdots\ \vdots\ \vdots & \ddots \vdots & \vdots & \vdots \\ 0\ 0 \ldots\ldots I & & \alpha^{j-1} & \ldots \alpha^{(j-1)(k-j)} \end{bmatrix}$$

where *0* is the $p \times p$ null matrix. The LDPC codes defined by H have a codeword length $N = kp$, $M = jp$ parity checks, and an information block length $K = (k - j)p$. Also, H is 4-cycle free. Note that the triangular form induces a slight irregularity in the initially regular LDPC code. Indeed, the symbol-node and check-node degree distribution polynomials are given by

$$\lambda(x) = \sum_{i=1}^{j} \lambda_j x^{i-1}; \lambda_i = \begin{cases} \frac{1}{k} & 1 \le i \le j-1 \\ 1 - \frac{(j-1)}{k} & i = j \end{cases}$$

and

$$\rho(x) = \sum_{i=k-(j-1)}^{k} \rho_i x^{i-1}; \rho_i = \frac{1}{j}, k - (j-1) \le i \le k$$

respectively.

The typical performance that can be achieved by the multilevel LDPC-coding scheme and the deterministic array-based code construction mentioned above is now briefly illustrated by means of simulations. The telephone-twisted-pair channel introduces frequency-dependent signal distortion as well as several other forms of disturbances, of which crosstalk is the most important. In the simulations, disturbance by AWGN only will be assumed. The reason for this is that if each DMT subchannel has a sufficiently narrow bandwidth, then each one independently approximates an AWGN channel with a particular SNR value. Clearly, impulse noise and narrowband interference of various origins, e.g., AM radio signals, also affect the reliability of communications in DSLs. Performance should ultimately be assessed using actual test-loop conditions.

Fig. 4.12 shows the SER performance of three array-code-based LDPC codes of block lengths $N = 529$, 2209, and 4489 bits, and code rates 0.870, 0.915 and 0.940, respectively. For 16-QAM symbol mapping, all bits are coded. For 256 and 4096-QAM, there are three coded bits along each dimension, and the remaining bits are uncoded.

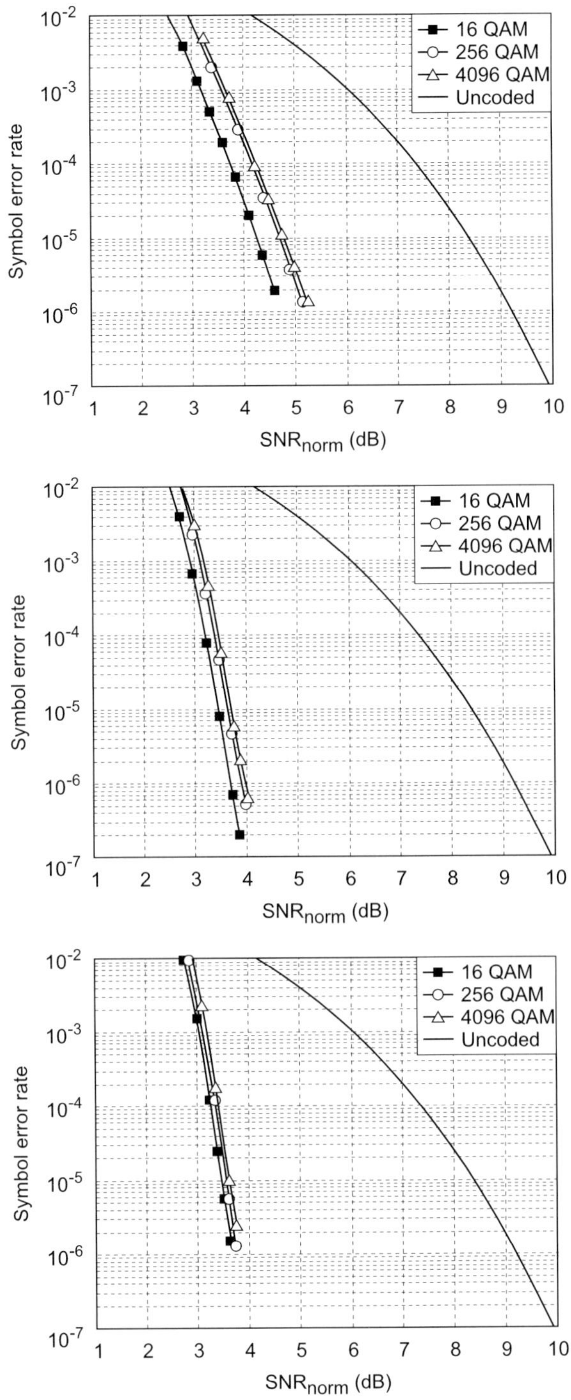

Fig. 4.12. Performance of three LDPC codes: (a) (529, 460), (b) (2209, 2021), and (c) (4489, 4221), for transmission over the AWGN channel using 16, 256, and 4096-QAM.

As mentioned earlier, latency is an important parameter in DSL systems. "Voice" applications are known to demand rather low latency whereas other applications, such as video streaming, tolerate larger delays but need stronger error-correction capability. Thus, in studying new coding techniques for DSLs, trade-offs between coding gain and latency have to be clearly established. Generally speaking, if higher latencies can be tolerated, then longer–and hence more powerful–codes can be employed. Conversely, lower coding gains are imposed by small latencies. Recall that DMT frames in ADSL systems are generated at the rate of 4000 Hz. Therefore if one information block is encoded into one DMT frame, the encoding and decoding functions introduce a latency of 250 μs each, resulting in a total latency of 0.5 ms. When a codeword spans more than a single frame, latency increases accordingly.

Finally, similar to turbo codes, LDPC codes offer significant gains over TCM, see e.g., [41]. Notice, however, that care should be exercised in attempting a comparison between turbo and LDPC codes because they mostly operate at different spectral efficiencies and discrepancies usually exist between code rate, latency, implementation complexity, etc. For this reason, such a comparison was not included in this chapter.

4.7 Final Remarks

In the wake of the invention of the turbo codes, an intense research effort has been undertaken in the information-theory and coding community in order to improve the understanding of capacity-approaching codes as well as to design tools and techniques for constructing codes with practical utility. While important work is still ongoing in this field, the state of the art has reached a level of maturity that encourages engineers in the communications and recording industries to port these coding methods into the realm of practical realizations. This chapter has described, at some level of detail, possible approaches for incorporating turbo and LDPC coding into ADSL technologies.

For both DSL subscribers and network operators, the benefits of capacity-approaching coding will be measured in terms of the achievable increase in data rates, loop reach and service penetration. By employing turbo or turbo-like coding, it will be possible to operate DSL links with greater robustness and reliability under difficult transmission conditions due to spectral crowding, loop unbundling, etc. It is indeed conjectured that the 'turbo principle' will be applied, in various forms, in all future high-performance receiver designs in order to maximize performance.

Other techniques, such as the use of selectable masks for signal transmission, dynamic spectrum management, multi-input multi-output transmission, have partly found and will find their way into the definition of future DSL modems. Capacity-approaching coding is complementary to these techniques, as it can be combined with essentially anyone of them to achieve close to Shannon-limit performance in future generations of DSL systems.

References

1. C. Berrou, A. Glavieux, and P. Thitimajshima, "Near Shannon limit error-correcting coding and decoding : Turbo codes," in *P*roc. IEEE International Conference in Communications, pp. 1064–1070, June 1993.
2. J. Terry and J. Heiskala, *OFDM Wireless LANs: A Theoretical And Practical Guide*, Sams Publisher, 1st edition, December 2001.
3. ANSI *Network and Customer Installation Interfaces — Asymmetric Digital Subscriber Line (ADSL) Metallic Interface*, American National Standard for Telecomm., vol. T1.413, 1995.
4. R.G. Gallager, *Low-density Parity-check Codes*, IRE Transaction on Information Theory, vol. IT-8, pp. 21–28, January 1962.
5. R.G. Gallager, *Low Density Parity Check Codes*, Cambridge, MA: MIT Press, 1963.
6. N. Wiberg, *Codes and Decoding on General Graphs*, Ph.D. Thesis, Linkping University, Sweden, 1996.
7. D.J.C. MacKay, "Good error-correcting codes based on very sparse matrices," *IEEE Trans. Inform. Theory*, vol. 45, pp. 399–431, March 1999.
8. H.-A. Loeliger, "An introduction to factor graphs," *IEEE Signal Processing Magazine*, vol. 45, pp. 28–41, January 2004.
9. J. Tu and J. Cioffi, "A loading algorithm for the concatenation of coset codes with multichannel modulation methods," *Proc. IEEE Globecom 1990*, pp. 1183–1187.
10. R. Sonalkar and H.R. Sadjadpour, "A New Initialization Method for Minimization of Near End Cross Talk due to DMT Transmission in Cable Binders," *Proc. ICCS2000*, Singapore, November 20-24, 2000.
11. S. Lin and D.J. Costello Jr., *Error Control Coding: Fundamentals and Applications*, Prentice-Hall Publisher, 1983.
12. L. Bahl, J. Cocke, F. Jelinek, and J. Raviv, "Optimum decoding of linear codes for minimizing symbol error rate," *IEEE Trans. on Inf. Theory*, vol. IT-20, pp. 284–287, March 1974.
13. T.A. Summers and S.G. Wilson, "SNR Mismatch and Online Estimation in Turbo Decoding," *IEEE Trans. on Comm.* vol. 46, no. 4, pp. 421–424, April 1998.
14. P. Robertson, P. Hoeher, and E. Villebrun, "Optimal and Sub-Optimal Maximum A Posteriori Algorithms Suitable for Turbo Decoding," *European Trans. on Telecomm.* vol. 8, no. 2, pp. 119–126, March-April 1997.
15. P. Robertson, E. Villebrun, and P. Hoeher, "A Comparison of Optimal and Sub-optimal MAP Decoding Algorithms Operating in the Log Domain," *Proc. International Conference on Communications*, pp. 1009–1013, June 1995.
16. S. Benedetto, G. Montorsi, D. Divsalr, and F. Pollara "Soft-Output Decoding Algorithm in Iterative Decoding of Turbo Codes," *TDA Progress Report 42-124*, pp. 63–87, February 15, 1996.
17. J. Hagenauer and P. Hoeher, "A Viterbi Algorithm with Soft-Decision Outputs and Its applications," *Proc. of GLOBECOM*, pp. 1680–1686, November 1989.
18. J. Hagenauer, "Source-Controlled Channel Decoding," *IEEE Transaction on Communications*, vol. 43, No. 9, pp. 2449–2457, September 1995.
19. W.J. Gross and P.G. Gulak, "Simplified MAP algorithm suitables for implementation of turbo decoders," *Electronics Letters*, vol. 34, no. 16, August 6, 1998.

20. V.M. Eyuboglu and G.D. Forney, "Trellis Precoding: Combined Coding, Precoding and Shaping for intersymbol Interference Channels," *IEEE Transaction on Information Theory*, pp. 301–314, March 1992.
21. J. Hokfelt, O. Edfors, and T. Maseng, "Turbo Codes: Correlated Extrinsic Information and its Impact on Iterative Decoding Performance," *Proc. IEEE VTC '99*, Houston, Texas.
22. A.K. Khandani, "Group Structure of Turbo Codes with Applications to the Interleaver Design," *Proc. International Symposium on Information Theory*, pp. 421, August 1998.
23. O.Y. Takeshita and D.J. Costello Jr., "New Classes of Algebraic Interleavers for Turbo Codes," *Proc. International Symposium on Information Theory*, pp. 419, August 1998.
24. H. Herzberg, "Multilevel Turbo Coding with Short Interleavers," *IEEE Journal on selected areas in Communications*, vol. 16, no. 2, pp. 303–309, February 1998.
25. S. Dolinar and D. Divsalar, "Weight Distribution for Turbo codes Using Random and Nonrandom Permutations," *JPL Progress report 42-122*, pp. 56–65, August 15, 1995.
26. B. Kreko, *Linear programming,* translated by Ahrens and Safe, Sir Isaac Pitman & Sons Ltd., 1968.
27. J. Hokfelt, O. Edfors, and T. Maseng, "Interleaver Design for Turbo Codes Based on the Performance of Iterative Decoding," *Proc. IEEE ICC '99*, Vancouver, Canada.
28. H.R. Sadjadpour, N.J.A. Sloane, M. Salehi, G. Nebe, "Interleaver Design for Turbo Codes," *IEEE Selected Areas in Communications*, pp. 831–837, May 2001, special issue on Turbo codes.
29. J. Yuan, B. Vucetic, and W. Feng, "Combined Turbo Codes and Interleaver Design," *IEEE Trans. on Comm.*, vol. 47, no. 4, pp. 484–487, April 1999.
30. R.M. Tanner, "A recursive approach to low complexity codes," *IEEE Trans. Inform. Theory*, vol. 27, pp. 533–548, September 1981.
31. M.G. Luby, M. Mitzenmacher, M.A. Shokrollahi, and D.A. Spielman, "Improved low-density-parity-check codes using irregular graphs," *IEEE Trans. Inform. Theory*, vol. 47, No. 2, pp. 585–598, February 2001.
32. T. J. Richardson, M. A. Shokrollahi, and R. Urbanke, "Design of capacity-aproaching irregular low-density-parity-check codes," *IEEE Trans. Inform. Theory*, vol. 47, No. 2, pp. 619–637, February 2001.
33. M.C. Davey and D.J.C. MacKay, "Low density parity check codes over GF(q)," *IEEE Commun. Letters*, June 1998.
34. E. Eleftheriou and S. Ölçer, "Low-density parity-check codes for multilevel modulation," in *Proc. 2002 IEEE International Symp. on Information Theory*, ISIT 2002, Lausanne, Switzerland, p. 442, June 30-July 5, 2002.
35. T. Starr, J.M. Cioffi, and P.J. Silverman, *Digital Subscriber Line Technology*, Upper Saddle River, NJ: Prentice Hall, 1999.
36. G. Caire, G. Taricco, and E. Biglieri, "Bit-interleaved coded modulation," *IEEE Trans. Inform. Theory*, vol. 44, No. 3, pp. 927–946, May 1998.
37. E. Eleftheriou and S. Ölçer, *Advanced coding techniques for Digital Subscriber Lines*, in DSL Handbook, CRC Press, to be published.
38. J. Chen, A. Dholakia, E. Eleftheriou, M. Fossorier, and X.Y. Hu, "Reduced-Complexity Decoding of LDPC Codes," *IBM Research Report*, RZ 3498, June 2003.

39. J.L. Fan, "Array codes as low-density parity-check codes," in *Proc. 2nd Int'l Symposium on Turbo Codes and Related Topics*, Brest, France, pp. 543–546, September 2000.
40. E. Eleftheriou and S. Ölçer, "Low-density parity-check codes for digital subscriber lines," in *Proc. High-Speed Networks Symp.* ICC 2002, New York, NY, paper D21-3, April-May 2002.
41. E. Eleftheriou, S. Ölçer, and H. Sadjadpour, "Application of capacity-approaching techniques to digital subscriber lines," *IEEE Communications Magazine*, vol. 42, No. 4, pp. 88–94, April 2004.

Chapter 5

Turbo Codes for Single-Mode and Multimode Fiber Optic Communications

Cenk Argon[1] and Steven W. McLaughlin[2]

[1] Seagate Technology, USA
[2] Georgia Institute of Technology, USA

In this chapter, the application of turbo product codes in both single-mode and multimode fiber optic links is reviewed. In the case of single-mode fiber links, it is shown that turbo product codes outperform the current standard forward error correction approach which is using a hard-decision Reed-Solomon code. Furthermore, for multimode fiber links, it is demonstrated that turbo product codes can be combined with spatial equalization techniques to enable significant enhancement of fiber reach.

5.1 Forward Error Correction in Fiber Optic Links

Fiber optic links are now widely deployed for high-speed communications. When compared to their electrical counterparts, i.e., copper cables (twisted pair or coax), optical fiber has the following advantages [1]: (i) Higher bandwidth, (ii) smaller diameter and light weight, (iii) no crosstalk between parallel fiber cables, (iv) immunity to inductive interference, (v) security, (vi) safety, (vii) long life span, etc. Today's optical fiber systems are mainly of intensity modulation / direct detection (IM/DD) type [2]. In these systems, either lasers or light emitting diodes (LED's) are employed as optical transmitters [3]. LED's are usually used for short-distance applications, whereas long-haul (or long-distance) systems use lasers. Typical transmission wavelengths are about 850, 1300, and 1550 nm, where optical fiber has low attenuation. Furthermore, the major portion of optical receivers are of direct detection type, where incident optical power is converted to a proportional electrical current. A direct detection receiver consists usually of a photodetector [PIN (positive-intrinsic-negative) or Avalanche photodiode] followed by a transimpedance amplifier (TIA), which can be viewed as a current-to-voltage converter [3].

Receiver noise is one of the important design parameters of optical fiber links. Major noise sources at the receiver side are [4] : (i) Quantum or shot noise, (ii) dark-current noise, (iii) thermal noise, and (iv) amplifier noise. Besides these receiver noise factors, the main impairment in optical fiber links is

K. Sripmanwat (ed.), Turbo Code Applications: a journey from a paper to realization, 95–119.

dispersion. Dispersion is basically the broadening of a light pulse as it travels along the fiber. At high data rates, this introduces intersymbol interference (ISI) [5] that must be compensated for. Dispersion is also the dominant limiting factor in multimode fiber (MMF) links, since each mode inside the fiber travels at different speeds (this is also termed as intermodal distortion [4]).

The application of turbo codes [6] can reduce the negative effects of noise and dispersion in fiber optic communications by many orders of magnitude. Nevertheless, the turbo implementation is challenging, since especially the extraction of soft-information from the optical channel can be costly. Furthermore, although forward error control (FEC) codes in general have been applied to many types of communications systems, their application to optical communications can be considered as recent. The main reason for the late consideration of FEC codes for optical systems is due to the outstanding properties and low bit error probabilities of optical devices and subsystems.

On the other hand, with the increasing demand for higher capacity for dense wavelength division multiplexing (DWDM) systems, and with the demand for fiber links with longer repeater or amplifier spacing, FEC codes attract much attention. In general, FEC codes can provide increased system margins and relaxed component tolerances. FEC codes for optical networks are classified as in-band and out-of-band [9]. Out-of-band FEC codes can provide unrestricted error correcting capability and are protocol independent. However, out-of-band FEC codes also increase the data rate due to code redundancy. On the other hand, in-band FEC codes use idle periods of the transmission protocol; therefore, in-band FEC codes preserve the transmission data rate. For example, for the SONET (synchronous optical network) protocol, a shortened version of a triple error correcting (8191,8152) Bose, Chaudhuri, and Hocquenghem (BCH) code is applied for in-band FEC. This code is shortened so that the number of information bits is equal to 4320, and the code length is 4359. The 39 redundant bits of a codeword are sent in unused slots of the SONET protocol. Hence, the line rate is preserved. When compared to out-of-band codes, the main disadvantage of in-band codes is that they have limited strength because of the limited amount of idle periods depending on the applied protocol. As an example for out-of-band codes, a (255,239) Reed-Solomon (RS) code with hard-decision decoding is now widely applied in most modern trans-oceanic optical transmission systems [7][8]. In addition, critical aspects for the application of FEC codes (and especially turbo codes) in high-speed optical communications systems are reframe time and decoding latency. Reframe time is the time it takes for the FEC decoder to obtain all information before it can start decoding of the received codewords. Decoding latency is the delay until the decoder outputs the corrected codeword.

As the requirements for fiber link lengths and noise margins increase, soft-decision decoding will likely replace the current hard-decision decoding approach, and the application of turbo codes, and in particular turbo product codes (TPC) [10], also called block turbo codes (BTC), will be possible. In this chapter, the application of TPC's in optical fiber networks for both long-haul applications using single-mode fibers, and for short-haul multimode fiber links is investigated. For single-mode fiber links, the performance of TPC's is compared with that of the current standard RS code. Furthermore, it is demonstrated that for multimode fiber links, TPC's can be combined with a spatially resolved equalizer to improve fiber reach drastically.

5.2 Turbo Product Codes

5.2.1 Background

Turbo product codes (TPC's) [10], or block turbo codes (BTC's), form a subclass of turbo codes. Different than regular turbo codes that are formed with convolutional codes, turbo product codes are typically formed via linear block codes. Therefore, TPC's can be processed with algebraic decoders. Furthermore, efficient decoding algorithms for TPC's have been suggested recently [11][12][13][14]. Due to their low implementation complexity and high performance, TPC's are attractive FEC codes for future optical fiber links. These are basically the main motivations for the application of TPC's in fiber links as considered here.

A two-dimensional TPC is implemented as follows: To encode, the information bits (or symbols) are initially arranged in a two-dimensional $k_1 \times k_2$ array. Afterwards, the columns are encoded using an (n_1, k_1, δ_1) linear block code C_1, and the resulting n_1 rows are encoded using an (n_2, k_2, δ_2) linear block code C_2. Here, n_w, k_w, and δ_w $(w = 1, 2)$ are the codeword length, number of information bits, and minimum Hamming distance of the FEC code, respectively. Furthermore, codes C_1 and C_2 are called the component (or constituent) codes. The resultant product code has parameters $(n_1 n_2, k_1 k_2, \delta_1 \delta_2)$. The order of row and column encoding does not matter; i.e., encoding first the rows and then the columns yields the same product code.

At the TPC decoder, the Chase algorithm [15] is applied iteratively on the rows and columns of the product codeword. At each iteration, a reliability information, or log-likelihood ratio (LLR), is calculated for all decoded bit positions. For a stationary additive white Gaussian noise (AWGN) channel and a communications system using binary phase shift keying (BPSK), the LLR for bit position d_l can be approximated by [11][12]

$$\Lambda(d_l) \approx \left(\frac{R \cdot D - R \cdot \hat{D}}{2} \right) d_l. \tag{5.1}$$

Here, $l \in \{0, 1, \ldots, n-1\}$, n is the code length (typically $n = n_1 = n_2$), $d_l \in \{-1, +1\}$, $D = d_0 d_1 \ldots d_{n-1}$ is the decided codeword after Chase decoding, $\hat{D}$ (if it exists) is a competing codeword that has a different bit value at its lth bit position when compared to D, $R = r_0 r_1 \ldots r_{n-1}$ is the received noisy sequence, and $A \cdot B$ denotes the inner product [11] of vectors $A = a_0 a_1 \ldots a_{n-1}$ and $B = b_0 b_1 \ldots b_{n-1}$ defined as

$$A \cdot B = \sum_{l=0}^{n-1} a_l b_l. \tag{5.2}$$

The above described algorithm can be easily generalized to N-dimensional product codes, and it can be shown that TPC decoding can achieve near-optimum performance levels at low signal-to-noise ratios (SNR's). For further details on TPC's, the interested reader is referred to [10][16].

5.2.2 Finite Bit Precision Effects on TPC Decoding

Under the assumption that soft information is available from the optical channel, the effects of finite bit precision on the performance of TPC's have to be investigated. This is an important issue, since analog-to-digital converters (ADC's) are used after photodetection, and quantized data is passed to the FEC decoder, rather than the original received analog signal.

Integrated electronic ADC's form the bottleneck in today's high-speed optical communications systems at gigabit data rates. Current electronic ADC's have a limiting speed of about 1-2 Gbps [17]. Hence, the current trend is to investigate the design of electro-optical [18] or all-optical ADC's [19][17]. The design of these converters is critical for the application of FEC soft decoding in high-speed optical communications. To circumvent the need for electronic ADC's, alternative approaches propose to perform FEC decoding in the optical domain instead in the electrical domain. For example, all-optical FEC decoding of low-density parity-check (LDPC) codes is proposed in [20], where the message passing algorithm is implemented via optical splitters and adders. In case of TPC's, the incoming optical signal needs to be converted to an electrical equivalent. Hence, it is desirable to assess the quantization effects during electro-optical conversion on the TPC performance.

Fig. 5.1, 5.2, and 5.3 show the quantization effects on the performance of an eBCH$(64,57)^2$ TPC, eBCH$(128,120)^2$ TPC, and eBCH$(256,247)^2$ TPC, respectively. Here, eBCH$(n,k)^2$ TPC denotes that both the row and column component codes of the product code are extended BCH codes [i.e., the row and column codes are eBCH(n,k) codes]. Furthermore, the Chase decoder has parameter p equal to four; i.e., for each row or column decoding of the product code, 2^p candidate codewords are formed. For all three code types, it is noticed that about $q = 5$ quantization bits (or $2^q = 32$ quantization levels) are required to achieve performances close to the performance with no quantization.

Therefore, in the following discussions, it is assumed that the ADC's have at least $q = 5$ or $q = 6$ bits quantization, so that the TPC performance degradation due to quantization can be ignored. However, it must be always kept in mind that the electronic analog-to-digital conversion is one of the

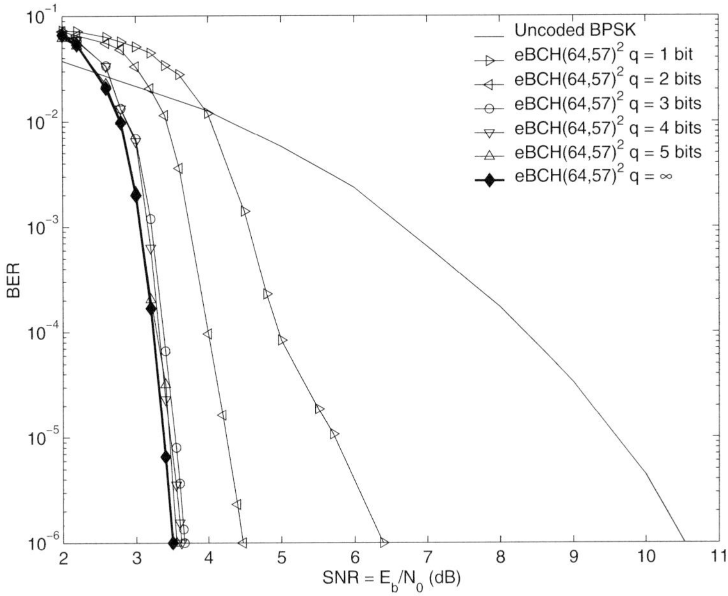

Fig. 5.1. Quantization effects on eBCH $(64{,}57)^2$ TPC decoding ($p = 4$, four iterations, AWGN channel).

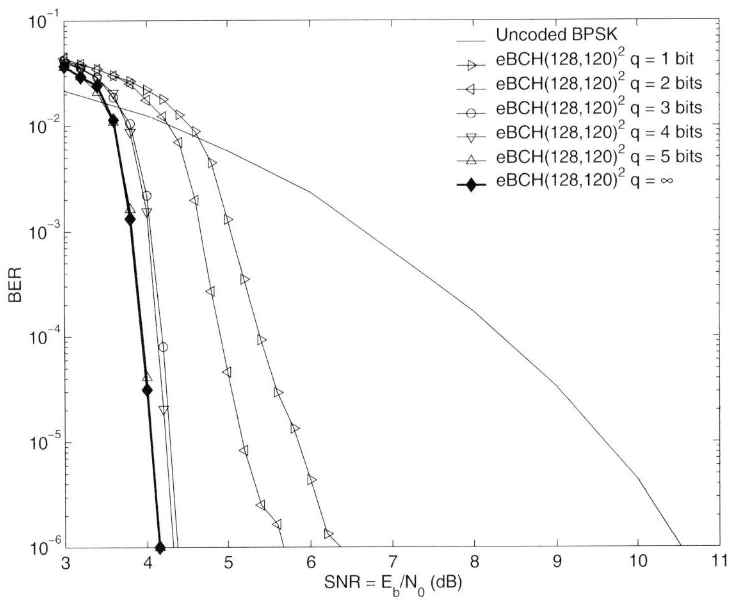

Fig. 5.2. Quantization effects on eBCH $(128{,}120)^2$ TPC decoding ($p = 4$, four iterations, AWGN channel).

noise is modeled as an AWGN signal $v(t)$ added to the photodetector output $s_e(t)$. This is a valid assumption for practical direct detection receivers, where noise in the electrical signal is mainly because of thermal noise [21].

The noisy photodetector output is the input signal to an ADC with sufficient levels of quantization. In case no FEC is used, or a simple hard-decision FEC is employed, the ADC is simply replaced by a threshold device that, depending on the signal intensity, determines which bit (0 or 1) has been received. Data is finally recovered after the FEC decoder.

Since the input of the system is an optical signal while the dominant noise is in the receiver electronics, it is necessary to define the optical signal-to-noise ratio, $\mathrm{SNR}_{\mathrm{o}}$, as

$$\mathrm{SNR}_{\mathrm{o}} = \frac{P_{\mathrm{avg}}\Re}{\sigma_n}, \tag{5.11}$$

where P_{avg} is the average received optical power, $\Re$ is the responsivity of the detector, and σ_n is the standard deviation [or root-mean-square (RMS)] value of the noise current of the pre-amplifier referenced at the output of the photodetector. By doing this, at all times a relative measurement is maintained with the exception of the responsivity $\Re$, which is normalized to 1 Ampere/Watt. Since square-law detectors are employed, the equivalent electrical SNR is defined as

$$\mathrm{SNR}_{\mathrm{e}} = (\mathrm{SNR}_{\mathrm{o}})^2. \tag{5.12}$$

For the remainder of this chapter, the electrical SNR is considered while evaluating the performance of optical fiber links.

5.3.2 FEC Performance

Using the system model in Fig. 5.5, the noise-free eye-diagrams for an SMF link at a data rate of 10 Gbps and for fiber lengths $L = 50$ km and $L = 120$ km are shown in Fig. 5.6 and 5.7, respectively. The transceiver characteristics $p(t)$ is modeled with a raised cosine pulse with 80% rise and fall times, and full-width half-maximum (FWHM) equal to the bit interval. Furthermore, a binary communications system using on-off keying (OOK) and nonreturn-to-zero (NRZ) line coding is assumed, and the transmission wavelength is set to $\lambda = 1550$ nm. As stated before, at this wavelength, the dispersion coefficient $D(\lambda)$ for a standard SMF is 17 ps/nm/km [23].

Fig. 5.6 and 5.7 reveal that as the SMF link distance is increased, the effects of chromatic dispersion become severe and result in strong ISI at the SMF output. This reduces the eye opening in both the vertical and horizontal directions at the optimum sampling time.

Especially for higher data rates and longer lengths of fiber, the system sensitivity to noise increases due to the decrease in eye opening. To enhance the link performance, i.e., to increase the electro-optical repeater (or all-optical amplifier) spacing, and enable error-free signal detection at lower SNR's, FEC techniques play an important role in SMF links. As mentioned before, the

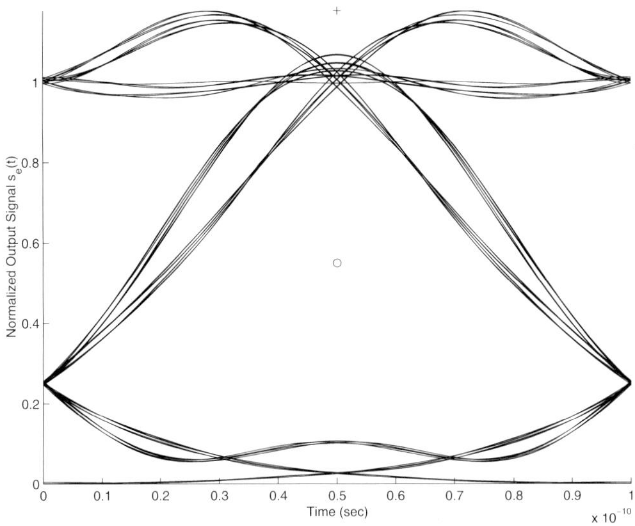

Fig. 5.6. SMF link eye-diagram at 10 Gbps, $L = 50$ km, $\lambda = 1550$ nm (circle shows threshold at optimum sampling time).

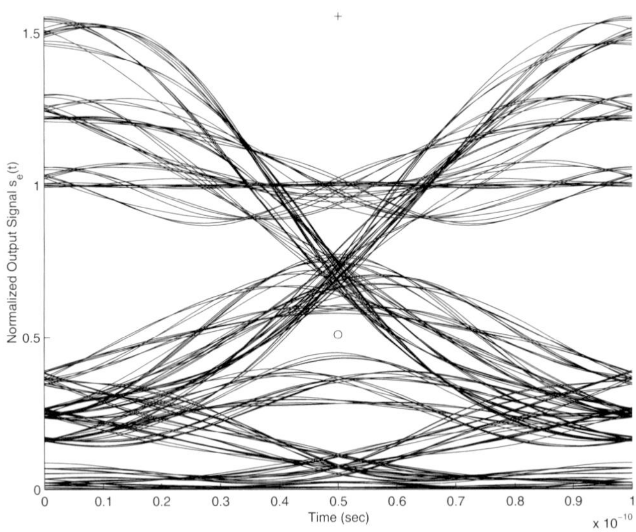

Fig. 5.7. SMF link eye-diagram at 10 Gbps, $L = 120$ km, $\lambda = 1550$ nm (circle shows threshold at optimum sampling time).

Table 5.1. Line rates for various FEC codes.

Information Data Rate (Gbps)	1.25	2.5	5	10	40
RS(255,239)	1.333	2.667	5.334	10.669	42.677
eBCH$(256,247)^2$ TPC	1.342	2.685	5.371	10.742	42.968
eBCH$(128,120)^2$ TPC	1.422	2.844	5.688	11.377	45.511
eBCH$(64,57)^2$ TPC	1.575	3.151	6.303	12.606	50.427

RS(255,239) FEC code is the standard for long-haul SMF links [7]. However, with the increasing demand for denser wavelengths and increased data rates, more advanced coding techniques need to be applied. For this reason, TPC's seem to be the perfect advanced FEC candidates for SMF links, since TPC's provide high performance at a relatively low decoding complexity. Therefore, a performance comparison is needed for FEC codes at various data rates and fiber lengths. In this context, out-of-band codes are considered where the line rate is higher than the information data rate due to the FEC redundancy.

The line rates for various FEC codes are shown in Table 5.1. It can be observed that especially for low code rates, the line rate has to be increased significantly so that the net information data rate is maintained. This has the effect that more ISI is present for low code rate FEC codes.

The simulation results for an SMF link at a data rate of 5 Gbps and length of 200 km are shown in Fig. 5.8. Here, all TPC's are implemented with four iterations and 16 test patterns. It is observed that a TPC coded system can outperform the standard RS(255,239) code by more than 2 dB at BER $= 10^{-6}$. It is also observed that the eBCH$(256,247)^2$ TPC outperforms the eBCH$(128,120)^2$ and eBCH$(64,57)^2$ TPC's. As stated before, this is due to the fact that more ISI is present in out-of-band FEC with low code rates.

The same phenomenon is seen when the fiber length is increased to $L = 300$ km (Fig. 5.9). In this case, the ISI for the low code rate eBCH$(64,57)^2$ TPC is so severe that up to an SNR of 16 dB, the uncoded system performs better.

Similar results are obtained when the data rate is increased to 10 Gbps. Fig. 5.10 and 5.11 show the BER versus SNR curves for $L = 50$ km and $L = 75$ km, respectively.

A close inspection of these figures reveal that the performance curves for 5 Gbps and $L = 200$ km are very close to the curves obtained for 10 Gbps and $L = 50$ km. Similarly, the performance curves for 5 Gbps and $L = 300$ km are almost overlapping with the curves obtained for 10 Gbps and $L = 75$ km. Hence, it can be concluded that doubling the data rate results in about $4\times$ decrease in the fiber reach. However, it is noted that these performance curves depend strongly on the transceiver characteristics, which in simulations is typically approximated by a raised cosine function with 80% rise and fall times.

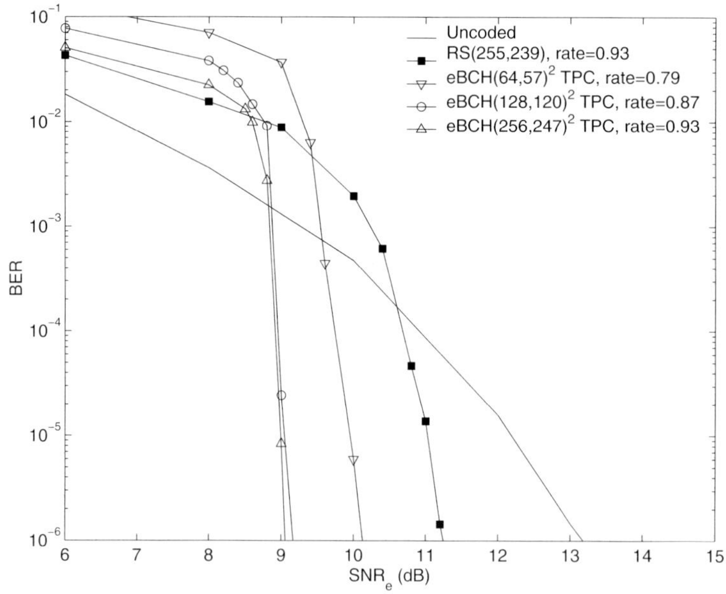

Fig. 5.8. SMF performance (data rate = 5 Gbps, $\lambda = 1550$ nm, $L = 200$ km).

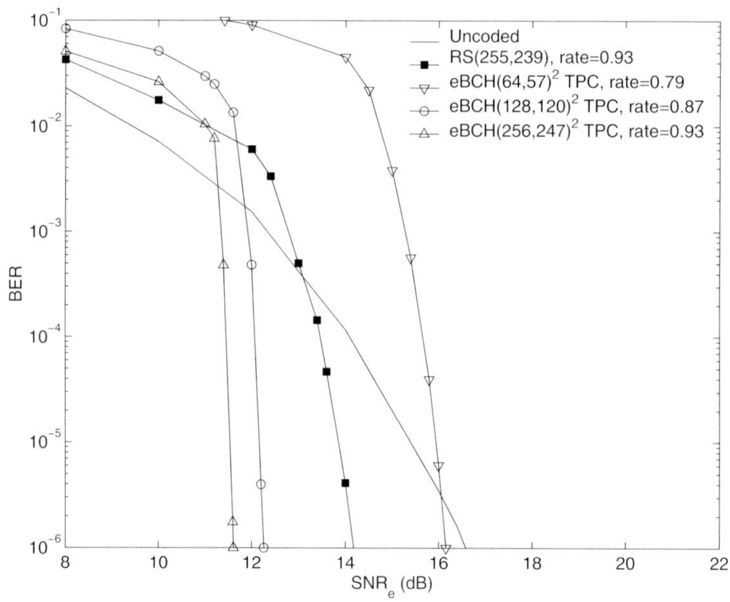

Fig. 5.9. SMF performance (data rate = 5 Gbps, $\lambda = 1550$ nm, $L = 300$ km).

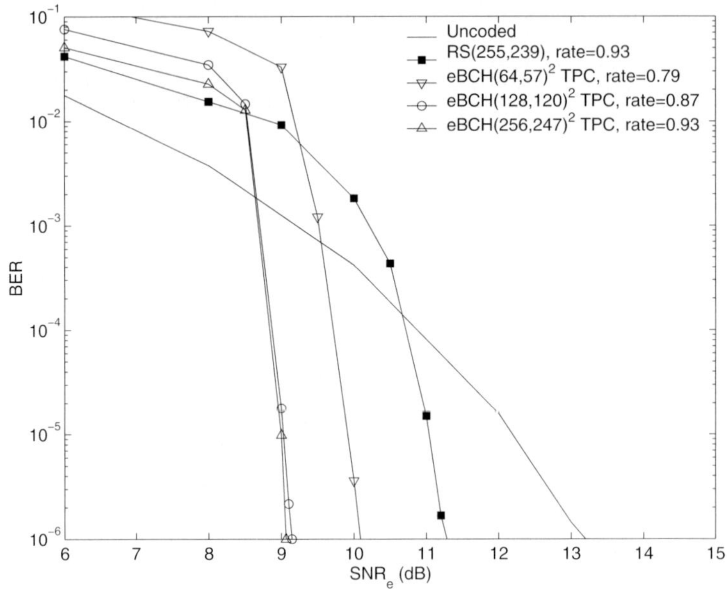

Fig. 5.10. SMF performance (data rate = 10 Gbps, $\lambda = 1550$ nm, $L = 50$ km).

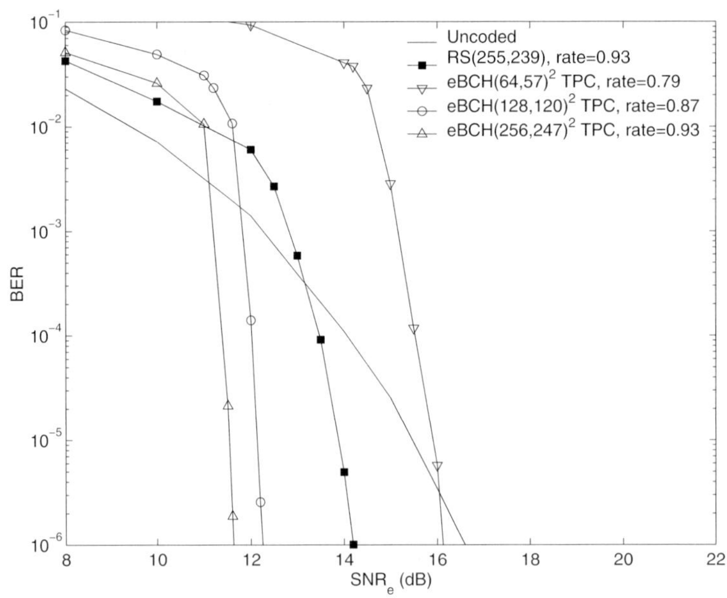

Fig. 5.11. SMF performance (data rate = 10 Gbps, $\lambda = 1550$ nm, $L = 75$ km).

As an overall result, it can be stated that TPC's outperform the standard RS codes for SMF links. This is an expected result, because typically, on an ISI-free AWGN channel, a soft-decision FEC performs about 3 dB better than hard-decision FEC. However, one interesting result obtained is that TPC's with low code rates do not perform as well as expected due to the increase in line rate resulting in strong ISI. To achieve better performances with TPC's with lower code rates, equalization techniques have to be used prior to FEC decoding. One suggestion for equalization for SMF links is to use a decision feedback equalizer (DFE) [5] as outlined by Otte and Rosenkranz [23], who propose a squared-DFE structure that takes the square-law detection at the photodetector into account. Other proposals, such as in [21], treat the overall fiber channel as near-linear, and apply well-known equalization techniques. Critical aspects in applying electrical equalization techniques in high-speed optical communications systems, as it is the case for FEC implementations, are again the design of ADC's and other high-speed circuitry.

5.4 Multimode Fiber Links

The increased demand in local area networks (LAN's) has led to the deployment of Gigabit Ethernet optical LAN's, which will of course lead to deployment of 10 Gigabit Ethernet LAN's. As discussed earlier, Non-LAN (or long-haul) optical links consist of SMF, whereas LAN's are primarily comprised of multimode fiber (MMF). The current installed base of MMF dramatically limits the bandwidth-distance product of the link. Specifically, the different propagation velocities of the distinct optical modes in MMF lead to large differential mode delay (DMD) resulting in intersymbol interference (ISI). The severity of ISI can limit MMF links to bandwidth-distance products of a few 500 MHz-kilometers. Since the ISI in MMF is caused by multiple modes (or rays) traveling inside the fiber, this phenomenon is called intermodal dispersion (IMD). IMD significantly limits the achievable data transmission rates on today's installed base of MMF links.

Looking towards 10 Gigabit Ethernet LANs, there are several possibilities including:

(i) reusing the current installed base of MMF and augmenting the transmission equipment to include new modulation and/or receiver signal processing techniques,

(ii) developing and deploying new MMF with better dispersion properties to upgrade LANs, and

(iii) deploying single-mode fiber in the LAN.

Options (ii) and (iii) are extremely costly because they require deploying new fiber. Given the huge installed base of MMF in buildings and campuses, option (i) is an attractive from an overall cost perspective. In the following we will adopt this strategy and propose our own unique solution. It should be noted that there are number of other novel solutions, most include some kind of signal equalization at the receiver, FEC and possibly alternative modulation schemes. It should be noted that in SMF, FEC is sufficient for getting an increase in coding gain and increased reach. In MMF, ISI is the limiting factor and FEC alone is not sufficient for increasing reach and/or data rate.

To compensate for the ISI, either electrical equalization, optical equalization, or a combination of equalization and FEC techniques need to be applied. For example, Kasper [24] suggests to implement electrical equalization techniques, such as a linear equalizer or a DFE for MMF systems. However, there are also optical domain approaches to increase the bandwidth of MMF. One method is based on exciting only the higher order modes by using an offset launch scheme (OLS) [25] . OLS is basically a technique where light is initially injected into an SMF and then into an MMF. The launch from the SMF into the MMF is performed with an offset of about 14-20 μm from the core center of the MMF, and it is shown that this technique can increase the bandwidth. One major drawback of this technique is the need for a single-mode laser source. Furthermore, splicing of the SMF to the MMF might cause difficulties that presents a controversy to the economics of MMF links. Hence, optical domain equalization is a more attractive method when compared to OLS.

Optical domain compensation of ISI is complicated by the large number of modes (channels) propagating in MMF, and the difficulty in separating the modes at the receiver. However, a simple and effective multisegment detector (MSD) [26] can be employed to obtain diversity information which in turn can be used to compensate for ISI. An MSD is based on spatially resolving the optical energy at the fiber output. In the following we shall show MSD together with FEC can provide a relatively low cost option for augmenting the installed fiber base to handle significantly higher data rates and/or reach.

In general, an MMF link with an MSD can be viewed as a single-input multiple-output (SIMO) system . Diversity combining and equalization for SIMO systems are well studied techniques for multiple antenna receivers employed in wireless communications [27][28] . On the other hand, a novel approach for performance improvement of MMF links, i.e., the combination of an optical domain MSD and FEC techniques is presented in [29]. In particular, the effectiveness of using optical domain equalization and TPC's is demonstrated. Furthermore, an MSD can be efficiently combined with electrical equalization techniques, such as a DFE or a Viterbi detector as investigated in [30], [31], and [32].

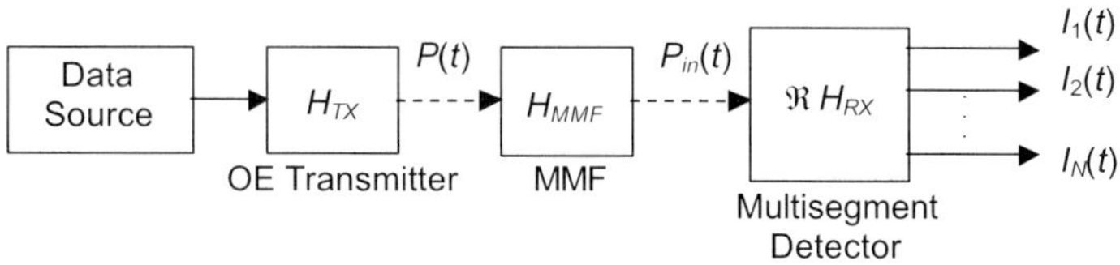

Fig. 5.12. Short-haul MMF link with multisegment detector (MSD) (solid arrows indicate electrical signals, dashed arrows indicate optical signals).

5.4.1 MMF System with MSD

The block diagram of an MMF system with a multisegment detector (MSD) is shown in Fig. 5.12.

At the transmitter side, information bits (or symbols) are fed into an opto-electronic (OE) transmitter (i.e. an LED or a laser with appropriate drive electronics) with transfer function $\mathrm{H}_{\mathrm{TX}}(f)$. The optical output power $P(t)$ of the OE transmitter is then coupled into the short-haul MMF link characterized by transfer function $\mathrm{H}_{\mathrm{MMF}}(f)$. At the receiver side, instead of a standard photodetector with a single segment, it is proposed to use an MSD with $\Re\mathrm{H}_{\mathrm{RX}}(f)$, where $\Re$ denotes, as before, the responsivity of the photodetector. $\mathrm{H}_{\mathrm{RX}}(f)$ may include equalization effects of the MSD to compensate for the ISI. Due to the spatial diversity present in the received optical power $P_{in}(t)$, an N-segment MSD is able to provide N output currents, denoted by $I_m(t)$, where $m = 1, 2, \ldots, N$. Using this diversity information, electrical equalization can be applied later to compensate for the ISI present in the received optical signal.

As stated before, photodetectors are square-law detectors. Therefore, the output current of the mth segment of the MSD is equal to

$$I_m(t) = \Re P_m(t), \tag{5.13}$$

where $P_m(t)$ is the input optical power to the mth detector segment. This fiber channel is treated as a linear system with respect to optical power. This is possible assuming spatially and temporally incoherent interaction among the guided modes in the fiber. It should be noted that this is not only a convenient assumption, but also necessary due to potential concerns with modal noise. Hence, the output power $P_m(t)$ in terms of input power $P(t)$ can be obtained using the convolution integral defined as

$$P_m(t) = P(t) * h_m(t) = \int_{-\infty}^{+\infty} P(\tau) h_m(t - \tau) d\tau, \tag{5.14}$$

where $h_m(t)$ is the impulse response of the MMF link as received by the mth segment. In this system model, the dispersion limited channel is considered

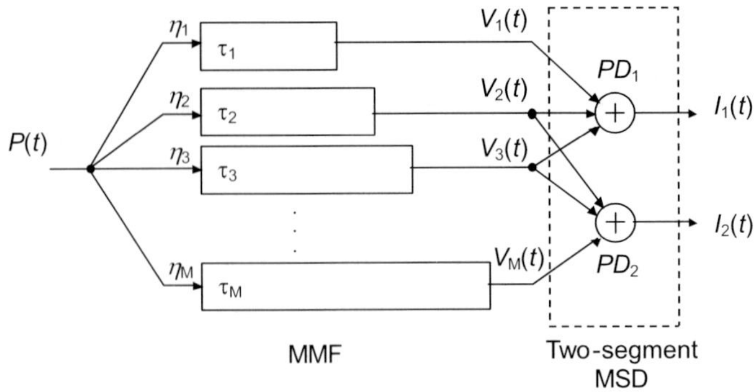

Fig. 5.13. MMF link model with a two-segment MSD.

and it is assumed that white Gaussian noise is added by electronics after photodetection.

An MSD is a very promising alternative to standard single-segment photodetection. The focus in this section is on two-segment multisegment detectors; however, the ideas and principles presented here can be easily generalized to N-segment detectors. To demonstrate the efficiency of an MSD, a two-segment device has been actually designed and fabricated at the Georgia Institute of Technology [26]. This device consists basically of two concentric photodetectors, i.e., an inner segment, and an outer segment. In general, for an N-segment MSD, there would be N concentric photodetectors.

In a typical MMF used for communications, the number of modes, or channels, in the fiber can number in the hundreds. As shown in Fig. 5.13, each received mode power $V_i(t)$ can be represented as the input power $P(t)$ scaled by η_i and delayed by τ_i seconds, i.e.,

$$V_i(t) = \eta_i P(t - \tau_i), \tag{5.15}$$

where $i = 1, \ldots, M$ denotes the mode number, and

$$\sum_{i=1}^{M} \eta_i = 1. \tag{5.16}$$

Here, the assumption is made that the MMF exhibits minimal intermodal coupling, and that the impulse response of each mode is dominated by its group-delay.

In Fig. 5.13, the two concentric segments are denoted by PD_1 (inner segment) and PD_2 (outer segment). In an MMF, modes of propagation can have significant spatial overlap making their isolation and separation difficult. That is, a fraction of mode power $V_i(t)$ can be detected by the first segment, while the remaining fraction can be detected by the second segment. However,

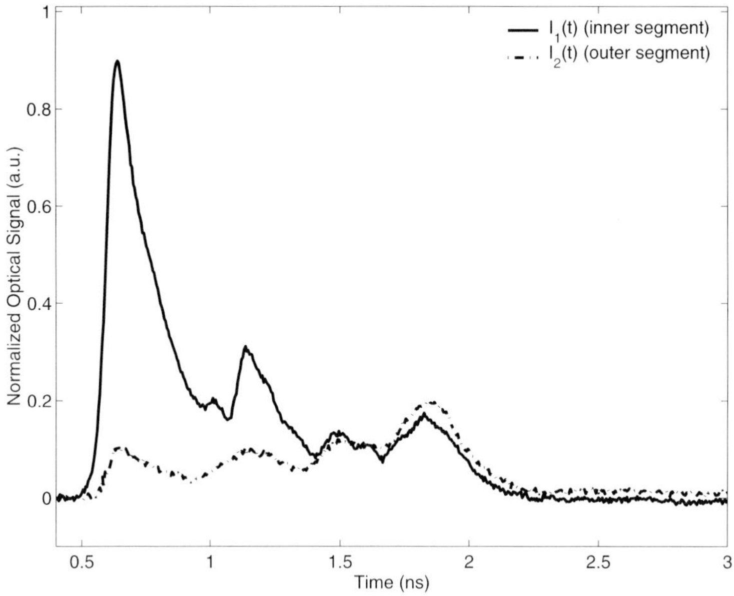

Fig. 5.14. Measured impulse responses for inner and outer segments of two-segment MSD.

despite the large spatial overlap among the fiber modes, there still exists significant diversity in the signals detected by the MSD. In fact, as stated earlier, using an MSD at the MMF output is in some ways comparable to the use of multiple antennas in wireless communications systems.

In the simplest case of spatial equalization, the outer segment signal $I_2(t)$ can be subtracted from the inner segment signal $I_1(t)$ within the MSD via alternating bias polarity. This configuration results in the overall output signal

$$I_{OE}(t) = I_1(t) - I_2(t). \tag{5.17}$$

Throughout the remainder of the following discussions, a two-segment MSD with inherent subtraction of signals is denoted by MSD-i.

In earlier studies [24], Gaussian approximations have been used to model the impulse response of an MMF link. To obtain a more realistic model of the MMF, it is suggested to use measured impulse responses. Fig. 5.14 shows the temporal impulse response[3] for the inner and outer segments of a two-segment MSD. For this measurement, a 1.1-km MMF optical link has been excited with a 1 ps optical pulse at 1550 nm wavelength.

Fig. 5.14 justifies the subtraction of the outer signal from the inner signal in Eq. 5.17 to achieve a narrower overall impulse response. Using this simple

[3] MMF impulse response measurements provided by Ketan M. Patel of Georgia Institute of Technology.

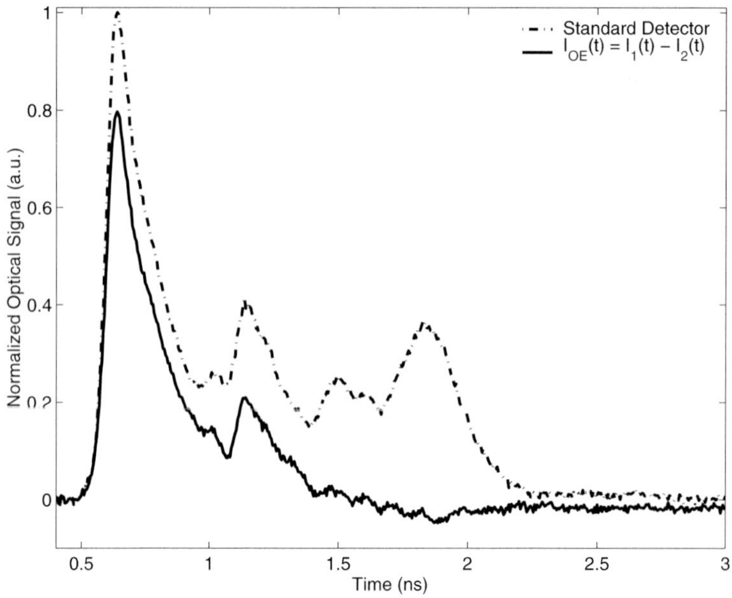

Fig. 5.15. Measured impulse responses for standard photodetection and two-segment MSD-i.

but effective approach, it is possible to compare the simple MSD configuration output to that of a standard single-segment photodetector in Fig. 5.15.

Fig. 5.15 reveals that standard detection is severely affected by DMD. On the other hand, with the two-segment MSD-i, most of the interference is eliminated. It can be shown that a fixed MSD-i geometry can result in a 2× increase in bandwidth as measured by the 3-dB cutoff. Nevertheless, it is also observed in Fig. 5.15 that a power penalty (about 4 dB) is introduced when compared to standard single-segment detection. As will be shown next, TPC's are suitable to reduce the effects of this power penalty.

5.4.2 MSD and TPC for MMF Links

The two-segment MSD-i presented in [26] is a very efficient method to enhance MMF performance and fiber reach. However, as indicated earlier, due to the inherent subtraction of inner and outer signals, an unavoidable power penalty is present in the overall output signal of this device. To recover this power penalty, the combination of this device with FEC codes is considered; in particular, the combination with advanced codes such as TPC's. The system model employing this technique is shown in Fig. 5.16.

The system model considered in Fig. 5.16 is similar to the model used for SMF links, with the exception that there is an MSD-i instead of a single segment photodetector at the receiving end. The optical and electrical SNR's

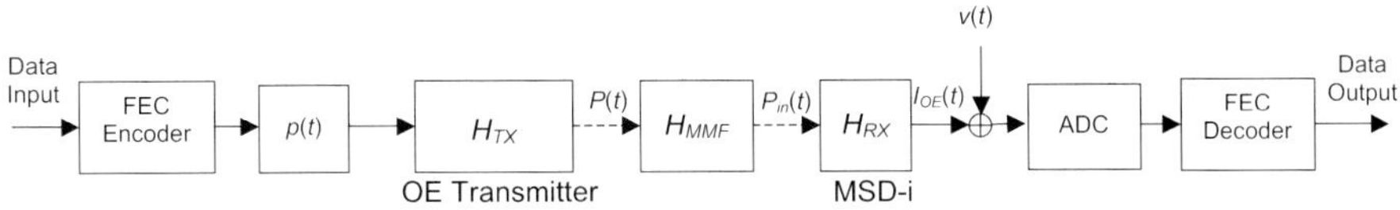

Fig. 5.16. MMF link model with MSD-i and FEC.

are also defined as before. However, in contrast to the SMF case where the frequency response has been approximated by $H_{SMF}(f)$, for the MMF case, measured impulse responses are used for both standard single-segment detection and MSD-i detection. Hence, more realistic models of the MMF link are achieved.

As in the SMF case, the transceiver characteristics $p(t)$ is modeled by a raised cosine function with 80% rise and fall times, and FWHM equal to the bit interval. Furthermore, binary communications with OOK and NRZ are assumed. It should be pointed out that the following results are actually quasi-simulated, because, except the transceiver characteristics and noise model, measured real data (as shown in Fig. 5.14 and 5.15) is used for the impulse responses for standard and MSD-i detection of the MMF link output.

As in the SMF case, out-of-band FEC is considered, i.e., the line rate is increased so that the information data rate is preserved for all codes. This again causes more ISI for codes with lower code rates. First, a 1.1 km MMF link at a data rate of 1.25 Gbps is considered. The obtained BER versus SNR curves are shown in Fig. 5.17.

For an information data rate of 1.25 Gbps, standard detection, and an eBCH$(64,57)^2$ TPC, the line rate has to be increased to about 1.575 Gbps. However, at this line rate, the ISI is too severe. Therefore, standard detection is not applicable and an eBCH$(64,57)^2$ TPC cannot be used at this data rate. On the other hand, an eBCH$(128,120)^2$ TPC and eBCH$(256,247)^2$ TPC are applicable and achieve superior BER levels when compared to the standard RS(255,239) code. Considering system complexity, the number of iterations for TPC decoding is equal to four and the number of test patterns used by the Chase algorithm equal to 16. Similar to the SMF case, the eBCH$(256,247)^2$ TPC outperforms the eBCH$(128,120)^2$ TPC. On the other hand, an MSD-i with no FEC achieves a coding gain of about 7 dB (BER $= 10^{-6}$) when compared to standard detection with no FEC. In fact, MSD-i with no coding outperforms the standard detection and FEC combinations. This shows the effectiveness of this simple spatially resolved equalization technique. To increase the performance level, it is of course possible to combine the MSD-i with FEC codes. Fig. 5.17 shows that the simple RS code with an MSD-i can perform about 3.4 dB better than the MSD-i with no coding. Furthermore, more gain can be achieved when MSD-i is combined with TPC's. Here, all

References

1. J.H. Franz and V.K. Jain, *Optical Communications: Components and Systems.* Narosa Publishing House, 2000.
2. V.W.S. Chan, "Coding and error correction in optical fiber communication systems," in *Optical Fiber Telecommunications IIIA*, ed. I.P. Kaminow and T.L. Koch, Academic Press, 1997, pp. 42-62.
3. T.E. Stern and K. Bala, *Multiwavelength Optical Networks.* Addison-Wesley, 1999.
4. G. Keiser, *Optical Fiber Communications.* 3rd Ed., New York, NY: McGraw-Hill, 2000.
5. J.G. Proakis, *Digital Communications. 3rd Ed.* New York, NY: McGraw-Hill, 1995.
6. C. Berrou, A. Glavieux, and P. Thitimajshima, "Near Shannon limit error-correcting coding and decoding: turbo-codes," in *Proc. IEEE ICC 1993,* (Geneva, Switzerland, May 1993), pp. 1064-1070.
7. International Telecommunication Union, Telecommunication Standardization Sector, "Forward error correction for submarine systems," *Tech. Recommendation* G. 975.
8. O. Aitsab, "FEC techniques in submarine transmission systems," in *Proc. OFC 2001*, pp. TuF1-1 - TuF1-3, 2001
9. A. Schmitt, "Improving optical networks with forward error correction," *Electronic Engineering*, pp. 87-90, March 2000.
10. R. Pyndiah, "Near-optimum decoding of product codes: block turbo codes," *IEEE Trans. Commun.*, vol. 46, no. 8, pp. 1003-1010, Aug. 1998.
11. C. Argon and S.W. McLaughlin, "Efficient decoding of turbo product codes," in *Proc. 39th Annual Allerton Conference on Communications, Control and Computing*, (Monticello, Illinois, USA, Oct. 2001).
12. C. Argon and S.W. McLaughlin, "An efficient Chase decoder for turbo product codes," *IEEE Trans. Commun.*, vol. 52, no. 6, pp. 896-898, June 2004.
13. S.A. Hirst, B. Honary, and G. Markarian, "Fast Chase algorithm with an application in turbo decoding," *IEEE Trans. Commun.*, vol. 49, no. 10, pp. 1693-1699, Oct. 2001.
14. S. Dave, J. Kim, and S.C. Kwatra, "An efficient decoding algorithm for block turbo codes," *IEEE Trans. Commun.*, vol. 49, no. 1, pp. 41-46, Jan. 2001.
15. D. Chase, "A class of algorithms for decoding block codes with channel measurement information," *IEEE Trans. Inform. Theory*, vol. IT-18, no. 1, pp. 170-182, Jan. 1972.
16. J. Hagenauer, E. Offer, and L. Papke, "Iterative decoding of binary block and convolutional codes," *IEEE Trans. Inform. Theory*, vol. 42, pp. 429-445, Mar. 1996.
17. E. Dunkor, P.D. Kumavor, S. Summers, M. Hayduk, and R. Bussjager, "A 5 gigabit/sec all-optical parallel analog-to-digital converter," in *Proc. IEEE CLEO 2001*, (Baltimore, MD, May 2001), pp. 182-183.
18. C. Pala, L. Thylen, M. Mokhtari, and U. Westergren, "A high-speed electro-optical analog-to-digital converter principle," in *Proc. IEEE ISCAS 2001*, (Sydney, NSW, Australia, May 2001), pp. 432-435.
19. L. Brzozowski and E.H. Sargent, "All-optical analog-to-digital converters, hardlimiters, and logic gates," *IEEE J. Lightw. Techn.*, vol. 19, no.1, pp. 114-119, Jan. 2001.

20. A. Banihashemi and S. Hemati, "Decoding in Optics," in *Proc. IEEE ISIT 2002*, (Lausanne, Switzerland, July 2002), pp.231.
21. J.H. Winters and R.D. Gitlin, "Electrical signal processing techniques in long-haul fiber-optic systems," *IEEE Trans. Commun.*, vol. 38, no. 9, pp. 1439-1453, Sept. 1990.
22. H. Bulow, D. Schlump, J. Weber, B. Wedding, and R. Heidemann, "Electronic equalization of fiber PMD-induced distortion at 10 Gbit/s," in *Proc. OFC '98*, 1998, pp. 151-152.
23. S. Otte and W. Rosenkranz, "A decision feedback equalizer for dispersion compensation in high speed optical transmission systems," in *Proc. IEEE ICTON '99*, 1999, pp. 19-22.
24. B.L. Kasper, "Equalization of multimode optical fiber systems," Bell Syst. Tech. J., vol. 61, pp. 1367-1388, Sept 1982.
25. L. Raddatz, I.H. White, D.G. Cunningham, and M.C. Nowell, "Influence of restricted mode excitation on bandwidth of multimode fiber links," *IEEE Photon. Tech. Lett.*, vol. 10, no. 4, pp. 534-536, April 1998.
26. K.M. Patel and S.E. Ralph, "Enhanced multimode fiber link performance using a spatially resolved receiver," *IEEE Photon. Tech. Lett.*, vol. 14, no. 3, pp. 393-395, Mar. 2002
27. P. Balaban and J. Salz, "Optimum diversity combining and equalization in digital data transmission with applications to cellular mobile radio - Part I: Theoretical considerations," *IEEE Trans. Commun.*, vol. 40, no. 5, pp. 885-894, May 1992.
28. Y. Li and Z. Ding, "A simplified approach to optimum diversity combining and equalization in digital data transmission," *IEEE Trans. Commun.*, vol. 43, no. 8, pp. 2285-2288, Aug. 1995.
29. C. Argon, K.M. Patel, S.W. McLaughlin, and S.E. Ralph, "Spatially resolved equalization and forward error correction for multimode fiber links," in *Proc. IEEE ICC 2002*, (New York, NY, USA, April 2002).
30. C. Argon, K.M. Patel, S.W. McLaughlin, and S.E. Ralph, "Exploiting diversity in multimode fiber communications links via multisegment detectors and equalization," *IEEE Commun. Letters*, vol. 7, no.8, pp. 400-402, Aug. 2003.
31. C. Argon, K.M. Patel, S.W. McLaughlin, and S.E. Ralph, "Spatially resolved equalization and decision feedback equalization for multimode fiber links," in *Proc. IEEE/LEOS Summer Topical Meetings 2002*, (Mont Tremblant, Quebec, Canada, July 2002).
32. S.E. Ralph, K.M. Patel, C. Argon, A. Polley, and S.W. McLaughlin, "Intelligent receivers for multimode fiber: Optical and electronic equalization of differential modal delay," in Proc. *IEEE/LEOS Annual Meeting 2002*.

Part III

Wireless Communications

Chapter 6

Iterative Demodulation and Decoding

Christian Schlegel

University of Alberta, Canada

6.1 Information Theoretic Communications

6.1.1 The Shannon Capacity

Among the many contributions of Claude Elwood Shannon, the channel coding theorem is probably one of the most far reaching results [1]. Shannon posed the communications problem as that of "reproducing at a point, either exactly or approximately, a message chosen at another point". The theoretical solution of this problem led to his celebrated source-channel coding theorem [2]. This theorem essentially establishes an upper bound on the maximum transmission rate at which digital data can be moved between two points such that the original message can be restored reliably. This maximum transmission rate is the capacity of the channel, and depends on many channel properties, foremost its bandwidth and received average signal power-to-noise power ratio.

Transmission above the channel capacity is theoretically impossible without the loss of a reliable message reconstruction at the receiver. The theorem also proves that transmission right up to capacity is possible with arbitrary reliability. However, the proof methodology used then and now is one of averages using the law of large numbers. This proof gives no descriptive way of constructing good communications systems.

However, the general structure of the solution is known and requires the use of error control codes. Fig. 6.1 shows the basic configuration of a point-to-point digital communications link using error control coding. Digital data to be transmitted over this link typically consists of a string of binary symbols, ones and zeros. These symbols enter the *encoder/modulator* whose function it is to condition them for transmission over the channel. The encoder accepts the input digital data and introduces controlled redundancy. The modulator converts discrete symbols from the encoder into waveforms which are suitable for transmission through the channel. On the receiver side the demodulator reconverts the waveforms back into a discrete sequence of received symbols,

K. Sripmanwat (ed.), Turbo Code Applications: a journey from a paper to realization, 123–155.

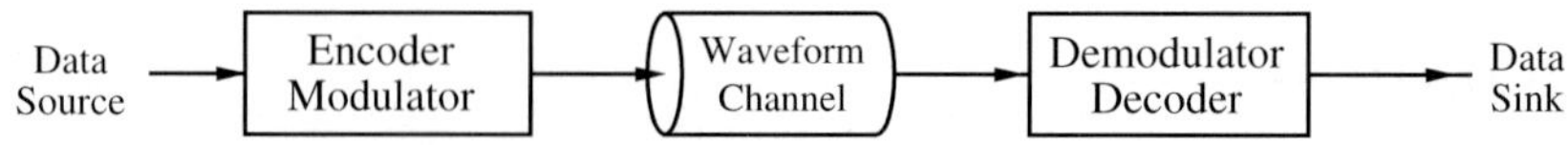

Fig. 6.1. Capacity achieving system for a point-to-point communications link.

and the decoder reproduces an estimate of the digital input data sequence, which is forwarded to the data sink. The purpose of the error control functions is to maximize data reliability when transmitted over an unreliable channel.

The most fundamental channel is the *additive white Gaussian noise* (AWGN) channel, whose only impairment is the addition of Gaussian noise to the transmitted signal. This channel model is as simple as it profound. Without the presence of noise, the communications problem would not exist, since one could reconstruct a signal arbitrarily accurately after transmission, and hence, in theory, transmit an infinite amount of information with a finite amount of resources.

For the AWGN channel with a limited bandwidth W, Shannon's capacity evaluates to

$$C = W \log_2 (1 + S/N) \quad \text{[bits/second]}. \tag{6.1}$$

In this formula C is the channel capacity, measured in bits per second, W is the bandwidth of the channel, and S/N is the signal-to-noise power ratio at the receiver.

6.1.2 Spectral and Power Efficiency

The parameter which characterizes how efficiently a system uses its allotted bandwidth is the *spectral efficiency* η, defined as

$$\eta = \frac{\text{Bit Rate}}{\text{Channel Bandwidth } W} \quad \text{[bits/}s\text{/Hz]}. \tag{6.2}$$

Using (6.1) and dividing by W we obtain the maximum spectral efficiency for an additive white Gaussian noise channel, the *Shannon limit*, as

$$\eta_{\text{max}} = \log_2 \left(1 + \frac{S}{N}\right) \quad [\text{bits}/s/Hz]. \tag{6.3}$$

To calculate η we must suitably define the channel bandwidth W. One commonly used definition is the 99% bandwidth definition, i.e., W is defined such that 99% of the transmitted signal power falls within the band of width W. This 99% bandwidth corresponds to an out-of-band power of -20dB.

The average signal power S can be expressed as

$$S = \frac{kE_b}{T} = RE_b, \tag{6.4}$$

where E_b is the energy per received bit, k is the number of bits transmitted per symbol, and T is the duration of that symbol. The parameter $R = k/T$ is the transmission rate of the system in bits/s. Rewriting the signal-to-noise power ratio S/N, where $N = WN_0$, i.e., total noise power simply equals the *noise power spectral density* N_0 multiplied by the width of the transmission band, we obtain

$$\eta_{\max} = \log_2\left(1 + \frac{RE_b}{WN_0}\right) = \log_2\left(1 + \eta\frac{E_b}{N_0}\right). \tag{6.5}$$

Since $R/W = \eta_{\max}$ is the limiting spectral efficiency, we obtain a bound from (6.5) on the minimum bit energy required for reliable transmission, given by

$$\frac{E_b}{N_0} \geq \frac{2^{\eta_{\max}} - 1}{\eta_{\max}}, \tag{6.6}$$

also called the *Shannon bound.*

In the limit as we allow the signal to occupy an infinite amount of bandwith, i.e., $\eta_{\max} \to 0$, we obtain

$$\frac{E_b}{N_0} \geq \lim_{\eta_{\max}\to 0} \frac{2^{\eta_{\max}} - 1}{\eta_{\max}} = \ln(2) = -1.59\text{dB}, \tag{6.7}$$

the minimum bit energy to noise power spectral density required for reliable transmission. It is remarkable that such a limit exists, stating that as soon as the signal-to-noise ratio exceeds (6.7), completely reliable communication is possible.

6.1.3 Discrete-Time Communications

In order to relate the continuous communications problem to a time-discrete equivalent problem, we apply a fundamental signal decomposition theorem. Nyquist showed in 1928 [3] that a channel of bandwidth W (in Hz) is capable of supporting $2W$ independent signal dimensions per second. If two carriers ($\sin(2\pi f_c)$ and $\cos(2\pi f_c)$) are used in quadrature, as in double side-band suppressed carrier (DSB-SC) amplitude modulation [4], we alternatively have W pairs of dimensions (or complex dimensions) per second, leading to the popular QAM constellations, represented by points in 2-dimensional space. Some popular constellations for DSB-SC modulation are shown in Fig. 6.2.

With these assumptions, the equivalent discrete channel is a memoryless channel given by

$$r_i = c_i + z_i \tag{6.8}$$

where c_i is a complex signal point (Fig. 6.2), z_i is a complex Gaussian random variable with variance $N_0/2$ in each independent component, and r_i is the received complex symbol at time i. The channel is memoryless, i.e., there is no correlation between n_i and $n_{i+j}, j \neq i$. (For more details see [4, 5]).

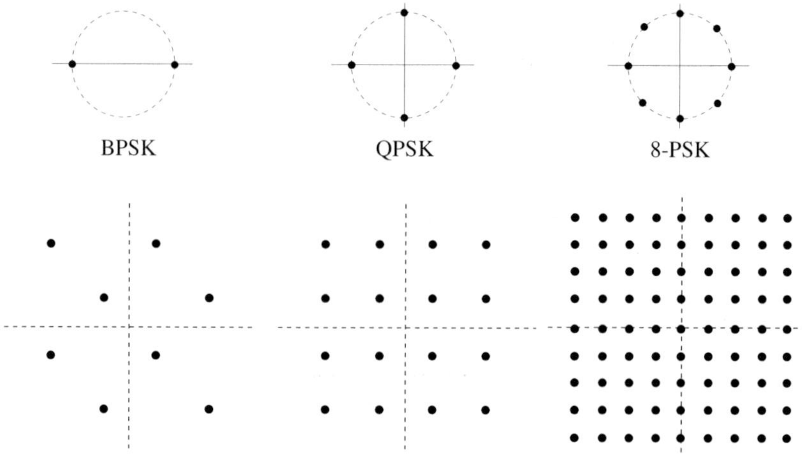

Fig. 6.2. Popular 2-dimensional signal constellations used for digital radio systems.

Fig. 6.3 shows the power and bandwidth efficiencies of some popular uncoded quadrature constellations as well as that of a number of coded transmission schemes, comparing them to the Shannon bound. Preferably, one would want to operate a system in the top left hand corner of this graph, but the Shannon bound draws a sharp line between what is and what is not possible. As a performance point approaches the Shannon bound, more resources in terms of computational complexity have to be expended. However, the figure clearly demonstrates the advantages of error control coding.

6.1.4 Low-Density Parity-Check and Turbo Codes

As we mentioned above, the theoretical capacity of a channel is approached by strong error control coding methods. Among the most accomplished such methods we find the *low-density parity-check* (LDPC) codes, proposed by Gallager in 1962 [6], and the *turbo codes* proposed by Berrou in 1993 [7, 8]. Shortly after their discovery, LDPC codes were forgotten, then rediscovered [9, 10] in the wake of the storm that swept the communications field after the discovery of turbo codes [11]. Nowadays both codes are known as graph-based codes, which includes a variety of powerful codes that approach the Shannon capacity [4]. With the discovery of turbo codes, a novel decoding method, iterative graph-based decoding pioneered by Tanner [12], became fashionable. Iterative decoding of graph-based codes is today considered to be the "magic bullet" of error control coding, since it allows the Shannon capacity to be approached virtually arbitrarily closely. The current record is

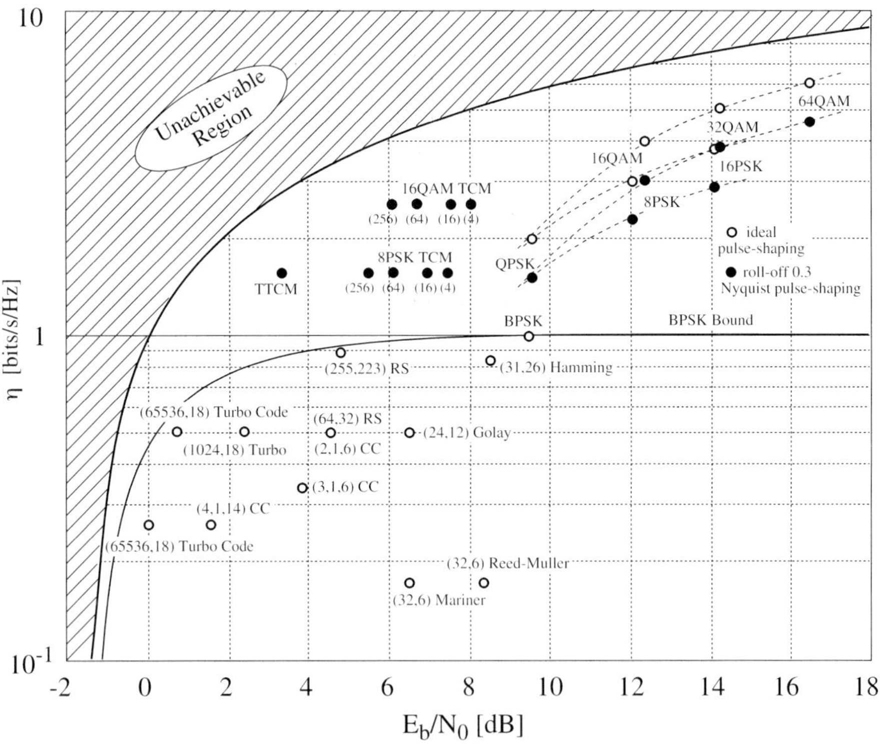

Fig. 6.3. Spectral and power efficiencies achieved by various coded and uncoded transmission methods. The open circles assume ideal Nyquist-rate transmission, while the solid circles are for a more practical Nyquist pulse with excess bandwidth of 0.3 (see [4]).

held by an LDPC code that obtains an error rate of below 10^{-6} at an SNR of only 0.04dB above capacity [13].

6.2 Large-Constellation Channels

6.2.1 The Demodulation Problem

A close inspection of Fig. 6.3 shows quickly that at higher values of the signal-top-noise ratio large constellations are required in order to exploit a channel's capacity. At 20dB a single channel can support 10 bits per complex dimension. The signal constellation therefore contains 1024 signal points. In the demodulator, a reliability measure for each of the symbols has to be generated. We call this, somewhat inappropriately, demodulation of a complex received signal point. In its direct form then, a demodulator generates the reliability measure $m_i = |r_i - s_i|^2$ for each signal point s_i. This measure is directly related to the probability that signal point s_i was transmitted, and is used by the error control decoder. As long as the signal points are nicely arranged on a rectangular grid, as in Fig. 6.2 this task is relatively simple to accomplish. Even with points on a more general lattice, efficient demodulation structures can be found [4].

However, in many circumstances, the channel itself will correlate transmitted signal points and generate a constellation with very difficult disorderly geometrical arrangements. In this case it is often not possible to generate the list of reliabilities m_i in an efficient manner. In other words, the generation of these lists would require the computation of all distances $|r_i - s_i|^2$. Clearly, if the constellation becomes large, this enumeration becomes computationally impossible. Channels that fall into this category are the *code-division multiple access* (CDMA) channel and the multiple antenna channel, both discussed below. Another example is the multi-wire crosstalk channel that arises in high-speed chip-to-chip communications.

The channels mentioned above are all linear channels, composed of linear passive elements. We therefore model a general *multi-input multi-output* (MIMO) channel by the following complex linear algebraic equation

$$\mathbf{r}_i = \mathbf{R}\boldsymbol{c}_i + \mathbf{z}_i, \tag{6.9}$$

where $\mathbf{r} = (r_{1i}, \cdots, r_{Ni})$ is the N-dimensional received complex vector at time i, $\boldsymbol{s} = (s_{1i}, \cdots, s_{Mi})$ is the M-dimensional vector of complex transmit symbols, $\mathbf{R}$ is the $N \times M$ matrix of correlation gains. The transmission is embedded in noise, and $\boldsymbol{n}$ is a N-dimensional complex Gaussian noise vector with variance N_0 for each complex entry.

The Shannon capacity of this channel can be calculated by applying the multi-dimensional capacity formula [2], and is given by

$$C = \log_2 \left[\det \left(\boldsymbol{I}_N + \frac{E_s}{N_0 M} \mathbf{R}\mathbf{R}^+ \right) \right]; \quad \text{[bits/channel use]}, \tag{6.10}$$

which is a generalization of the capacity formula (Eq. 6.1). E_s is the energy of the entire M-dimensional symbol, and $\boldsymbol{I}_N$ is the identity matrix of dimension N.

If the channel $\mathbf{R}$ is known at the transmitter, a reductionist approach can be used via decomposition into the eigenmodes of the channel. This is accomplished by applying the singular-value decomposition (SVD) [14] to $\mathbf{R}$ to obtain

$$\mathbf{R} = \boldsymbol{U}\mathbf{D}\boldsymbol{V}^+ \tag{6.11}$$

where both $\boldsymbol{U}$ and $\boldsymbol{V}$ are unitary matrices of size $N \times N$ and $M \times M$, and therefore invertible, *i.e.*, $\boldsymbol{U}^{-1} = \boldsymbol{U}^+$. The beauty of the SVD is that the matrix $\mathbf{D}$ is diagonal and contains the *singular values* $d_1 \geq \cdots \geq d_n \geq 0; n = \min(M, N)$ of $\mathbf{R}$, which are the square roots of the non-zero eigenvalues of $\mathbf{R}\mathbf{R}^+$.

Simple manipulations now lead to a decomposed equivalent channel form:

$$\begin{aligned}\mathbf{r} &= \mathbf{R}\boldsymbol{c} + \mathbf{z} \\ &= \boldsymbol{U}\mathbf{D}\boldsymbol{V}^{+}\boldsymbol{c} + \mathbf{z} \\ \tilde{\boldsymbol{y}} = \boldsymbol{U}^{+}\mathbf{r} &= \mathbf{D}\tilde{\mathbf{c}} + \tilde{\mathbf{z}}.\end{aligned} \tag{6.12}$$

Due to the unitary nature of $\boldsymbol{U}$, $\mathbf{z}$ and $\tilde{\mathbf{z}}$ have the same statistics.

Eq. 6.12 is now simply an aggregate of $n = \min(M, N)$ parallel channels. Each channel has power d_j^2, corresponding to the j-th singular value of $\mathbf{R}$. That is, simple linear pre-, and postprocessing by multiplying the transmitted and received symbol vectors by unitary matrices decomposes the channel into parallel channels, and conventional coding techniques can be used on these individual channels. The real problem arises when the channel is not known at the transmitter, which is often the case.

6.2.2 The Code-Division Multiple Access (CDMA) Channel

CDMA as an accessing method [15, 16] grew out of the military's direct sequence spread spectrum program. Spread spectrum signals have a number of desirable advantages, such as resistance to interference, inherent frequency diversity, and low spectral power levels which lets them co-exist with other transmission in the same frequency band. Direct sequence spreading of a given narrow-band signals is accomplished simply by multiplying the signal with a large-bandwidth *signature sequence*, which is typically a sequence of rapid and random positive and negative pulses. The application of spread spectrum techniques to multiple access channels is relatively recent, and creates a very interesting large-constellation channel.

The basic model of asynchronous CDMA is shown in Fig. 6.4. K independent transmitters generate binary signal vectors $\boldsymbol{u}_k$, which are encoded by K error control encoders each with code rate R. Equal rates are chosen for model simiplicity. Random interleavers may separate the error control encoders from the spreading operation, where each signal stream is multiplied by the user individual signature sequence $a_{k,l}(t)$, which may vary with each symbol l, as well as with each user. The particular properties of these signature sequences shall not interest us here, other than that we assume that they behave like completely random sequences, an assumption that can be supported well by practical sequences used [15]. We also assume that one-dimensional BPSK modulation is chosen as the modulation format, but extensions are quite straightforward.

The signal from the kth spreader is given by

$$c_k(t) = \sum_{l=0}^{L-1} \sqrt{P_k} d_{k,l} a_{k,l}(t - lT - \tau_k) \tag{6.13}$$

where L is the number code symbols per user per frame, $d_{k,l}$ is the l-th symbol in the data stream of user, and P_k is the power of user k. Each signal may

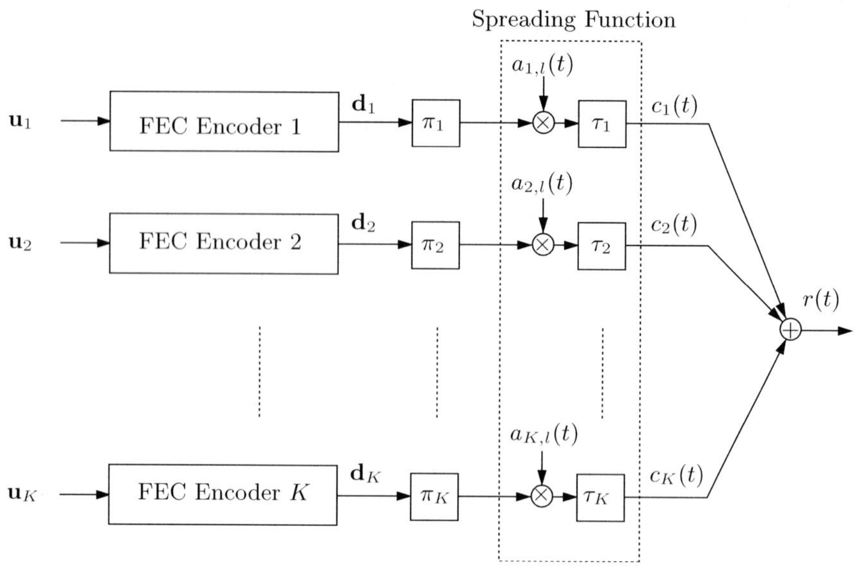

Fig. 6.4. Asynchronous CDMA with error control system model. The spreading sequences randomly change for each symbol period l.

experience a random delay τ_k as it passes through the channel, causing the asynchronicity of system. The spreading waveform of user k during symbol time l, $a_{k,l}(t)$ is supported on the interval $[0, T]$ and may be a random pulse sequence as illustrated in Fig. 6.5.

In this paper we consider an AWGN channel multiple access channel and the received signal is

$$r(t) = \sum_{k=1}^{K} c_k(t) + z(t), \tag{6.14}$$

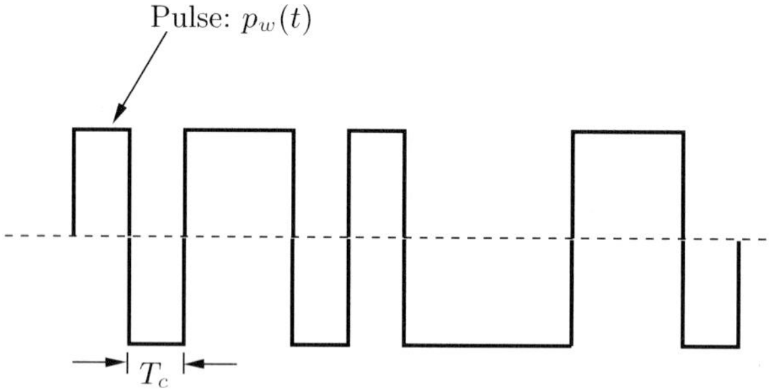

The signature sequence is the pulse train:

$$a_{k,l}(t) = \sum_{n=0}^{N-1} a_{k,l,n} p_w(t - nT_c)$$

Fig. 6.5. Example signature sequence using random binary pulses.

where $z(t)$ is zero mean white Gaussian noise with double-sided noise power spectral density $\sigma^2 = N_0/2$.

Assuming that timing and phase references have been established, the received signal sampled by chip matched filters can be written in the convenient discrete chip-based matrix model

$$\mathbf{r} = \mathbf{RAd} + \mathbf{z} \tag{6.15}$$

where $\mathbf{z}$ is an $(L+1)N$ vector of sampled white noise with variance σ^2, $\mathbf{R}$ is an $(L+1)N \times LK$ matrix whose j-th column is $\mathbf{a}_{k,l} = [\mathbf{0}_{lN+\tau_k/T_c}, a_{k,l,0}, \cdots, a_{k,l,N-1}, \mathbf{0}_{(L-l)N-\tau_k/T_c}]^T$; where $l = \lfloor j/K \rfloor$ is the symbol time index, $\mathbf{0}_i$ is a length-i all-zero column vector, $\mathbf{d} = [d_{1,0}, \ldots, d_{K,0}, d_{1,1}, \ldots, d_{K,L-1}]^T$ is the KL-sized vector of encoded (BPSK) symbols, and $\mathbf{A}$ is a diagonal matrix containing the amplitudes $\sqrt{P_k}$ of each user.

6.2.3 The Multiple Antenna Channel

Another popular representative of a large-constellation channel is the multiple antenna channel. This wireless radio channel uses an array of transmit and an array of receive antennas shown in Fig. 6.6.

Each transmit antenna is fed with a DSB-SC modulated signal, represented by a complex number, and therefore each channel from transmit antenna i to receive antenna j is conveniently described by a complex *path gain* h_{ij}. This constitutes the most simple model which does not suffer from intersymbol interference, given that the symbol rate is slower than the channel dispersion [17]. The composite channel can again be described by the complex vector equation

$$\mathbf{r}_i = \mathbf{R}\boldsymbol{c}_i + \mathbf{z}_i, \tag{6.16}$$

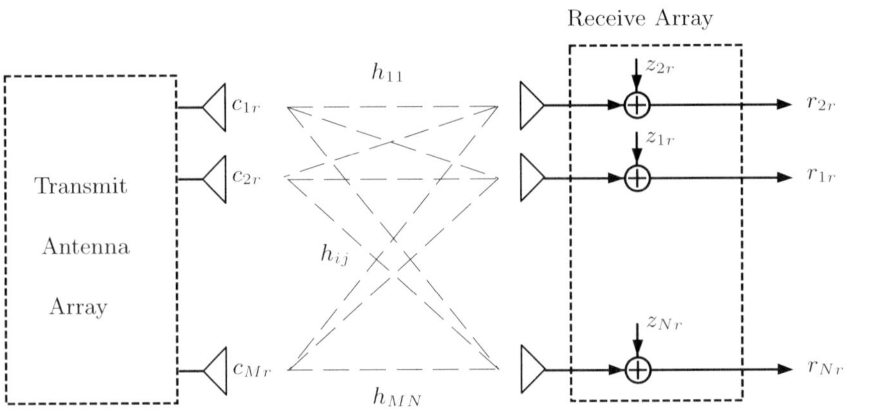

Fig. 6.6. Multiple antenna wireless transmission creates a multiple-input multiple-output (MIMO) channel.

where $\boldsymbol{c}_i = (c_{1i}, \cdots, c_{Mi})$ is the M-dimensional vector of complex transmit symbols, called the *space-time symbol.*

The capacity of this wireless channel has been studied in a number of recent papers [18–20] and analyzed as a function of the transmission environment [21, 22]. Under favorable conditions it is possible that this channel increases the Shannon capacity by a factor of $n = \min(M, N)$, i.e., essentially by the number of antennas in the array. Measurements in real indoor and outdoor environments confirm these conclusions [23–25].

The multiple antenna and CDMA channels are formally equivalent, however, in practice there are some important differences. The multiple antenna channel is synchronous, since all transmissions originate at the same location. It is also possible to jointly code over the different signal streams, while in the CDMA channel only independent individual coding is practical. Interestingly, these advantages of the multiple antenna channel do not produce a higher information theoretic capacity, which is given in both cases by Eq. 6.10.

In many applications the channel gains h_{ij} are subject to fading, modeled by Rayleigh fading [17]. For the case where these channels are uncorrelated, Telatar [18] has calculated the capacity by averaging Eq. 6.10 over the random entries of $\mathbf{R}$ using random matrix theory arguments. In the case of uncorrelated fading, the primary conclusion that capacity grows with n remains valid. Severe correlation among the channel path h_{ij} will degrade this linear capacity growth. More detailed information on the information theoretic behavior of this channel, as well as the CDMA channel, can also be found in [26].

6.3 Layering of Large-Constellation Channels

6.3.1 Back to Single-Stream Channels

If the cardinality of the large-constellation channel (Eq. 6.9) is large, direct demodulation becomes infeasible due to the large number of signal points. For example, a multiple antenna channel with 10 transmit antennas using 8-PSK on each will generate $2^{30} = 1,073,741,824$ signal vectors $\boldsymbol{c}_i$. A maximum-likelihood search through all of these is going to dominate the complexity of the receiver, and, in fact, is quite unnecessary as we will demonstrate below.

The generic approach to solving this demodulation problem is called *signal layering*, first proposed for the multiple antenna channel by Foschini [19, 27]. The basic idea is to decouple the channel by judicious processing into parallel or layered single-stream channels, each with smaller constellation sizes.

Using a matched filter bank on the received signal $\mathbf{r}_i$ generates a *sufficient statistic* for the detection of the symbols $\boldsymbol{c}_i$, that is,

$$\mathbf{r}_{\mathrm{MF},i} = \mathbf{R}^*\mathbf{r}_i = \mathbf{R}^*\mathbf{R}\mathbf{A}\boldsymbol{c}_i + \mathbf{R}^*\mathbf{z}_i = \mathbf{C}\mathbf{A}\boldsymbol{c}_i + \tilde{\mathbf{z}}_i, \tag{6.17}$$

where the noise $\tilde{\mathbf{z}}$ is multi-variate, zero-mean Gaussian with correlation $E\left[\tilde{\mathbf{z}}\tilde{\mathbf{z}}^*\right] = \mathbf{C}$. Invertible linear preprocessing does not diminish the capacity of

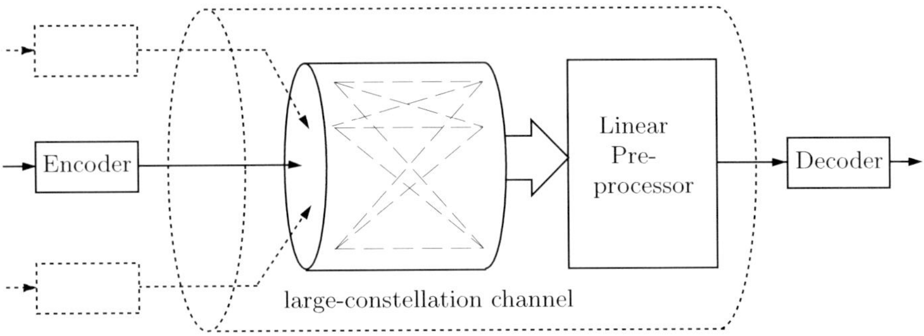

Fig. 6.7. Reduction of a large-constellation channel by linear processing.

the original channel, but conditions the channel for improved single-stream detection. The goal is to create single-stream channels whose aggregate capacity is as large as possible, ideally identical to that of the original channel. This principle is shown in Fig. 6.7.

Appropriate encoding and decoding operations can now efficiently exploit the capacity of this layered channel. Hence it is very important to understand the capacity of the layered channels.

6.3.2 The Zero-Forcing Filter (Decorrelation)

The conceptually simplest way to layer the channel is to simply invert the channel matrix $\mathbf{C}$, or equivalently to pseudo-invert $\mathbf{R}$ in Eq. 6.17. If the columns of $\mathbf{R}$ are linearly independent, then the correlation matrix $\mathbf{C}$ is invertible and we proceed to calculate

$$\tilde{\boldsymbol{c}}_i = \mathbf{C}^{-1}\mathbf{r}_{\mathrm{MF},i} = \boldsymbol{c}_i + \mathbf{C}^{-1}\mathbf{R}^*\mathbf{z}_i \tag{6.18}$$

The elements of $\tilde{\boldsymbol{c}}_i$ are an estimate for $\mathbf{A}\boldsymbol{c}_i$. If the symbols $\boldsymbol{c}$ are antipodal binary, a decision can be made by a sign detector, also known as the *decorrelator* [28]:

$$\boldsymbol{c}_{\mathrm{DEC},i} = \mathrm{sign}\left(\mathbf{C}^{-1}\mathbf{R}^*\mathbf{r}_i\right). \tag{6.19}$$

A fairly sophisticated analysis assuming random signature sequences and signal projection [29], or using random matrix theory [30] as in [31], can be used to show that the per-dimension capacity of the zero-forcing receiver with random matrix entries is given by

$$C_{\mathrm{dec}} = \frac{\beta}{2}\log\left(1 + 2\frac{RE_b}{N_0}\left(1 - \beta + \frac{1}{N}\right)\right) \quad \text{bits/dimension} \tag{6.20}$$

In this equation the load $\beta = K/N$ is the ratio of users or data streams to number of complex received dimensions. A system with $\beta > 1$ is called *oversaturated.* In this cased the decorrelator equation becomes indeterminate since $\mathbf{C}$ has no inverse.

Eq. 6.20 in turn can be converted into a Shannon Bound for the zero-forced layered sub-channel given by

$$\frac{E_b}{N_0} \geq \frac{2^{2C/\beta} - 1}{2C(1-\beta)}, \tag{6.21}$$

and shown in Fig. 6.8 for a few different loads with comparisons to a channel without interference, i.e., the AWGN Shannon bound (Eq. 6.6). Optimal loads are $0.5 \leq \beta \leq 0.75$ for most of the range of interest. It is instructive to note that there is a significant loss in capacity that has to be paid as a prize for the simplicity of the layering.

6.3.3 Minimum-Mean Square Error (MMSE) Layering

The MMSE detector is an estimation theory concept transplanted to the field of detection. Functionally it is completely analogous to the decorrelator, and the sub-channels are created by inversion also, as

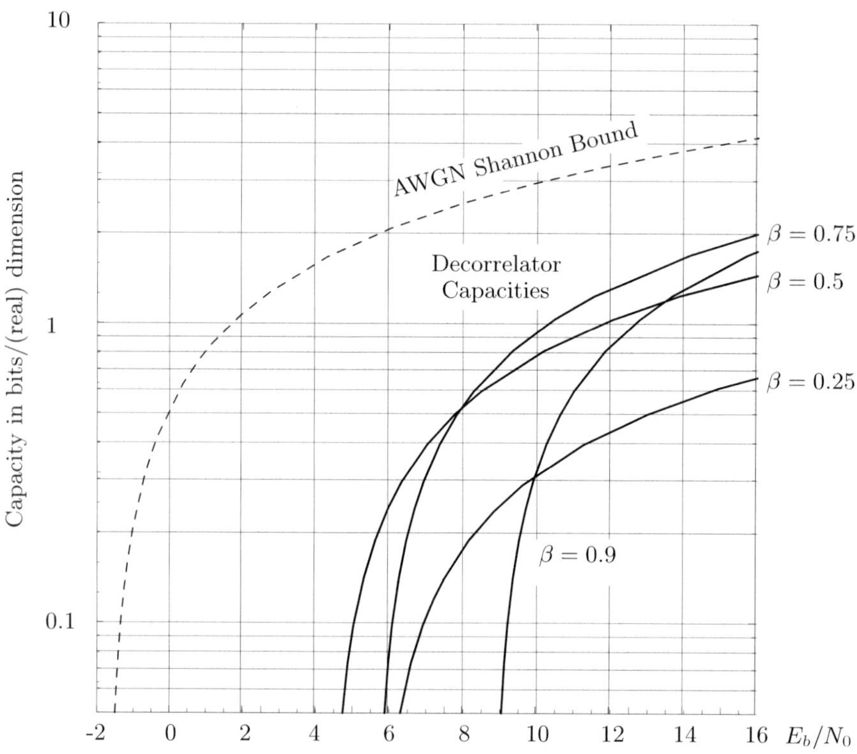

Fig. 6.8. Spectral efficiencies achievable with a zero-forcing layering filter.

$$\tilde{\boldsymbol{c}}_i = \left(\mathbf{C} + \sigma^2 \mathbf{P}^{-1}\right)^{-1} \mathbf{r}_{\mathrm{MF},i}; \qquad \mathbf{P} = \mathbf{A}^T \mathbf{A}. \tag{6.22}$$

The difference lies in the kernel matrix that is being inverted. The MMSE filter is defined as the filter $\mathbf{M}$ which minimizes the squared norm from the data vector, that is,

$$\mathbf{M}_{\mathrm{mmse}} = \arg\min_{\mathbf{M}} \mathrm{E}\left[\|\boldsymbol{c} - \mathbf{Mr}\|^2\right], \tag{6.23}$$

whose solution leads to Eq. 6.22. The derivation proceeds via the well-known *orthogonality principle* and can be found in [26]. This filter has been proposed for joint detection of CDMA by [32], then been further developed in [33, 34].

A sophisticated random analysis similar to that for the decorrelator lets us calculate spectral efficiencies of this filter for the case where the signature sequences are random via the distribution of eigenvalues of random matrices [26, 31]. The resulting layered channel capacity is given by

$$C_{\mathrm{mmse}} = \frac{\beta}{2} \log\left[1 + 2\frac{RE_b}{N_0} - \frac{1}{4}\mathcal{F}\left(2\frac{RE_b}{N_0}, \beta\right)\right], \tag{6.24}$$

where the function [31]

$$\mathcal{F}(x, z) = \left[\sqrt{x\left(1+\sqrt{z}\right)^2 + 1} - \sqrt{x\left(1-\sqrt{z}\right)^2 + 1}\right]^2. \tag{6.25}$$

Setting $\beta R = C_{\mathrm{mmse}}$ leads to an implicit equation for the maximum spectral efficiency, i.e., the Shannon bound for the MMSE filter, which is shown in Fig. 6.9. It is worth noting that the MMSE filter is quite close to capacity for low values of E_b/N_0, however, at these low values noise is dominant, and simple matched filtering, i.e., $\tilde{\boldsymbol{c}}_i = \mathbf{R}^* \mathbf{r}_i$ will do very well also.

In either case, both filters will fail for oversaturated systems with $\beta > 1$; the decorrelator because no inverse can be found, the MMSE filter because it will strongly suppress the desired channel also. Nontheless, the spectral efficiencies achievable with the MMSE filter are strictly larger than those achievable with the decorrelator. The difference is most marked for low SNR values or for large loads.

6.3.4 Iterative Filter Implementations

Despite these drawbacks of linear layering structures, they are very powerful for the amount of processing complexity that is required. This is particularly the case for *iterative implementations* of these filters. Iterative filter implementations are based on iterative matrix solution methods which were known since the days of Gauss [14, 35].

Consider the linear system $\mathbf{Mx} = \boldsymbol{v}$, and let $\mathbf{M} = \boldsymbol{S} - \mathbf{T}$ be a arbitrary *splitting* of $\mathbf{M}$. Now

$$\boldsymbol{S}\mathbf{x} = \mathbf{Tx} + \boldsymbol{v} \Longrightarrow \boldsymbol{S}\mathbf{x}^{(j+1)} = \mathbf{Tx}^{(j)} + \boldsymbol{v}, \tag{6.26}$$

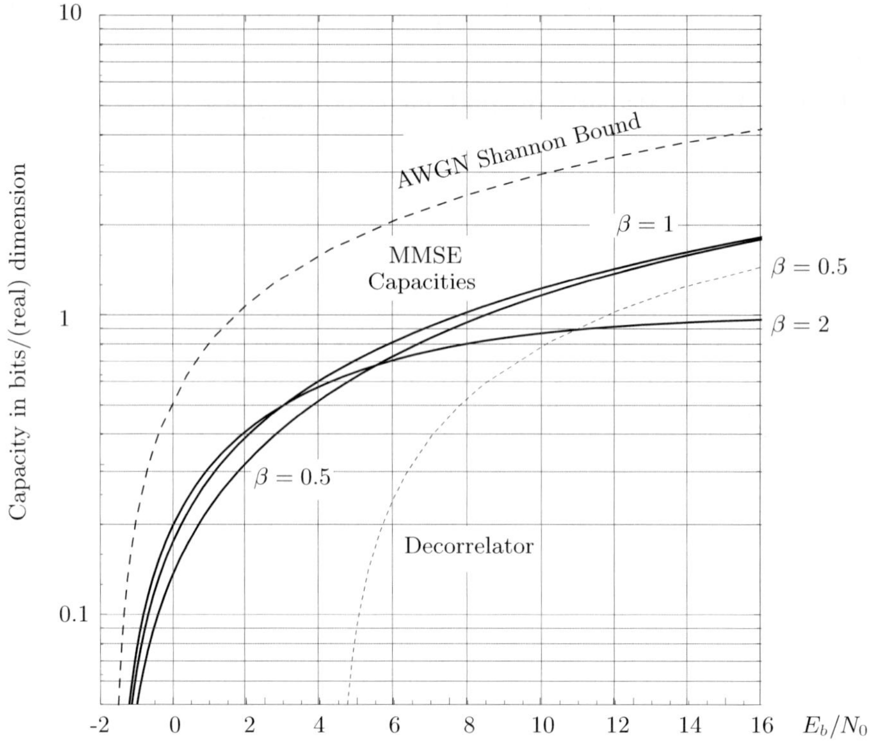

Fig. 6.9. Spectral efficiency of a channel layered with the MMSE filter.

and we have generated an iterative equation for $\mathbf{x}$:

$$\mathbf{x}^{(j+1)} = \boldsymbol{S}^{-1}\left(\mathbf{T}\mathbf{x}^{(j)} + \boldsymbol{v}\right). \tag{6.27}$$

The question is now simply if, and under what conditions, this equation converges to the solution of $\mathbf{M}\mathbf{x} = \boldsymbol{v}$. Convergence of Eq. 6.27 is understood by studying the error. Let $\mathbf{x}^{(j)} = \mathbf{x} + \mathbf{e}^{(j)}$, where $\mathbf{e}^{(j)}$ is the error at iteration j.

Consider then

$$\begin{aligned} \boldsymbol{S}(\mathbf{x} + \mathbf{e}^{(j+1)}) &= \mathbf{T}(\mathbf{x} + \mathbf{e}^{(j)}) + \boldsymbol{v} \\ \boldsymbol{S}\mathbf{e}^{(j+1)} &= \mathbf{T}\mathbf{e}(j) \\ \mathbf{e}^{(j+1)} &= \boldsymbol{S}^{-1}\mathbf{T}\mathbf{e}^{(j)} = (\boldsymbol{S}^{-1}\mathbf{T})^{j}\,\mathbf{e}^{(0)}. \end{aligned} \tag{6.28}$$

We can use the *spectral decomposition* [14] to write $\boldsymbol{S}^{-1}\mathbf{T} = \mathbf{Q}\boldsymbol{\Lambda}\mathbf{Q}^{*}$, where $\boldsymbol{\Lambda}$ is a diagonal matrix of the eigenvalues of $\boldsymbol{S}^{-1}\mathbf{T}$, and hence convergence always occurs if

$$\begin{aligned} \left\|\mathbf{e}^{(j+1)}\right\| = \left\|\mathbf{Q}\boldsymbol{\Lambda}^{j}\mathbf{Q}^{*}\mathbf{e}^{(0)}\right\| &< \left\|\mathbf{e}^{(0)}\right\| \\ &\longrightarrow \rho(\boldsymbol{S}^{-1}\mathbf{T}) < 1, \end{aligned} \tag{6.29}$$

where $\rho(\mathbf{M})$ is the spectral radius of $\mathbf{M}$, i.e., the absolute value of the largest eigenvalue [36]. If Eq. 6.29 is fullfilled, then Eq. 6.27 will converge to the true solution of the linear equation irrespective of the starting point $\mathbf{x}^{(0)} \neq \mathbf{0}$.

The only question to be resolved now is to find an efficient splitting, naturally one such that $\boldsymbol{S}^{-1}$ is easy to compute, or nothing is gained by this procedure. The first such splitting is called the *Jacobi iteration*, where we simply select $\boldsymbol{S} = \text{diag}\mathbf{M}$, which is trivially invertible.

Using either $\mathbf{C}$ or $\mathbf{C}+\sigma^2 P^{-1}$ as matrix $\mathbf{M}$ we can iteratively solve Eq. 6.19 or Eq. 6.22. For the first case, the update formulas are given by the *Jacobi decorrelator* as

$$\boldsymbol{c}^{(j+1)} = \mathbf{r}_{\text{MF}} - (\mathbf{C} - \boldsymbol{I})\,\boldsymbol{c}^{(j)} \Longrightarrow c_k^{(j)} = r_k - \sum_{\substack{i=1\\ i\neq k}}^{K} R_{ki} c_i^{(j-1)}, \tag{6.30}$$

where we have used as convenient initial estimate $\boldsymbol{c}^{(0)} = \mathbf{r}_{\text{MF}}$. This application of iterative matrix inversion to decoding CDMA systems was first proposed by [37], but early versions of cancellation receivers following the format of Eq. 6.30 were proposed in [38] without exploring the connection to matrix theory. These used the the iterative feedback equation

$$c_k^{(j)} = r_k - \sum_{\substack{i=1\\ i\neq k}}^{K} R_{ki}\text{sign}\left(c_i^{(j-1)}\right), \tag{6.31}$$

realizing that $c_i \in \pm 1$.

A thorough analysis of iterative cancellation receivers in [39] revealed that the Jacobi decorrelator suffers from a serious drawback in that, for large systems using random signature sequences, it converges to the decorrelator if and only if $K < N(\sqrt{2}-1)^2$, i.e., for a system load of only 17% or less. An illustration of this behavior is shown in Fig. 6.10, where the divergence for loads $\beta > 0.17$ becomes evident, even for the hard-decision iterative canceler proposed in [38].

There are many other iterative matrix solution methods which more or less can all be used in cancellation receivers [35, 39]. Their performance does not typically differ by much, however, arguably the most efficient structure is the *Gauss-Seidel canceler*. It is based on the matrix splitting

$$\mathbf{M} = \text{diag}(\mathbf{M}) - \omega \boldsymbol{L} - (1-\omega)\boldsymbol{L} - \boldsymbol{L}^*, \tag{6.32}$$

where $-\boldsymbol{L}$ is the lower triangular part of $\mathbf{M}$. Eq. 6.32 is the splitting leading to the general form of over-relaxation iterative inversion. Choosing the relaxation parameter $\omega = 1$ we obtain the *Gauss-Seidel* iteration equation

$$\left(\text{diag}(\mathbf{M}) - \boldsymbol{L}\right)\boldsymbol{c}^{(j+1)} = \mathbf{r}_{\text{MF}} + \boldsymbol{L}^*\boldsymbol{c}^{(j)}. \tag{6.33}$$

The major consequence for an interference canceler is that computed values are used in the next iteration as soon as they are available, unlike the Jacobi

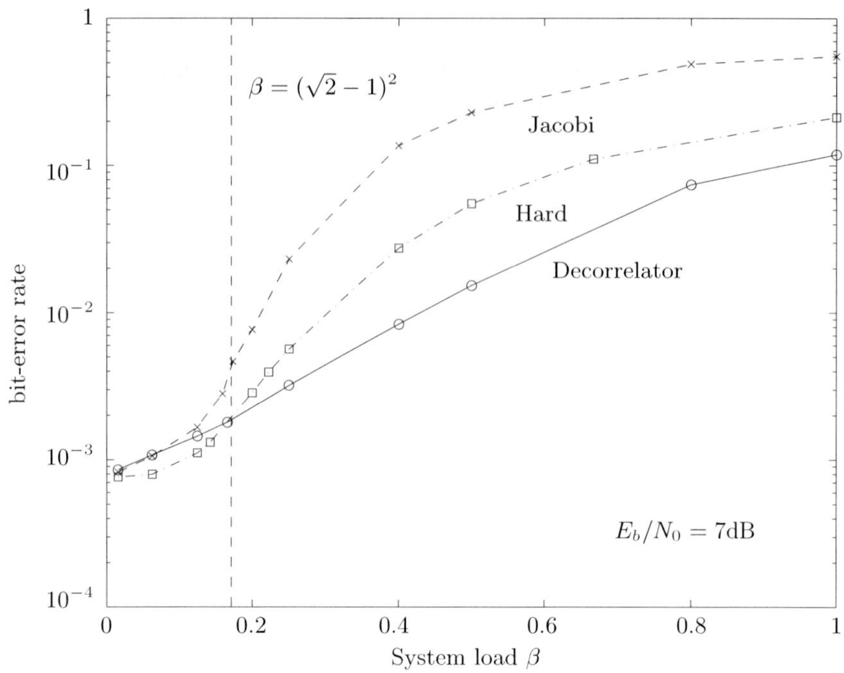

Fig. 6.10. BER performance versus load of Jacobi-type decorrelators.

canceler, which performs a complete block update before new values are used. The iteration equations are therefore given by

$$c_k^{(j)} = r_k - \sum_{i=1}^{k-1} R_{ki} c_i^{(j)} - \sum_{i=k+1}^{K} R_{ki} c_i^{(j-1)} \tag{6.34}$$

6.4 Iterative Decoding

Iterative Decoding involves the error control decoder in the iterative decoding loop. It results from interpreting the joint multiple access system with error control coding in Fig. 6.4 as a serially concatenated coding system [4, 26], in which the error control codes take the role of outer codes and the multiple access channel takes the role of the inner code. With this interpretation, a soft information exchange decoding algorithm suggests itself, similar to the turbo decoding algorithm, and has proven very effective. A diagram of such an iterative multiuser decoder is shown in Fig. 6.11.

The error control decoders are soft-output a posteriori probabiltiy (APP) decoders [4] which generate the a posteriori probabilities of the individual transmitted coded bits, i.e.,

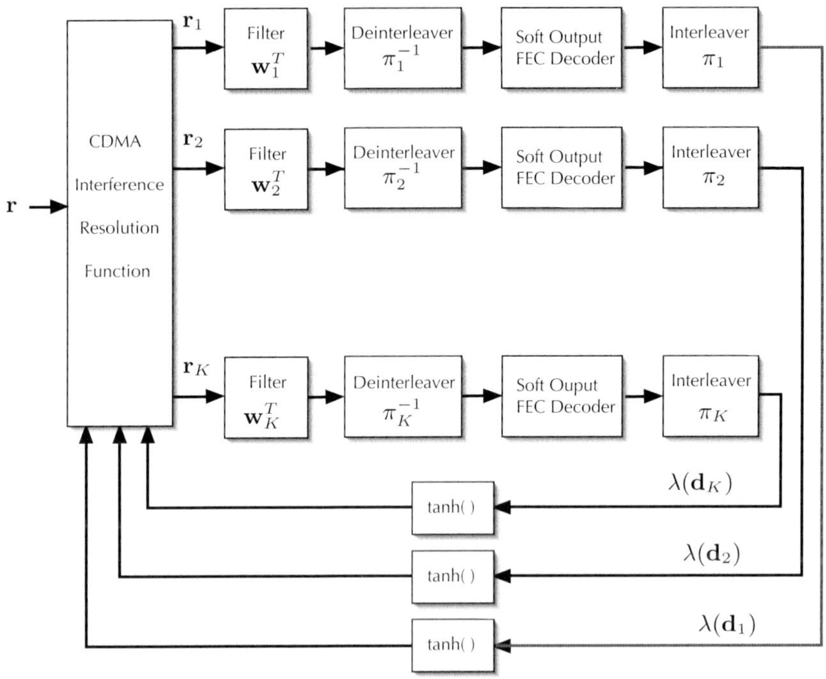

Fig. 6.11. Iterative multiuser decoder with soft information exchange.

$$\Pr(d_{k,i} = 1|\mathbf{r}_k); \Pr(d_{k,i} = -1|\mathbf{r}_k), \tag{6.35}$$

which is typically calculated as the symbol-wise *log-likelihood ratio (LLR)*

$$\lambda(d_{k,i}) = \log\left(\frac{\Pr(d_{k,i} = 1|\mathbf{r}_k)}{\Pr(d_{k,i} = -1|\mathbf{r}_k)}\right). \tag{6.36}$$

The interleavers $\pi_1, \cdots, \pi_K$ are assumed to be large enough to destroy any correlation between different LLR values. This *independance assumption* is very important for the statistical analysis of the detector, as well as for its successful operation.

Iterative decoders mainly differ in the type of FEC codes used, in the soft-output decoding algorithms applied, in the CDMA interference resolution function employed, and, in the loop filters $\mathbf{w}_k^T$ used. We will see in the next sections that a cancellation front-end is not only an attractive low-complexity solution, but, in conjunction with different received power levels or code rates, can approach the capacity of the multiple access channel [40].

6.4.1 Signal Cancellation

The interference resolution block has the function of extracting soft estimates from the received vector $\mathbf{r}$ using as input soft estimates of the coded symbols from the soft-output error control decoders. A possible way of doing this is by calculating the APP probability of the coded symbols using the channel constraints [26], but this leads typically to a receiver with prohibitive complexity. Such an APP front-end would calculate

$$\hat{d}_{k,i} = \arg \max_{d \in \mathcal{D}^{KN}} \sum_{\mathbf{d}:d_{k,i}=d} \Pr(\mathbf{r}|\mathbf{d})\Pr(\mathbf{d}), \tag{6.37}$$

where $\Pr(\mathbf{r}|\mathbf{d})$ is the multiple access channel transition probability and is Gaussian for CDMA, and $\Pr(\mathbf{d})$ is the vector of a priori probabilities of the transmitted symbols learned from a previous iteration.

Eq. 6.37 represents the "optimal" APP detector proposed by Moher [41]. Unfortunately, even a cursory analysis shows that the sum has 2^K terms and becomes thus quickly unmanageable, and is therefore not of much practical interest.

Guided by information theoretic thinking, or simple intuition, one can argue that the next best thing to do is to cancel the interference for each of the individual users by subtracting an appropriate estimate $\sqrt{P_k}\tilde{d}_{k,j}\mathbf{s}_{i,j}$ of the interference caused by other users. This operation generates a received sequence $\mathbf{r}_k$ for user k as

$$\begin{aligned} \mathbf{r}_k &= \sum_{i=1}^{N} \sqrt{P_k} d_{k,i}\mathbf{s}_{i,k} + \overbrace{\sum_{i=1}^{N} \sum_{\substack{j=1 \\ (j\neq k)}}^{K} \sqrt{P_j}\left(d_{j,i} - \tilde{d}_{j,i}\right)\mathbf{s}_{i,j}}^{I_k} + \mathbf{z} \\ &= \sum_{i=1}^{N} \sqrt{P_k} d_{k,i}\mathbf{s}_{i,k} + I_k + \mathbf{z}. \end{aligned} \tag{6.38}$$

That is, $\mathbf{r}_k$ is the signal of the k-th user distorted by additive noise and the interference I_k. Central limit theorem arguments quickly establish that this interference, which results from incomplete cancellation of the interference users' signals, is Gaussian. The filters $\mathbf{w}_1^T, \cdots, \mathbf{w}_K^T$ are receiver loop filters inserted in the processing chain of each of the users. In the absence of multiple access interference these filters would simply be signal matched filters.

The question now turns to how to generate the estimate of interference. Clearly, if we know precisely all of the interfering symbols, perfect cancellation is theoretically possible. However, the other APP decoders are not, initially, capable of generating exact estimates. We therefore strive to minimize the error of the estimate $d_{j,i} - \tilde{d}_{j,i}$ between actual and estimated symbols in the mean square sense. This is accomplished by using as estimated symbols the conditional expectations

$$\tilde{d}_{j,i} = E\left[d_{j,i}\middle|\lambda(d_{j,i})\right] = \tanh\left(\frac{\lambda(d_{j,i})}{2}\right), \tag{6.39}$$

which leads to the $\tanh(\cdot)$ functions to generate the *soft-bits* $\tilde{d}_{j,i}$. These soft-bits will minimize the variance of I_k given the locally probability estimates from the code APP decoders.

The received signal $\mathbf{r}_k$ are now considered to by distorted only by Gaussian noise, and matched filters $\mathbf{w}_k^T = \mathbf{s}_{i,k}^T$ are therefore the optimal receiver filters.

6.4.2 Convergence – Variance Transfer Analysis

If considering a CDMA channel with random spreading sequences, i.e., the "chips" $a_{k,l,n}$ which make up the signature sequences shown in Fig. 6.5, are chosen randomly with probability $p(a_{k,l,n} = 1) = p(a_{k,l,n} = -1) = 1/2$, then the interference is proportional to the ratio of users to processing gain. Additionally, the interference is also proportional to the variance of the estimation error

$$\sigma_d^2 = E\left[\left(d_{j,i} - \tilde{d}_{j,i}\right)^2\right]. \tag{6.40}$$

Assuming equal power for all users $P_k = P$, and matched receiver filters, *i.e.*, $\mathbf{w}_k^T = \mathbf{s}_{k,i}^T$, leads to a particularly simple expression for the variance of the residual interference plus noise, given by

$$\sigma_{\mathrm{IC}}^2(\sigma_d^2) = \sigma^2 + \frac{K-1}{N} P\sigma_d^2; \quad \stackrel{K,N\to\infty}{\Longrightarrow} \sigma^2 + \beta P \sigma_d^2, \tag{6.41}$$

which is intuitive: there are $K-1$ interferers for user k, each being suppressed by the processing gain N. The APP decoder estimation error variance σ_d^2 is the average power of the residual interference.

Therefore, given an average residual symbol interference variance σ_d^2, Eq. 6.41 describes the noise variance in the signal $\mathbf{r}_k$ for the next iteration through the FEC decoders. This *variance transfer curve (VT)* is a linear function in the system load and the residual symbol variance, and be conveniently pictured by the following Fig. 6.4.2.

The more accurate the symbol estimates of the interferers, the smaller the residual noise that the error control decoder has to overcome. Note that even if the interfering symbols are exactly known, $\sigma_d^2 = 0$, there is still the noise variance which the error control decoder has to overcome.

The idea of iterative decoding is, that through iterations, the variance on all the single-user channels can be improved until complete cancellation of all interference is achieved. In order to understand this process, we need to gain understanding of the operation of the soft-output FEC APP decoders.

The input signal $\mathbf{r}_k$ to the k-th APP decoder has an additive Gaussian noise distortion with variance σ_{IC}^2, and the decoder analysis can be handled analogously to an AWGN channel. The output of the decoder are the soft-bits (Eq. 6.39), and the primary measure of their reliability is the variance

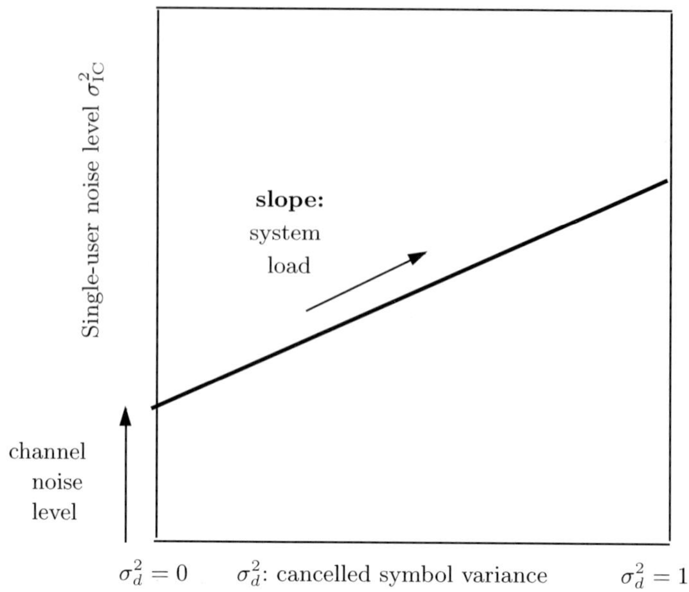

Fig. 6.12. Soft cancellation variance transfer curve.

σ^2_d. While much numerical analysis [42, 43] seems to indicate that the LLRs (Eq. 6.36) are Gaussian distributed with good accuracy, the soft-bits are not Gaussian distributed. However, since in the limit of many users, the number of terms in I_k is large, I_k is still asymptotically Gaussian by central limit theorem arguments, and therefore σ^2_d is a sufficient measure of the reliability of $\tilde{d}_{j,i}$.

The FEC code has a similar VT curve which translates input noise variance σ^2_{IC} into soft-output symbol variance values σ^2_d, and the operation of an iterative receiver now carries out in the following way: Cancelled signals $\mathbf{r}_k$ for each user are generated according to Eq. 6.38 and passed to the FEC APP decoders. They, in turn, generate soft estimates of the coded signals which are used to generate interference estimates to be used in a new cancellation iteration. If the interleavers used in the system are large enough to ensure that the implicit assumption of independence between signals is accurate, the process of iterative cancellation and decoding can be captured by a variance transfer chart shown in Fig. 6.13 for the rate $R = 1/3, \nu = 3$ convolutional code and a system load of $\beta = 3$.

Decoding starts at point (0) where the FEC decoders have to work with the full interference and noise. Nonetheless, after one iteration, the system manages to reduce the interference by subtracting estimates of the interfering signals, this leads to point (1), whose vertical component is the now reduced noise variance at the input of the soft-output error control decoder. In the next

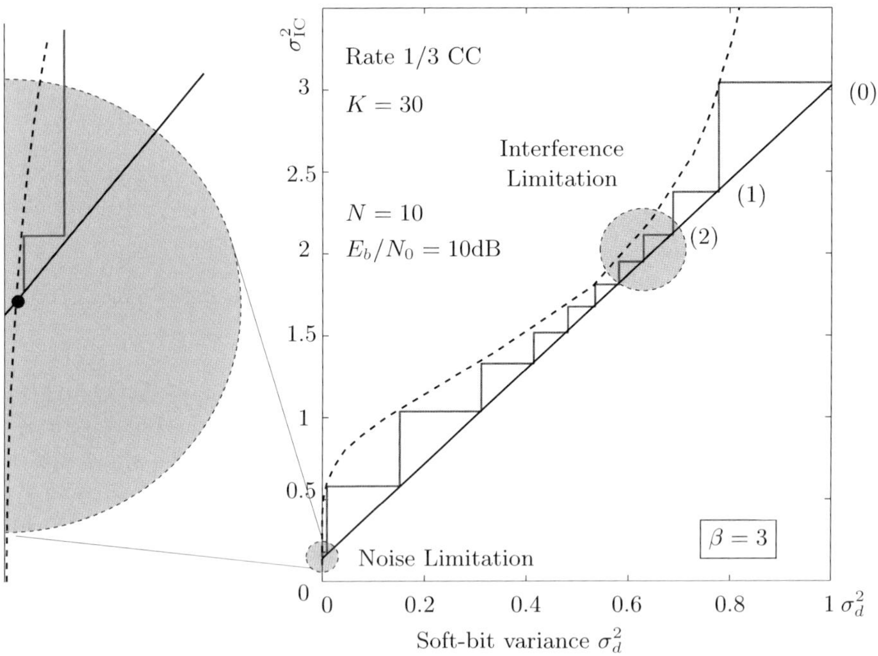

Fig. 6.13. VT transfer chart and iteration example for a highly loaded CDMA system. FEC code VT is dashed, the cancellation VT curve is the solid curve.

iteration, the noise variance can be further reduced to point (2), and so on, until the iterations reach the intersection point between the two curves, the solid point in the enlargement. At this point virtually all the interference has been cancelled, and only channel noise is left. This intersection point is a fix point of a one-dimensional iterated variance transfer map, which models the behavior of the iterative decoder. The performance of the individual decoders at that point is essentially that of the decoders in Gaussian noise, and we call it the noise limitation of the iterative decoder.

Note that along the iterative trajectory there is an section which forms a narrow channel through which the trajectory progresses in small steps. This area is the interference limitation. A small increase in the system load will create another intersection point between the two VT curves, and the variance will stop improving at that new fix point, rather than at the noise limitation fix point. The decoder does not function due to an excessive system load. As the load decreases, or the signal-to-noise ratio increases, the channel opens up and convergence to the noise limitation fix point is suddenly enabled. This effect happens at a very sharp signal-to-noise threshold, and gives rise to the abrupt, cliff-like behavior of the error rates in such systems, which is illustrated by the simulations shown in Fig. 6.18. Clearly visible is the sharp onset of the drop in error rates as the signal-to-noise ratio is increased sufficiently to open

Table 6.1. Serial concatenated codes of total rate $R = 1/3$ used in Fig. 6.14.

	Inner Code	Outer Code	Rate
Serial Concatenated Code 1	$g_1 = \begin{bmatrix} 1\ 0\ \frac{1+D^2}{1+D+D^2} \\ 0\ 1\ \frac{1+D}{1+D+D^2} \end{bmatrix}$	$g_2 = [1\ \frac{1+D^2}{1+D+D^2}]$	$\frac{1}{3}$
Serial Concatenated Code 2	$g_1 = \begin{bmatrix} 1\ 0\ \frac{1+D^2}{1+D+D^2} \\ 0\ 1\ \frac{1+D}{1+D+D^2} \end{bmatrix}$	rate 1/2 repetition code	$\frac{1}{3}$

the convergence channel and allow the decoder system to progress beyond the interference limitation. At sufficient signal-to-noise ratios the bit error rates closely follow those of the FEC code operated in additive noise alone, this is the noise limitation and is only a function of the FEC code used.

The question now poses itself as to what FEC code should be used for efficient communication. Examining stronger error control codes, such as turbo codes or LDPC codes will be beneficial in the noise limitation area, where the single-user performance of the codes comes to bear. Fig. 6.14 show the VT curves of a selection of powerful error control codes, several rate $R = 1/3$ LDPC codes at their first iteration and at iteration 20, as well as two rate $R = 0.5$ serially concatenated turbo codes, whose rates and component codes are given in the following table. All of these codes have good to excellent performance on AWGN channels, and their VT transfer curves all show a general trend. The more powerful the code, the more abrupt the transition between unreliable and reliable code output behavior. Further analysis on the impact of the code and code choices are explored in [26, 40].

6.4.3 Filters in the Loop

A simple processing filter $\mathbf{w}_k^T$ in Fig. 6.11 is a filter matched to the spreading waveform of user k. Such a matched-filter canceler is both computationally efficient and has good performance. Nontheless, more sophisticated loop filters can eek out extra system gain. In this section we explore the potential of a local per-user minimum-mean square error (MMSE) filter, originally proposed by Wang and Poor [44]. From a (local) variance point of view the MMSE filter is the optimal choice. We therefore consider the filter

$$\mathbf{w}_{i,k}^T = \arg \min_{\mathbf{w} \in \mathcal{R}^N} \mathrm{E}\left(\left\| \mathbf{d}_{k,i} - \mathbf{w}^T \mathbf{r}_k \right\|^2 \right). \tag{6.42}$$

The advantage of the MMSE filter over the matched filter comes from its capability to include the correlation inherent in the residual interference signal. The derivation of the filter follows standard steps [26], and we obtain

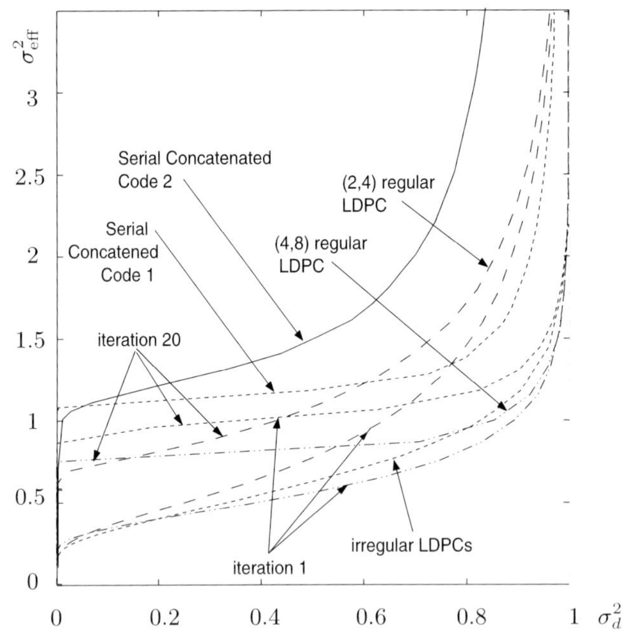

Fig. 6.14. Illustration of variance transfer curves of various powerful error control codes, such as practical-sized LDPC codes of lenght $N = 5000$ and code rate $R = 0.5$, as well as two serially concatenated turbo codes.

$$\mathbf{w}_{i,k}^T = \frac{\sqrt{P_k}\mathbf{s}_{i,k}^T\left(\mathbf{S}_k\mathbf{P}_k\mathbf{S}_k^T + \sigma^2\mathbf{I}\right)^{-1}}{P_k + \mathbf{s}_{i,k}^T\left(\mathbf{S}_k\mathbf{P}_k\mathbf{S}_k^T + \sigma^2\mathbf{I}\right)^{-1}\mathbf{s}_{i,k}^T}. \tag{6.43}$$

We see that this filter now depends not only on the spreading sequence of the interfering users, but also on their powers $\mathbf{P}_k$, where $\mathbf{P}_k$ equals $\mathbf{P}$ with the k-th entry removed. Furthermore, since these are residual powers, the matrix inverse in Eq. 6.43 has to be recomputed for every iteration – a truly tremendous effort.

In the case of a system with equal received power, a result by Tse and Hanly [45] allows us to calculate the signal-to-noise ratio after the MMSE filter in closed form as a function of the residual interference as

$$\beta_k = \frac{\sigma_d^2 P(\beta-1) - \sigma^2}{2\sigma^2} + \sqrt{\frac{(\sigma_d^2 P)^2(\beta-1)^2}{4\sigma^{2^2}} + \frac{\sigma_d^2 P(\beta-1) - \sigma^2}{2\sigma^2} + \frac{1}{4}}. \tag{6.44}$$

Schlegel et. al. [26, 40] prove the following limit result

$$\lim_{\frac{w}{\sigma^2}\to\infty} \frac{\sigma_{\text{IC}}^2}{\sigma_d^2} = \beta - 1, \tag{6.45}$$

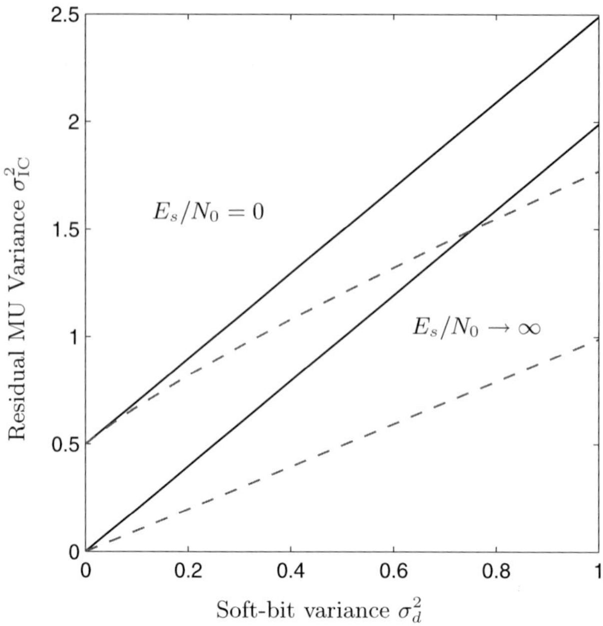

Fig. 6.15. Variance transfer curves for matched filter (simple) cancellation and per-user MMSE filter cancellation (dashed lines) for $\beta = 2$ and $E_s/N_0 = 0$dB and $E_s/N_0 \to \infty$.

which is illustrated by the canceler VT curves shown in Fig. 6.15. This leads to the following conclusion: If simple cancellation can support a load of β, the use of per-user MMSE filters instead of the matched filters allows a load of $\beta + 1$ for high signal-to-noise ratios. Therefore, an appreciable performance gain can be achieved for systems with a moderate level of loading.

6.4.4 Low-Complexity Loop Filters

While the MMSE loop filter can produce a marked advantage for a certain range of channel loads, its implementation is significantly more complex than that of a matched filter due to the inverse in Eq. 6.43. However, if we replace

$$\mathbf{K}_{i,k} = \mathbf{S}_{i,k}\mathbf{P}_k\mathbf{S}_{i,k}^T + \sigma^2\mathbf{I} \tag{6.46}$$

in Eq. 6.43 by the first-order stationary multi-stage approximation

$$\begin{aligned}\mathbf{K}_{i,k}^{-1} &= (\mathbf{I} - \tau\mathbf{K}_{i,k})^n + \sum_{l=0}^{n-1} \tau\,(\mathbf{I} - \tau\mathbf{K}_{i,k})^l \\ &= \phi_n\,(\mathbf{K}_{i,k})\end{aligned} \tag{6.47}$$

according to the principles explored in Sec. 6.3.4, the resulting system complexity is much reduced, depending only on the number of stages n. As shown in [26], this number does not depend on N or K, and only very few stages recover almost all of the difference between a matched-filter iterative decoder and an MMSE iterative decoder.

Instead of Eq. 6.43, the loop filter is now

$$\mathbf{w}_{i,k} = \phi_n\left(\mathbf{K}_{i,k}\right)\mathbf{s}_{i,k} \tag{6.48}$$

where we have dropped the normalization factor in Eq. 6.43 since its variance goes to zero for large systems with $K, N \to \infty$.

Fig. 6.16 below shows such VT curves for a number of stages and two values of the signal-to-noise ratio. It is remarkable that one or two stages are almost completely sufficient to achieve the VT characteristics of the MMSE loop filter over a wide range of input soft bit variance σ_d^2 and channel signal-to-noise ratios.

We use the following numerical example to illustrate the accuracy of the analysis as well as the power of the multi-stage loop filters. The test example is made up of a convolutional rate $R = 1/3$ four-state code with $G(D) = [1 + D^2, 1 + D + D^2, 1 + D + D^2]$ and a two-stage first order stationary loop

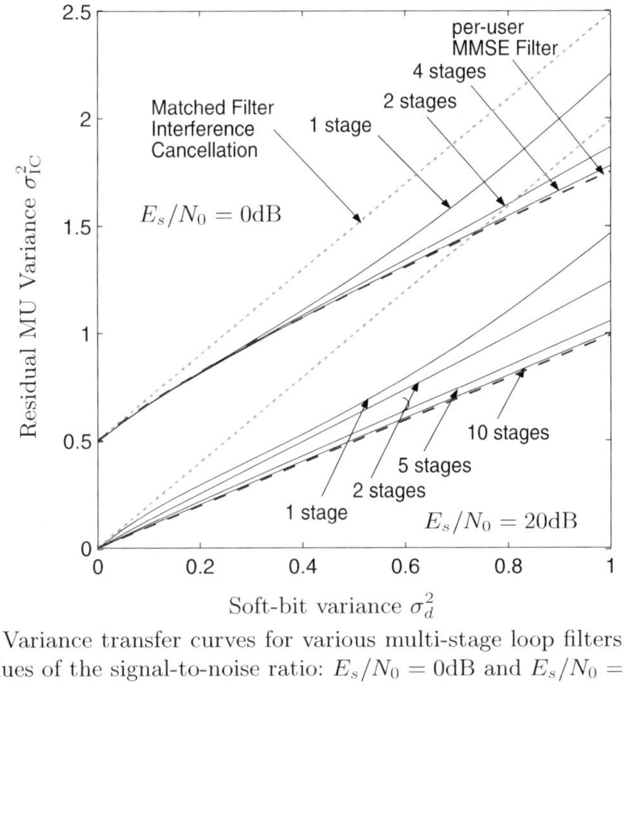

Fig. 6.16. Variance transfer curves for various multi-stage loop filters for a $\beta = 2$ and two values of the signal-to-noise ratio: $E_s/N_0 = 0$dB and $E_s/N_0 = 20$dB.

The conclusion we draw from this result is that power control is an important feature in CDMA systems, even if a linear joint detector is used.

For iterative decoders which include the error control code in the loop, the converse is true, i.e., the equal power case produces the worst case performance of the overall system in terms of average signal-energy to noise power ratio versus spectral efficiency. However, we are not aware of a rigorous proof of this statement. What is known is that, with properly adjusted power levels, the *capacity* of the multiple access channel can be achieved with an iterative decoder. Unfortunately these power levels are exponentially distributed [40]. The same effect can be achieved with different coding rates, rather than different power levels, also achieving the capacity of the channel. It seems safe to conclude that in a scenario where the received powers are not controlled, and where large fluctuations in the received power are expected, an iterative decoder including FEC will fare significantly better than a mere linear filter front-end. On the other hand, in a power regulated scenario, where the received powers are constant, linear processing – iterative or not – achieves a very large portion of the theoretical capacity of the channels for system loads as large as $\beta = 1$.

Fig. 6.19 shows achievable information theoretic capacities for random CDMA systems using both a linear minimum mean-square error filter and optimal processing for two system loads. The two capacities for $\beta = 0.5$ show very little difference, and it is therefore "nearly optimal" to use a minimum mean square error filter as linear preprocessor which converts the CDMA channel into parallel single-user AWGN channels where well-known FEC methods can be used. This is the case for $\beta \overset{\approx}{\leq} 0.5$ in random CDMA channels. Even for $\beta = 1$ the improvement possible is only in order of about 50% in terms of capacity. As the system load increases, linear preprocessing becomes increasingly less efficient, and for highly loaded systems joint detection can provide significant extra gains. When and where joint detection provides a capacity advantage depends on the particular system. In general, systems with a small number of users per signal dimension do not profit significantly from joint processing.

This picture changes if powers and/or rates are unequal, where iterative decoding can make a large difference. A precise analysis depends on the particular distribution of the power levels, and will have to be evaluated case by case. Further results, an in-depth discussion of asymmetric operating conditions, and proofs of the results stated above can be found in [26].

6.6 Conclusions

The topic of iterative demodulation and decoding is fascinating. The method is extremely powerful in handling large constellations arising from the (linear) superposition of a large number of signals, which we wish to decode, either dependently, or independently. The turbo decoding principle, which was so

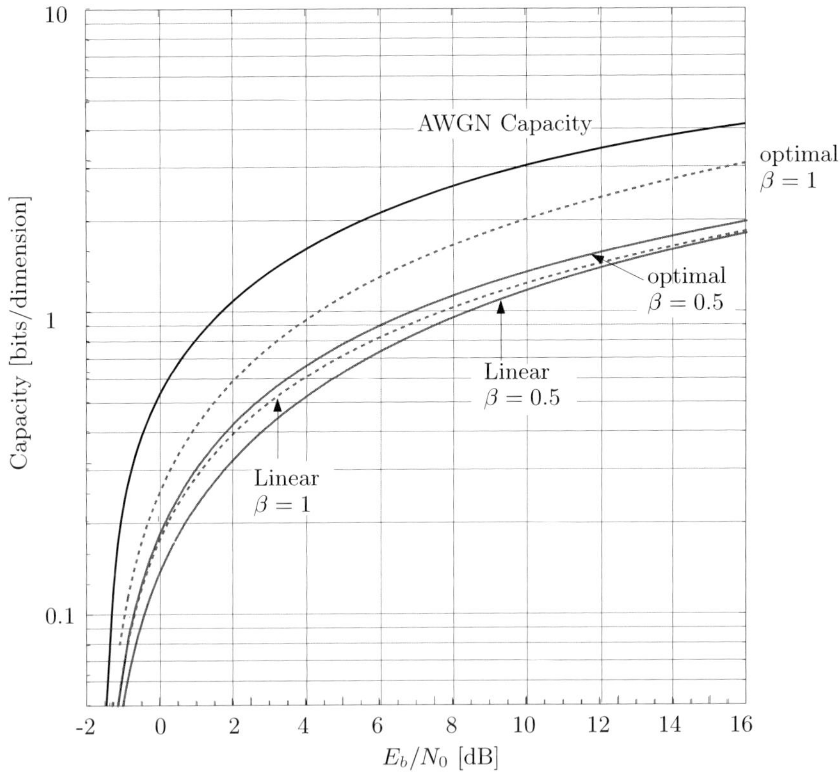

Fig. 6.19. Comparison of the per-dimension capacity of optimal and linearly processed random CDMA channels. The solid lines are for $\beta = 0.5$ for both linear and optimal processing, the dashed lines are for a full load $\beta = 1$.

successful in achieving the Shannon capacity on the AWGN channel [8], has also proven extremely useful in iterative demodulation and decoding.

We have shown that there are essentially two avenues to approach the problem. The first is linear filtering, which separates the superposed channels into independent channels via direct or iterative multistage linear operations. This method works very well for systems whose numbers of signal dimensions exceeds the number of signals – the typical case to date. Power control may be necessary to maintain a large system efficiency, even with use of these joint demodulators.

The second avenue is to consider a full iterative (turbo) decoder with soft information messages cycling back to an appropriate cancellation front-end. This approach is more powerful, benefits from different received power levels, and under judicious operating conditions, can be shown to achieve the capacity of the multiple access channel. It is also more complex, but not significantly so. Both approaches essentially scale linearly in the number of joint signal that need to be detected.

Future implementers will have to carefully weigh the advantages and costs of these two approaches in order to find optimal system configurations. Last, but not least, we wish to point out that this area is still undergoing significant research, and that hybrid systems are feasible. For example, simple soft-thresholding devices put into the iterative matrix inversion loops have proven to be very effective in improving performance of the linear demodulators. These non-linear devices can be viewed as simple symbol-level decoders.

In either way, iterative demodulation and decoding is a very powerful methodology to work with large numbers of interfering signals. Their complexity only linearly depends on the number of signals to be treated, and they are thus implementable even for large systems.

References

1. C.E. Shannon, "A mathematical theory of communications," *Bell Syst. Tech. J.*, vol.27, pp. 379-423, July 1948.
2. T. Cover and J. Thomas, *Elements of Information Theory*, John Wiley, New York, 1991.
3. H. Nyquist, "Certain topics in telegraph transmission theory," *AIEE Trans.*, pp. 617–644, 1928.
4. C. Schlegel and L. Perez, *Turbo and Trellis Coding*, IEEE/Wiley, Piscataway NJ, 2004.
5. J.M. Wozencraft and I.M. Jacobs, *Principles of Communication Engineering*, John Wiley and Sons, New York, 1965, reprinted by Waveland Press, 1993.
6. R. G. Gallager, "Low-density parity-check codes", *IRE Trans. on Inform. Theory*, pp. 21–28, Vol. 8, No. 1, January 1962.
7. C. Berrou, A. Glavieux, and P. Thitimajshima, "Near Shannon limit error-correcting coding and decoding: Turbo-codes," *Proc. 1993 IEEE ICC*, Geneva, Switzerland, pp. 1064–1070, 1993.
8. C. Berrou and A. Galvieux, "Near optimum error correcting coding and decoding: turbo-codes," *IEEE Trans. Commun.*, Vol. 44, No. 10,pp. 1261–1271, Oct. 1996.
9. D. J. C. MacKay and R. M. Neal, "Near Shannon limit performance of low density parity check codes", *IEE Electronic Letters*, vol. 32, no. 18, pp. 1645–1646, August 1996.
10. D. J. C. MacKay, "Good error-correcting codes based on very sparse matrices", *IEEE Trans. Inform. Theory*, vol. IT-45, No. 2, pp. 399–431, March, 1999.
11. E. Guizzo, "Closing in on the perfect code", *IEEE Spectrum*, Vol. 41, No. 3, pp. 36–42, Mar. 2004.
12. R.M. Tanner, "A recursive approach to low complexity codes," *IEEE Trans. Inform. Theory*, vol. IT-27, pp. 533–547, September 1981.
13. S. Y. Chung, G. D. Forney, T. J. Richardson and R. Urbanke, "On the design of low-density parity-check codes within 0.0045 dB of the Shannon limit", *IEEE Comm. Letters*, vol. COMM-5, no. 2, pp. 58–60, February, 2001.
14. R.A. Horn and J.C. Johnson, *Matrix Analysis*, Cambridge University Press, New York, 1990.
15. A.J. Viterbi, *CDMA : Principles of Spread Spectrum Communication*, Prentice Hall,1995.
16. R.E. Ziemer, R.L. Peterson, D.E. Borth *Introduction to Spread Spectrum Communications*, Pearson Education, 1995.
17. J.G. Proakis, *Digital Communications*, McGraw-Hill, Inc., 1995.
18. E. Telatar, "Capacity of Multi-antenna Gaussian Channels," *Europ. Trans. Telecommunications*, vol. 42, no. 2/3/4, pp. 1617–1627, Nov.-Dec. 1999.
19. G.J. Foschini, "Layered space-time architecture for wireless communication in a fading environment when using multi-element antennas," *Bell Labs Tech. J.*, vol. 1, Aug. 1996, pp. 41 – 59.
20. T.L. Marzetta and B.M. Hochwald, "Capacity of a mobile multiple-antenna communication link in Rayleigh flat fading," *IEEE Trans. Inform. Theory*, vol. 45, no. 1, pp. 139–157, Jan. 1999.
21. Z. Bagley and C. Schlegel, "Classification of correlated flat fading MIMO channels," *Proc. Third Canadian Workshop on Inform. Theory*, Vancouver, Canada, June 2001.

22. D. Gesbert, H. Bölcskei, D.A. Gore, and A.J. Paulraj, "Outdoor MIMO wireless channels: Models and performance prediction," *IEEE Trans. Commun.*, to appear.
23. A.L. Swindlehurst, G. German, J. Wallace, M. Jensen, "Experimental measurements of capacity for MIMO indoor wireless channels," *3rd IEEE Signal Proc. Wshp.*, Taiwan, Mar. 20–25, 2001.
24. P. Goud Jr, C. Schlegel, R. Hang, W. Krzymien, Z. Bagley, "MIMO Channel Measurements for an Indoor Office Environment," *IEEE Wireless '03*, July 7-9, Calgary, AB, Canada.
25. P. Goud Jr, C. Schlegel, W.A. Krzymien, R. Hang, "Multiple Antenna Communication Systems - An Emerging Technology," *Canadian Journal of Electrical and Computer Engineering*, accepted for publication.
26. C. Schlegel and A. Grant, *Coordinated Multiple User Communications*, Springer Publishers, 2005.
27. G.J. Foschini and M.J. Gans, "On limits of wireless communication in a fading environment when using multiple antennas," *Wireless Personal Commun.*, vol. 6, no. 3, pp. 311–355, Mar. 1998.
28. R. Lupas and S. Verdú, "Linear multiuser detectors for synchronous code–division multiple–access channels," *IEEE Trans Inform. Theory*, vol. 35, no. 1, pp. 123–136, January 1989.
29. P. D. Alexander, L. K. Rasmussen, and C. Schlegel, "A linear receiver for coded multiuser CDMA," *IEEE Trans. Commun.*, vol. 45, pp. 605–610, May 1997.
30. Z.D. Bai and Y.Q. Yin, "Limit of the smallest eigenvalue of large dimensional sample covariance matrix," *Ann. Probab.*,, vol. 21, pp. 1275–1294, 1993.
31. S. Verdú and S. Shamai, "Spectral Efficiency of CDMA with Random Spreading," *IEEE Trans. Inform. Theory*, vol. 45, pp. 622–640, March 1999.
32. Z. Xie, R. Short, and C. Rushforth, "A family of suboptimum detectors for coherent multiuser communications," *IEEE. J. Select. Areas Commun.*, vol. 8, no. 4, pp. 683–690, May 1990.
33. P. B. Rapajic and B. S. Vucetic, "Adaptive receiver structures for asynchronous CDMA systems," *IEEE J. Select. Areas Commun.*, vol. 12, no. 4, pp. 685–697, May 1994.
34. U. Madhow and M.L. Honig, "MMSE interference suppression for direct–sequence spread–spectrum CDMA," *IEEE Trans. Commun.*, vol. 42, no. 12, pp. 3178–3188, Dec. 1994.
35. O. Axelsson, *Iterative Solution Methods*, Cambridge University Press, 1994.
36. Gilbert Strang, *Linear Algebra and its Applications*, Harcourt Brace, 3 edition, 1988.
37. H. Elders-Boll, H-D. Schotten, and A. Busboom, "Efficient implementation of linear multiuser detectors for asynchronous CDMA systems by linear interference cancellation," *Europ. Trans. Telecommun*, vol. 9, pp. 427–438, Sept-Oct. 1998.
38. M.K. Varanasi and B. Aazhang, "Multistage detection in asynchronous code-division multiple-access communications," *IEEE Trans. Commun.*, vol. 38, No. 4, pp. 509–519,April 1990.
39. A. Grant and C. Schlegel "Iterative Implementations for Linear Multiuser Detectors" *IEEE Trans. Commun.*, vol. 49, pp. 1824-1834, October 2001.
40. C. Schlegel, Z. Shi, and M. Burnashev, "Asymptotically Optimal Power Allocation and Code Selection for Iterative Joint Detection of Coded Random CDMA," *IEEE Trans. Inform. Theory*, in revision.

41. M. Moher, "An Iterative Multiuser Decoder for Near-Capacity Communications", *IEEE Trans. Commun.*, vol. 47, pp. 870–880, July 1998.
42. S. ten Brink, "Convergence behavior of iteratively decoded parallel concatenated codes," *IEEE Trans. Commun.*, vol. 49, no. 10, October, 2001.
43. D. Divsalar, S. Dolinar, and F. Pollara, "Iterative Turbo decoder analysis based on density evolution," TMO Progress Report 42-144, February 15, 2001.
44. X. Wang and H.V. Poor, "Iterative (Turbo) Soft Interference Cancellation and Decoding for Coded CDMA", *IEEE Trans. Commun.*, vol. 47, no. 7, July 1999.
45. D. Tse and S. Hanly, "Linear Multiuser Receivers: Effective interference, effective bandwidth and user capacity," *IEEE Trans. Inform. Theory*, vol. 45, pp. 641–657, Mar. 1999.

Chapter 7

Turbo Receiver Techniques for Coded MIMO OFDM Systems *

Ben Lu and Xiaodong Wang

Columbia University, USA

In recent years, iterative ("turbo") processing techniques have received considerable attention stimulated by the discovery of the powerful turbo codes [2, 3]. The so called "turbo principle" can be successfully applied to many detection/decoding problems such as serial concatenated decoding, equalization, coded modulation, multiuser detection and joint source and channel decoding [16, 37, 44]. We start the discussion of this chapter by illustrating the general concept of turbo processing for concatenated systems using a simple example.

A typical communications system in general consists of a cascade of subsystems with different signal processing functionalities. Consider, for example, the simple communication system employing channel coding and signaling over an inter-symbol interference (ISI) channel, as shown in Fig. 7.1. In a "conventional" receiver, the demodulator makes hard decisions about the transmitted bits $\{b[i]\}$ based on the received signal $r(t)$, which are then passed to the channel decoder to decode the transmitted information. The problem with this approach is that by making hard decisions of the bits, information loss is incurred in each subsystem (i.e., demodulator and decoder). This is because while the subsystem only indicates whether it believes that a given bit is a "0" or a "1", it usually has sufficient information to estimate the degree of confidence in its decisions. One straightforward way to reduce the loss of information, and the resulting loss in performance, is to pass the confidence level along with the decision, i.e., to render soft decisions. This is often done when passing information from a demodulator to a channel decoder, which is known to result in approximately a 2dB performance gain in the additive white Gaussian noise (AWGN) channel [39]. However, even if optimal bit-by-bit soft decisions are passed between all the subsystems in the receiver, the overall performance can still be far from optimal. This is due to the fact that, while later stages (e.g., the channel decoder) can use the information

* This work was supported in part by the U.S. National Science Foundation (NSF) under grants CCR-0225826 and CCR-0207550, and in part by the U.S. Office of Navel Research (ONR) under grant N00014-03-1-0039.

K. Sripmanwat (ed.), Turbo Code Applications: a journey from a paper to realization, 157–191.

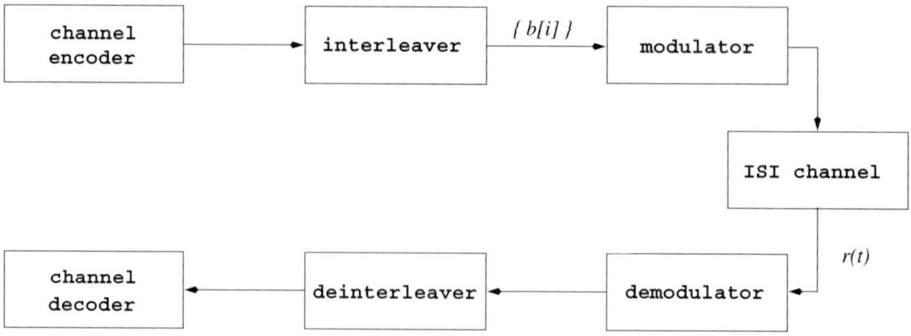

Fig. 7.1. A coded communication system signaling through an inter-symbol interference (ISI) channel.

gleaned from previous stages (e.g., the demodulator), the reverse is not generally true. While the optimal performance can be achieved by performing a joint detection, taking all receiver processing into account simultaneously (e.g., maximum likelihood detection based on the super-trellis of both the channel code and the ISI channel), the complexity of such a joint approach is usually prohibitive. This motivates an iterative (turbo) processing approach which allows earlier stages (e.g., the demodulator) to refine their processing based on information obtained from later stages (e.g., the channel decoder).

In order to employ turbo processing in the system shown in Fig. 7.1, both the demodulator and the channel decoder are of the maximum *a posteriori* probability (MAP) type. The function of the MAP demodulator is to produce soft decisions which reflect the probability that a given bit is a "0" or a "1". At the lth iteration, the information available to the MAP demodulator consists of the received signal $r(t)$ and the *a priori* probabilities of the input bits, the later of which are applied by the MAP channel decoder based on its output from the $(l-1)^{\text{th}}$ iteration. The MAP demodulator uses this information, combined with knowledge of the chosen modulation and of the channel structure, to produce the *a posteriori* probabilities (APPs) of the channel bits

$$P^l(b[i] = 1 \mid r(t)) = \frac{p(r(t) \mid b[i] = 1)\, P^{l-1}(b[i] = 1)}{p(r(t))}, \tag{7.1}$$

$$\text{and} \quad P^l(b[i] = 0 \mid r(t)) = \frac{p(r(t) \mid b[i] = 0)\, P^{l-1}(b[i] = 0)}{p(r(t))}, \tag{7.2}$$

for all $\{b[i]\}_i$. Consider the log-likelihood ratio (LLR) formed from the *a posteriori* probabilities of Eq. 7.1 and 7.2:

$$\begin{aligned} \Lambda_1^l(b[i]) &\triangleq \log \frac{P^l(b[i]=1 \mid r(t))}{P^l(b[i]=0 \mid r(t))} \\ &= \underbrace{\log \frac{p(r(t) \mid b[i]=1)}{p(r(t) \mid b[i]=0)}}_{\lambda_1^l(b[i])} + \underbrace{\log \frac{P^{l-1}(b[i]=1)}{P^{l-1}(b[i]=0)}}_{\lambda_2^{l-1}(b[i])} . \end{aligned} \tag{7.3}$$

It is seen from Eq. 7.3 that the LLR is the sum of two distinct quantities. The first term, $\lambda_1^l(b[i])$, is the so-called *extrinsic* information produced by the first stage subsystem in the receiver (i.e., the MAP demodulator), which is information that the MAP demodulator gleans about $b[i]$ from the received signal $r(t)$ and the *a priori* probabilities of the other transmitted bits, without utilizing the *a priori* probability of $b[i]$. The second term, $\lambda_2^{l-1}(b[i])$, contains the *a priori* probability of $b[i]$. Note that typically, for the first iteration ($l = 1$), we set $P^0(b[i]=1) = P^0(b[i]=0) = \frac{1}{2}$, i.e., $\lambda_2^0(b[i]) = 0$ for all i. The extrinsic information $\{\lambda_1^l(b[i])\}$ produced by the MAP demodulator, is sent to the second stage subsystem, i.e., the MAP channel decoder, as the *a priori* information for channel decoding.

Based on the *a priori* information provided by the MAP demodulator, and the channel code constraints, the MAP channel decoder computes the *a posteriori* LLR of each code bit

$$\begin{aligned} \Lambda_2^l(b[i]) &\triangleq \log \frac{P(b[i]=1 \mid \{\lambda_1^l(b[i])\}; \text{code structure})}{P(b[i]=0 \mid \{\lambda_2^l(b[i])\}; \text{code structure})} \\ &= \lambda_2^l(b[i]) + \lambda_1^l(b[i]). \end{aligned} \tag{7.4}$$

Here again we see that the output of the channel decoder is the sum of the extrinsic information $\lambda_2^l(b[i])$ obtained by the second stage subsystem (i.e., the MAP channel decoder), and the prior information $\lambda_1^l(b[i])$ delivered by the previous stage (i.e., the MAP demodulator). The extrinsic information $\lambda_2^l(b[i])$ is then fed back to the MAP demodulator as the *a priori* information in the next (i.e., $(l+1)^{\text{th}}$) iteration. It is important to note that Eq. 7.3 and 7.4 hold only if the inputs to the demodulator or the decoder are independent. Since both the ISI channel and the channel encoder have memory, this independence assumption will not be valid; therefore, interleaving (i.e., permutation of time order) must be present between the demodulator and the decoder in order to provide approximate independence. Finally, the turbo receiver structure for the coded ISI system is illustrated in Fig. 7.2. This scheme was first introduced in [10], and is termed a "turbo equalizer". The name "turbo" is justified because both the demodulator and the decoder use their processed output values as *a priori* input for the next iteration, similar to the operation of a turbo engine. The same turbo processing principle can be applied to coded CDMA systems, yielding the so-called turbo multiuser detection [35, 40, 53–55].

In this chapter, we introduce the applications of turbo principle in the receiver design and code optimization of low-density parity-check (LDPC)

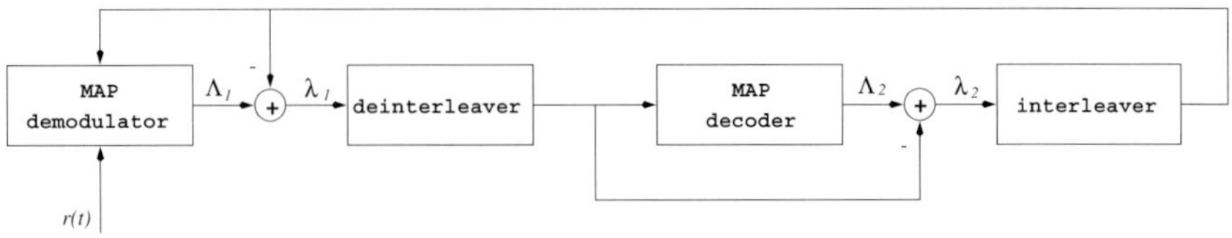

Fig. 7.2. A turbo receiver for coded communication over ISI channel.

coded multiple-input multiple-output (MIMO) orthogonal-frequency division-multiplexing (OFDM) systems.

7.1 LDPC-Coded MIMO OFDM Systems

Multiple-input multiple-output (MIMO) data transmission through sparsely-spaced antennas at both transmitter and receiver provides substantial increase in spectral efficiency of wireless links, both in theory and in practice. Technically, MIMO transmission can potentially accomplish both *multiplexing gain*, i.e., the information rate increase due to virtual multiple air-links, and *diversity gain*, i.e., the spatial diversity due to multiple antennas in addition to time-domain and frequency-domain diversity. Key to realizing such high spectral efficiency (i.e., high data rate) promised by MIMO systems is the practical coded modulation schemes. For small block size (e.g., smaller than ten), there are many solutions such as orthogonal space-time block codes [46], linear dispersion codes [17], threaded algebraic space-time codes [9] and lattice space-time codes [13]. When the block size is around several hundred, options include space-time trellis codes [47] and 'wrapped' space-time codes [5]. For moderate-to-large block size (e.g., larger than 1000) which is suitable for most data traffic in broadband communications, existing schemes are mostly based on powerful binary random codes (e.g., turbo codes or LDPC codes), and they include bit-interleaved coded modulation (BICM), [24, 27, 49, 56], [18, 19], stratified diagonal BLAST [43]. The LDPC-coded MIMO OFDM scheme proposed in this chapter falls into the category of BICM based MIMO with moderate-to-large block size.

7.1.1 MIMO OFDM Modulation

Consider a quasi-static block fading model for a generic MIMO OFDM modulation, as illustrated in Fig. 7.3. It is assumed that the fading channels remain static during each OFDM slot but vary independently from one OFDM slot to another. Assume proper cyclic insertion and sampling, the MIMO OFDM system with K subcarriers decouples frequency-selective channels into K correlated flat-fading channels with the following input-output relation

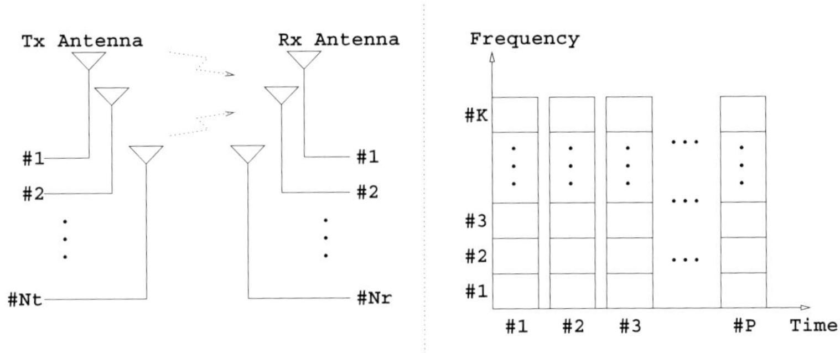

Fig. 7.3. System description of a MIMO OFDM system over correlated fading channels. Each transmit frame spans K subcarriers and P time slots; at a particular subcarrier and at a particular time slot, coded symbols are transmitted from N_t transmitter antennas and received by N_r receiver antennas.

$$\boldsymbol{y}[p,k] = \sqrt{\frac{\gamma}{N_t}}\boldsymbol{H}[p,k]\boldsymbol{x}[p,k] + \boldsymbol{n}[p,k] \; , \qquad k = 0,\ldots,K-1, \quad p = 0,\ldots,P-1, \tag{7.5}$$

where N_t (N_r) denotes the number of transmitter (receiver) antennas; $\boldsymbol{H}[p,k] \in \mathcal{C}^{N_r \times N_t}$ is the matrix of complex channel frequency responses to be defined later; $\boldsymbol{x}[p,k] \in \Omega^{N_t}$ and $\boldsymbol{y}[p,k] \in \mathcal{C}^{N_r}$ are respectively the transmitted signals and the received signals at the k-th subcarrier and the p-th slot; $\mathcal{C}$ denotes the set of complex number; Ω denotes quadrature amplitude modulation (QAM) constellation with unit average power; $\boldsymbol{n}[p,k] \in \mathcal{C}^{N_r}$ is the additive noise with independent and identically distributed (i.i.d.) entries $\boldsymbol{n}[p,k] \sim \mathcal{N}_c(\boldsymbol{0}, \boldsymbol{I})$; γ denotes the average signal-to-noise ratio (SNR) at each receiver antenna.

Consider the channel response between the j-th transmitter antenna and the i-th receiver antenna. Following [39], the time-domain channel impulse response can be modeled as a tapped-delay line. With only the non-zero taps considered, it can be expressed as

$$h_{i,j}(\tau;t) = \sum_{l=1}^{L_f} \alpha_{i,j}(l;t)\delta\left(\tau - \frac{n_l}{K\Delta_f}\right) \; , \tag{7.6}$$

where $\delta(\cdot)$ is the Dirac delta function; L_f denotes the number of non-zero taps; $\alpha_{i,j}(l;t)$ is the complex amplitude of the l-th non-zero tap, whose delay is $n_l/(K\Delta_f)$, where n_l is an integer and Δ_f is the tone spacing of the OFDM system. In mobile channels, for the particular (i,j)-th antenna pair, the time-variant tap coefficients $\alpha_{i,j}(l;t)$, $\forall l, \forall t$, can be modeled as wide-sense stationary random processes with uncorrelated scattering (WSSUS) and with band-limited Doppler power spectrum [39]. For the signal model in Eq. 7.5, we only need to consider the time responses of $\alpha_{i,j}(l;t)$ within the time interval $t \in [0, PT]$, where T is the total time duration of one OFDM word plus its cyclic extension, and PT is the total time involved in transmitting P adjacent OFDM words. Following [20], for the particular l-th tap of the (i,j)-th antenna pair, the dimension of the band- and time-limited random process $\alpha_{i,j}(l;t)$, $t \in [0, PT]$ (defined as the number of significant eigenvalues in the Karhunen-Loeve expansion of this random process), is approximately equal to $L_t \triangleq \lceil 2f_d PT + 1 \rceil$, where f_d is the maximum Doppler frequency. Hence, ignoring the edge effects, the time response of $\alpha_{i,j}(l;t)$ can be expressed in terms of the Fourier expansion as

$$\alpha_{i,j}(l;t) \simeq \sum_{n=-\lceil f_d PT \rceil}^{\lceil f_d PT \rceil} \beta_{i,j}(l,n) e^{\jmath 2\pi nt/(PT)} , \qquad (7.7)$$

where $\{\beta_{i,j}(l,n)\}_n$ is a set of independent circularly symmetric complex Gaussian random variables, indexed by n.

For OFDM systems with proper cyclic extension and sample timing, with tolerable leakage, the channel frequency response between the j-th transmitter antenna and the i-th receiver antenna at the p-th time slot and at the k-th subcarrier, which is exactly the (i,j)-th element of $\boldsymbol{H}[p,k]$ in Eq. 7.5, can be expressed as [28, 52]

$$H_{i,j}[p,k] \triangleq H_{i,j}(pT, k\Delta_f) = \sum_{l=1}^{L_f} \alpha_{i,j}(l;pT) e^{-\jmath 2\pi k n_l/K} = \boldsymbol{h}_{i,j}^{\mathrm{H}}(p) \boldsymbol{w}_f(k) , \qquad (7.8)$$

where $\boldsymbol{h}_{i,j}(p) \triangleq [\alpha_{i,j}(1;pT), \ldots, \alpha_{i,j}(L_f;pT)]^{\mathrm{H}}$ is the L_f-sized vector containing the time responses of all the non-zero taps; ^{H} denotes matrix or vector Hermitian transpose; and $\boldsymbol{w}_f(k) \triangleq \left[e^{-\jmath 2\pi k n_1/K}, \ldots, e^{-\jmath 2\pi k n_{L_f}/K}\right]^{\mathrm{T}}$ contains the corresponding DFT coefficients.

Using Eq. 7.7, $\alpha_{i,j}(l;pT)$ can be simplified as

$$\alpha_{i,j}(l;pT) = \sum_{n=-f_d PT}^{f_d PT} \beta_{i,j}(l,n) e^{\jmath 2\pi np/P} = \boldsymbol{\beta}_{i,j}^{\mathrm{H}}(l) \boldsymbol{w}_t(p) , \qquad (7.9)$$

where $\boldsymbol{\beta}_{i,j}(l) \triangleq [\beta_{i,j}(l, -f_d PT), \cdots, \beta_{i,j}(l,0), \cdots, \beta_{i,j}(l, f_d PT)]^{\mathrm{H}}$ is an L_t-sized vector; and $\boldsymbol{w}_t(p) \triangleq \left[e^{-\jmath 2\pi p f_d T}, \ldots, e^{\jmath 0}, \ldots, e^{\jmath 2\pi p f_d T}\right]^{\mathrm{T}}$ contains the corresponding inverse DFT coefficients. Substituting Eq. 7.9 into Eq. 7.8, we get

$$H_{i,j}[p,k] = \boldsymbol{g}_{i,j}^{\mathrm{H}} \boldsymbol{W}_t'(p) \boldsymbol{w}_f(k) , \qquad (7.10)$$

$$\text{with} \quad \boldsymbol{g}_{i,j} \triangleq \left[\boldsymbol{\beta}_{i,j}^{\mathrm{H}}(1), \ldots, \boldsymbol{\beta}_{i,j}^{\mathrm{H}}(L_f)\right]_{L \times 1}^{\mathrm{H}} ,$$

$$\boldsymbol{W}_t'(p) \triangleq \mathrm{diag}\{\boldsymbol{w}_t(p), \ldots, \boldsymbol{w}_t(p)\}_{L \times L_f} ,$$

where $\{\beta_{i,j}(l,n)\}_n$ are properly normalized as $\sum_l \sum_n \mathrm{Var}\{\beta_{i,j}(l,n)\}_n = 1, \forall i,j$. From Eq. 7.10, it is seen that due to the close spacing of OFDM subcarriers and the limited Doppler frequency, for a specific antenna pair (i,j), the channel responses $\{H_{i,j}[p,k]\}_{p,k}$ are different transformations [specified by $\boldsymbol{w}_t(p)$ and $\boldsymbol{w}_f(k)$] of the same random vector $\boldsymbol{g}_{i,j}$, and hence they are correlated in both frequency and time.

7.1.2 Channel Capacity

With no channel state information (CSI) at the transmitter side, the channel capacity for the above MIMO OFDM modulation has been studied in [11, 33]. Assume Gaussian signaling (i.e., $\Omega \to \mathcal{C}$), for MIMO OFDM channels with infinite fading channel observations (i.e, $P \to \infty$), the *ergodic capacity* is given by

$$C_{erg}(\gamma) \triangleq \mathrm{E}\Bigg\{ \underbrace{\frac{1}{KP}\sum_{k=0}^{K-1}\sum_{p=0}^{P-1}\left[\log_2 \det\left(\boldsymbol{I}_{N_r} + \frac{\gamma}{N_t}\boldsymbol{H}[p,k]\boldsymbol{H}^{\mathrm{H}}[p,k]\right)\right]}_{\mathcal{I}_{|\mathcal{H}}(\gamma)} \Bigg\} \text{ bits/Hz/sec,} \tag{7.11}$$

where the expectation is taken over random channel states $\mathcal{H}$, with $\mathcal{H} \triangleq \{\boldsymbol{H}[p,k]\}_{p,k}$; $\mathcal{I}_{|\mathcal{H}}(\gamma)$ is the instantaneous mutual information conditioned on $\mathcal{H}$. For MIMO OFDM channels with finite fading channel observations (i.e, $P \ll \infty$), the *outage capacity/probability* is a more sensible measure. For a target information rate R, the outage probability is given by

$$P_{\mathrm{out}}(R,\gamma) = \Pr(\mathcal{I}_{|\mathcal{H}}(\gamma) < R) \,. \tag{7.12}$$

However, in practice, the transmitted signals usually take values from constraint constellation, i.e., $\boldsymbol{x} \in \Omega^{N_t}$. In this case, following [51], the mutual information is computed instead as

$$\begin{aligned}\mathcal{I}_{|\mathcal{H}}(\gamma) = N_t \log_2 |\Omega| - &\frac{1}{KP|\Omega|^{N_t}} \sum_{k=0}^{K-1}\sum_{p=0}^{P-1}\sum_{j=0}^{|\Omega|^{N_t}-1} \\ &\mathrm{E}\left\{ \log_2 \sum_{i=0}^{|\Omega|^{N_t}-1} \exp\left[-\left\|\sqrt{\gamma/N_t}\,\boldsymbol{H}[p,k]\left(\boldsymbol{x}^j - \boldsymbol{x}^i\right) + \boldsymbol{z}\right\|^2 + \left\|\boldsymbol{z}\right\|^2\right] \right\},\end{aligned} \tag{7.13}$$

where the expectation is taken over random noise vector $\boldsymbol{z} \sim \mathcal{N}_c(\boldsymbol{0}, \boldsymbol{I})$.

7.1.3 Transmitter Structure

For an LDPC coded MIMO OFDM system with K subcarriers, N_t transmitter antennas and N_r receiver antennas, the transmitter structure is illustrated

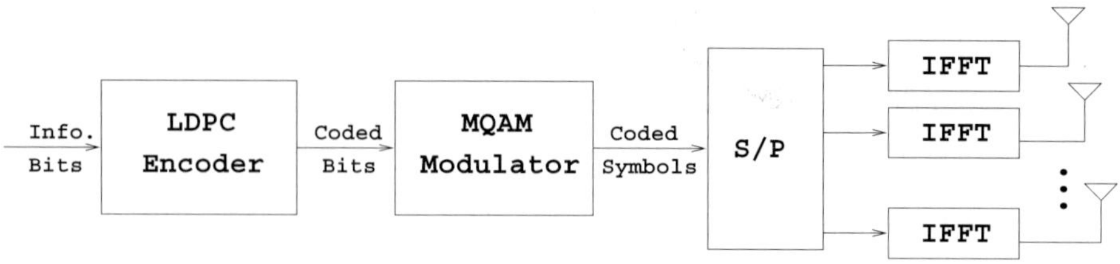

Fig. 7.4. Transmitter structure of an LDPC coded MIMO OFDM system.

in Fig. 7.4. A block of k bits of information data is encoded by a rate $r = k/n$ LDPC code. The output n coded bits are interleaved. The interleaved bits are modulated by QAM constellation Ω into a block of $n/\log_2|\Omega|$ QAM symbols. During each OFDM slot, N_tK out of the total $n/\log_2|\Omega|$ QAM symbols are transmitted from K OFDM subcarriers and N_t transmitter antennas simultaneously. Due to the inherent random structure of LDPC codes, the N_tK symbols can be mapped to K subcarriers and N_t transmitter antennas in any order. Without loss of generality, we assume $(n/\log_2|\Omega|)/(N_tK) = P$, i.e., the total block of QAM symbols is transmitted in P OFDM slots. Note that in Fig. 7.4, LDPC could also be replaced by other error-control codes such as turbo codes, however the relatively low and scalable decoding complexity and the freedom for code optimization make LDPC codes a more favorable candidate.

As in a typical data communication scenario, communication is carried out in a burst manner. A data burst is illustrated in Fig. 7.5. It spans $(Pq + 1)$ OFDM words, with the first OFDM word containing known pilot symbols. The rest (Pq) OFDM words contain q LDPC-based space-time code (STC) code words.

7.2 Turbo Receivers for LDPC-Coded MIMO OFDM

In this section, we consider receiver design for the proposed LDPC-coded MIMO OFDM system. Even with ideal CSI, the optimal decoding algorithm

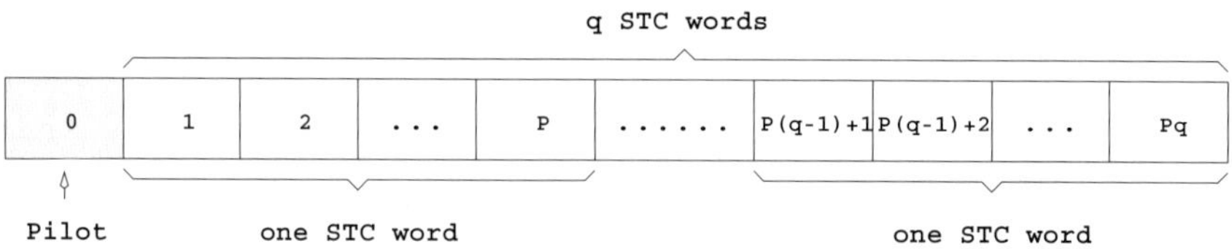

Fig. 7.5. OFDM time slots allocation in data burst transmission. A data burst consists of $(Pq + 1)$ OFDM words, with the first OFDM word containing known pilot symbols. The rest (Pq) OFDM words contain q STC code words.

for this system has an exponential complexity. Hence the near-optimal turbo receiver based on the turbo principle [16] becomes attractive.

7.2.1 Turbo Receiver with Ideal CSI

When ideal CSI is available at the receiver, we employ a serial concatenated turbo iterative receiver (as shown in Fig. 7.6) to approach the maximum likelihood (ML) receiver performance of joint MIMO OFDM demodulation and LDPC decoding. The extrinsic information of the LDPC coded bits is iteratively passed between a soft-input-soft-output (abbreviated as 'soft') demodulator and a soft belief-propagation LDPC decoder; in each demodulator-decoder iteration, a number of inner iterations is performed within the soft LDPC decoder during which extrinsic information is passed along the edges in the bipartite graph of LDPC code.

Demodulation of MIMO OFDM

It is clear from Eq. 7.5 that the demodulation of the received signals at a particular subcarrier and in a particular slot can be carried out independently. For notational convenience, in this subsection we temporarily drop the index $[p,k]$.

As illustrated in Fig. 7.6, at the l-th turbo iteration, the soft MIMO OFDM demodulator computes extrinsic information of the LDPC code bit b_i as

$$\lambda_1^l(b_i) = g(\boldsymbol{y}, \{\lambda_2^{l-1}(b_j)\}_j) \tag{7.14}$$

where $\boldsymbol{y}$ is the received data; $\{\lambda_2^{l-1}(b_j)\}_j$ is the extrinsic information computed by LDPC decoder in the previous turbo iteration, at the first turbo iteration $\lambda_2^{l-1}(b_j) \equiv 0, \forall j$; $g(\cdot)$ denotes the demodulation function, which is described below.

At a given subcarrier and time slot, N_t symbols or correspondingly $N_t \log_2 |\Omega|$ LDPC code bits are transmitted from N_t transmitter antennas. In maximum *a posterior* (MAP) MIMO OFDM demodulator, $\lambda_1^l(b_i)$ $(i = 1, \ldots, N_t \log_2 |\Omega|)$ is computed as

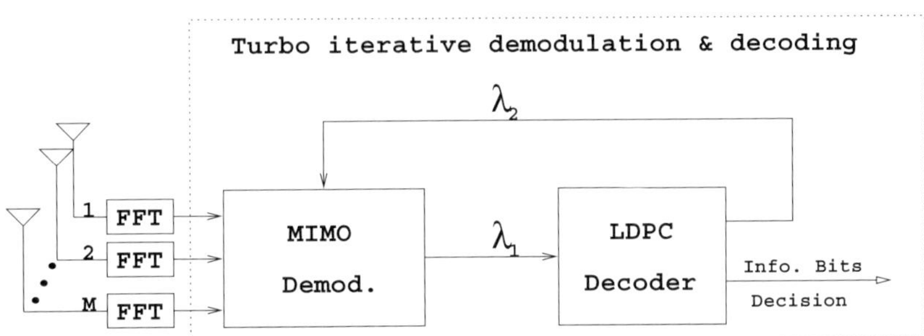

Fig. 7.6. A turbo receiver structure, which employs a soft MIMO demodulator and a soft LDPC decoder, for an LDPC coded MIMO OFDM system.

$$\lambda_1^l(b_i) \triangleq \log \frac{P(b_i = +1|\boldsymbol{y})}{P(b_i = -1|\boldsymbol{y})} - \underbrace{\log \frac{P(b_i = +1)}{P(b_i = -1)}}_{\lambda_2^{l-1}(b_i)}$$

$$= \log \frac{\sum_{\boldsymbol{x}^+ \in \mathcal{C}_i^+} \exp\left(-\|\boldsymbol{y} - \sqrt{\gamma/N_t}\boldsymbol{H}\boldsymbol{x}^+\|^2 + \sum_{j=1}^{N_t \log_2|\Omega|}\{\boldsymbol{x}^+\}_j \cdot \lambda_2^{l-1}(b_j)/2\right)}{\sum_{\boldsymbol{x}^- \in \mathcal{C}_i^-} \exp\left(-\|\boldsymbol{y} - \sqrt{\gamma/N_t}\boldsymbol{H}\boldsymbol{x}^-\|^2 + \sum_{j=1}^{N_t \log_2|\Omega|}\{\boldsymbol{x}^-\}_j \cdot \lambda_2^{l-1}(b_j)/2\right)}$$
$$-\lambda_2^{l-1}(b_i), \tag{7.15}$$

where $\mathcal{C}_i^+$ is the set of $\boldsymbol{x}$ for which the i-th LDPC coded bit b_i is "+1", and $\mathcal{C}_i^-$ is similarly defined; $\{\boldsymbol{x}^+\}_j$ denotes the corresponding j-th binary bit of the symbol $\boldsymbol{x}^+$, and similarly does $\{\boldsymbol{x}^-\}_j$. The soft MAP demodulator in Eq. 7.15 has a complexity at $\mathcal{O}(|\Omega|^{N_t})$, and can only be used in practice for small constellation size and small number of transmit antennas.

We next describe a suboptimal soft demodulator, which is based on the linear minimum-mean-square-error soft-interference-cancellation (LMMSE-SIC) techniques [54] and has a relatively low complexity at $\mathcal{O}(|\Omega|^3)$. Based on the *a priori* LLR of the code bits provided by the LDPC decoder, $\{\lambda_2^{l-1}(b_i)\}$, we first form soft estimates of the symbol transmitted from the j-th ($j = 1, 2, \ldots, N_t$) antenna as

$$\tilde{x}_j \triangleq \sum_{\hat{x}\in\Omega} \hat{x} P(x_j = \hat{x}) = \sum_{\hat{x}\in\Omega} \hat{x} \prod_{k=1}^{\log_2|\Omega|} \left[1 + \exp\left(-\{\hat{x}\}_k \cdot \lambda_2^{l-1}(\{x_j\}_k)\right)\right]^{-1}. \tag{7.16}$$

Denote $\tilde{\boldsymbol{x}}_j \triangleq [\tilde{x}_1, \ldots, \tilde{x}_{j-1}, 0, \tilde{x}_{j+1}, \ldots, \tilde{x}_{N_t}]^{\mathrm{T}}$, where ^{T} represents matrix or vector transpose. We then perform a soft interference cancelation $\boldsymbol{y}_j$ to obtain $\tilde{\boldsymbol{y}}_j \triangleq \boldsymbol{y} - \boldsymbol{H}\tilde{\boldsymbol{x}}_j = \boldsymbol{H}(\boldsymbol{x} - \tilde{\boldsymbol{x}}_j) + \boldsymbol{n}$. Next an instantaneous linear MMSE filter is applied to $\tilde{\boldsymbol{y}}_j$, to obtain $z_j = \boldsymbol{w}_j^{\mathrm{H}}\tilde{\boldsymbol{y}}_j$, where the filter $\boldsymbol{w}_j \in \mathcal{C}^{N_r}$ is chosen to minimize the mean-square error between the transmit symbol x_j and the filter output z_j, i.e.,

$$\boldsymbol{w}_j = \sqrt{\frac{\gamma}{N_t}}\left(\frac{\gamma}{N_t}\boldsymbol{H}\Delta_j\boldsymbol{H}^{\mathrm{H}} + \boldsymbol{I}\right)^{-1}\boldsymbol{H}\boldsymbol{e}_j \tag{7.17}$$

$$\text{with} \quad \Delta_j = \operatorname{diag}\left\{1 - |\tilde{x}_1|^2, \ldots, 1 - |\tilde{x}_{j-1}|^2, 1, 1 - |\tilde{x}_{j+1}|^2, \ldots, 1 - |\tilde{x}_{N_t}|^2\right\}; \tag{7.18}$$

and $\boldsymbol{e}_j$ denotes a N_r-sized vector with all-zero entries, except for the j-th entry being 1. The detailed derivation of Eq. 7.17 is further referred to [54].

As in [54], we approximate the soft instantaneous MMSE filter output z_j as Gaussian distributed, i.e., $p(z_j \mid x_j) \sim \mathcal{N}_c\left(\mu_j x_j, \eta_j^2\right)$.Conditioned on x_j, the mean and variance of z_j are given respectively by

$$\mu_j \triangleq E\{z_j x_j^*\} = \frac{\gamma}{N_t} \boldsymbol{e}_j^{\mathrm{T}} \boldsymbol{H}^{\mathrm{H}} \left(\frac{\gamma}{N_t} \boldsymbol{H} \Delta_j \boldsymbol{H}^{\mathrm{H}} + \boldsymbol{I} \right)^{-1} \boldsymbol{H} \boldsymbol{e}_j, \tag{7.19}$$

$$\eta_j^2 \triangleq \mathrm{var}\{z_j\} = E\{|z_j|^2\} - \mu_j^2 = \mu_j - \mu_j^2. \tag{7.20}$$

The extrinsic information $\lambda_1^l(b_{j,i})$ $(i = 1, \ldots, \log_2 |\Omega|)$ delivered by the LMMSE-SIC demodulator is calculated as

$$\begin{aligned}
\lambda_1^l(b_{j,i}) &\triangleq \log \frac{P(b_{j,i} = +1|z_j)}{P(b_{j,i} = -1|z_j)} - \underbrace{\log \frac{P(b_{j,i} = +1)}{P(b_{j,i} = -1)}}_{\lambda_2^{l-1}(b_{j,i})} \\
&= \log \frac{\sum_{x_j^+ \in \mathcal{S}_{i,j}^+} \exp\left(-\|z_j - \mu_j x_j^+\|^2/\eta_j^2 + \sum_{k=1}^{\log_2 |\Omega|} \{x_j^+\}_k \cdot \lambda_2^{l-1}(b_{j,k})/2 \right)}{\sum_{x_j^- \in \mathcal{S}_{i,j}^-} \exp\left(-\|z_j - \mu_j x_j^-\|^2/\eta_j^2 + \sum_{k=1}^{\log_2 |\Omega|} \{x_j^-\}_k \cdot \lambda_2^{l-1}(b_{j,k})/2 \right)} \\
&\quad -\lambda_2^{l-1}(b_{j,i}),
\end{aligned} \tag{7.21}$$

where $\mathcal{S}_{i,j}^+$ is the set of all possible values of x_i for which the i-th LDPC coded bit is "+1", and $\mathcal{S}_{i,j}^-$ is similarly defined; $\{x_j^+\}_k$ denotes the corresponding k-th binary bit of the symbol x_j^+, and similarly does $\{x_j^-\}_k$. Note that, LMMSE-SIC demodulator extracts the extrinsic LLR of code bit $b_{j,i}$ from z_j, the *scalar* output of the LMMSE filter, whereas MAP demodulator collects the extrinsic LLR from $\boldsymbol{y}$, the N_r-size *vector* of the received signals. The complexity of soft LMMSE-SIC demodulator hence is significantly lower than that of soft MAP demodulator, especially when N_t and $|\Omega|$ are large.

Decoding of LDPC Codes

The details of belief propagation LDPC decoding algorithm are referred to [12, 31].

7.2.2 Turbo Receiver without Ideal CSI

When ideal CSI need to be estimated, we employ a novel serial concatenated turbo iterative receiver, as illustrated in Fig. 7.7. As compared to the receiver with ideal CSI in Fig. 7.6, the major difference is that the soft MAP demodulator there is replaced by a soft maximum *a posteriori* expectation-maximization (MAP-EM) demodulator and its associated EM initialization circuit in Fig. 7.7. In addition, the data burst transmission frame defined in Fig. 7.5 is used. We next describe each component of the receiver in Fig. 7.7.

MAP-EM Demodulator

For notational simplicity, here we consider an LDPC-coded MIMO OFDM system with two transmitter antennas and one receiver antenna. The results can be easily extended to a system with N_t transmitter antennas and

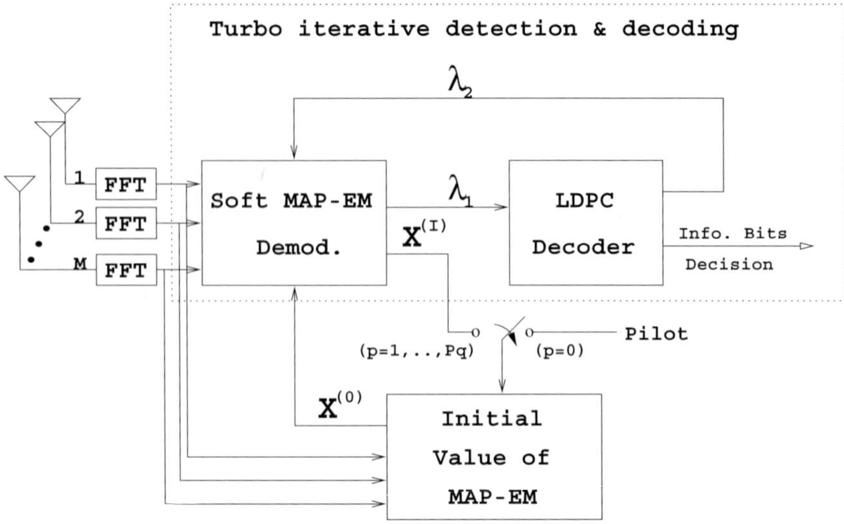

Fig. 7.7. The turbo receiver structure, which employs an MAP-EM demodulator and a soft LDPC decoder, for multiple-antenna LDPC-coded MIMO OFDM systems in unknown fading channels.

N_r receiver antennas. Note that for the purpose of performance analysis, the $\boldsymbol{h}_{i,j}(p)$ defined in Eq. 7.8 only contains the time responses of L_f non-zero taps; whereas for the purpose of receiver design, especially when the CSI is not available, the $\boldsymbol{h}_{i,j}(p)$ needs to be re-defined to contain the time responses of all the taps within the maximum multipath spread. That is, $\boldsymbol{h}_{i,j}(p) \triangleq \left[h_{i,j}[1;p], \ldots, h_{i,j}[L'_f;p]\right]^{\mathrm{T}}$, with $L'_f \triangleq \lceil \tau_m K \Delta_f + 1 \rceil \geq L_f$ and τ_m being the maximum multipath spread; and $\boldsymbol{w}_f(k)$ is correspondingly re-defined as $\boldsymbol{w}_f(k) \triangleq \left[e^{\jmath 0}, \ldots, e^{-\jmath 2\pi k (L'_f - 1)/K}\right]^{\mathrm{H}}$. The received signal during one data burst can be written as

$$\boldsymbol{y}[p] = \boldsymbol{X}[p]\boldsymbol{W}\boldsymbol{h}[p] + \boldsymbol{n}[p] \,, \qquad p = 0, 1, \ldots, Pq \,, \tag{7.22}$$

$$\begin{aligned}
\text{with} \qquad \boldsymbol{X}[p] &\triangleq [\boldsymbol{X}_1[p], \boldsymbol{X}_2[p]]_{K\times(2K)} \,, \\
\boldsymbol{X}_j[p] &\triangleq \mathrm{diag}\left\{x_j[p,0], x_j[p,1], \ldots, x_j[p,K-1]\right\}_{K\times K} \,, \quad j = 1, 2, \\
\boldsymbol{W} &\triangleq \mathrm{diag}[\boldsymbol{W}_f, \boldsymbol{W}_f]_{(2K)\times(2L'_f)} \,, \\
\boldsymbol{W}_f &\triangleq [\boldsymbol{w}_f(0), \boldsymbol{w}_f(1), \ldots, \boldsymbol{w}_f(K-1)]^{\mathrm{H}}_{K\times L'_f} \,, \\
\boldsymbol{h}[p] &\triangleq \left[\boldsymbol{h}^{\mathrm{H}}_{1,1}(p), \boldsymbol{h}^{\mathrm{H}}_{1,2}(p)\right]^{\mathrm{H}}_{(2L'_f)\times 1} \,,
\end{aligned}$$

where $\boldsymbol{y}[p]$ and $\boldsymbol{n}[p]$ are K-sized vectors which contain respectively the received signals and the ambient Gaussian noise at all K subcarriers and at the p-th time slot; the diagonal elements of $\boldsymbol{X}_j[p]$ are the K coded symbols transmitted from the j-th transmitter antenna and at the p-th time slot.

Without CSI, the maximum *a posteriori* (MAP) detection problem is written as,

$$\hat{\boldsymbol{X}}[p] = \arg\max_{\boldsymbol{X}[p]} \log p(\boldsymbol{X}[p]|\boldsymbol{y}[p]) , \quad p = 1, 2, \ldots, Pq . \tag{7.23}$$

(Recall that $\boldsymbol{X}[0]$ contains pilot symbols.) The optimal solution to Eq. 7.23 is of prohibitive complexity. We next discuss the application of the expectation-maximization (EM) algorithm [15] for solving Eq. 7.23.

The basic idea of the MAP-EM algorithm is to solve problem Eq. 7.23 iteratively according to the following two steps: (For notational convenience, we temporarily drop the time index p, with the understanding that the MAP-EM algorithm discussed below is applied to each OFDM word in the data burst.)

1. E-step: Compute $Q(\boldsymbol{X}|\boldsymbol{X}^{(i)}) = \mathrm{E}\left\{[\log p(\boldsymbol{y}|\boldsymbol{X}, \boldsymbol{h})] \,\Big|\, \boldsymbol{y}, \boldsymbol{X}^{(i)}\right\}$; (7.24)
2. M-step: Solve $\boldsymbol{X}^{(i+1)} = \arg\max_{\boldsymbol{X}} Q(\boldsymbol{X}|\boldsymbol{X}^{(i)}) + \log P(\boldsymbol{X})$; (7.25)

where $\boldsymbol{X}^{(i)}$ denotes hard decisions of the data symbols at the i-th EM iteration, and $P(\boldsymbol{X})$ represents the *a priori* probability of $\boldsymbol{X}$, which is fed back by the LDPC decoder from the previous turbo iteration. It is known that the likelihood function $Q(\boldsymbol{X}|\boldsymbol{X}^{(i)})$ is nondecreasing and under regularity conditions the EM algorithm converges to a local stationary point [34].

In the E-step, the expectation is taken with respect to the "hidden" channel response $\boldsymbol{h}$ conditioned on $\boldsymbol{y}$ and $\boldsymbol{X}^{(i)}$. It is easily seen that, conditioned on $\boldsymbol{y}$ and $\boldsymbol{X}^{(i)}$, $\boldsymbol{h}$ is complex Gaussian distributed as

$$\boldsymbol{h}|(\boldsymbol{y}, \boldsymbol{X}^{(i)}) \sim \mathcal{N}_c(\hat{\boldsymbol{h}}, \hat{\boldsymbol{\Sigma}}_h) , \tag{7.26}$$

$$\begin{aligned}
\text{with} \quad \hat{\boldsymbol{h}} &\triangleq (\boldsymbol{W}^{\mathrm{H}}\boldsymbol{X}^{(i)\mathrm{H}}\boldsymbol{\Sigma}_n^{-1}\boldsymbol{X}^{(i)}\boldsymbol{W} + \boldsymbol{\Sigma}_h^{\dagger})^{-1}\boldsymbol{W}^{\mathrm{H}}\boldsymbol{X}^{(i)\mathrm{H}}\boldsymbol{\Sigma}_n^{-1}\boldsymbol{y} \\
&= (\boldsymbol{W}^{\mathrm{H}}\boldsymbol{X}^{(i)\mathrm{H}}\boldsymbol{X}^{(i)}\boldsymbol{W} + \boldsymbol{\Sigma}_h^{\dagger})^{-1}\boldsymbol{W}^{\mathrm{H}}\boldsymbol{X}^{(i)\mathrm{H}}\boldsymbol{y} , \\
\hat{\boldsymbol{\Sigma}}_h &\triangleq \boldsymbol{\Sigma}_h - (\boldsymbol{W}^{\mathrm{H}}\boldsymbol{X}^{(i)\mathrm{H}}\boldsymbol{\Sigma}_n^{-1}\boldsymbol{X}^{(i)}\boldsymbol{W} + \boldsymbol{\Sigma}_h^{\dagger})^{-1}\boldsymbol{W}^{\mathrm{H}}\boldsymbol{X}^{(i)\mathrm{H}}\boldsymbol{\Sigma}_n^{-1}\boldsymbol{X}^{(i)}\boldsymbol{W}\boldsymbol{\Sigma}_h \\
&= \boldsymbol{\Sigma}_h - (\boldsymbol{W}^{\mathrm{H}}\boldsymbol{X}^{(i)\mathrm{H}}\boldsymbol{X}^{(i)}\boldsymbol{W} + \boldsymbol{\Sigma}_h^{\dagger})^{-1}\boldsymbol{W}^{\mathrm{H}}\boldsymbol{X}^{(i)\mathrm{H}}\boldsymbol{X}^{(i)}\boldsymbol{W}\boldsymbol{\Sigma}_h ,
\end{aligned}$$

where $\boldsymbol{\Sigma}_n$ and $\boldsymbol{\Sigma}_h$ denote respectively the covariance matrix of the ambient white Gaussian noise $\boldsymbol{n}$ and channel responses $\boldsymbol{h}$. According to the assumptions in Sec. 7.1.1, both of them are diagonal matrices as $\boldsymbol{\Sigma}_n \triangleq \mathrm{E}(\boldsymbol{n}\boldsymbol{n}^{\mathrm{H}}) = \boldsymbol{I}$ and $\boldsymbol{\Sigma}_h \triangleq \mathrm{E}(\boldsymbol{h}\boldsymbol{h}^{\mathrm{H}}) = \mathrm{diag}[\sigma_{1,1}^2, \ldots, \sigma_{1,L_f'}^2, \sigma_{2,1}^2, \ldots, \sigma_{2,L_f'}^2]$, where $\sigma_{j,l}^2$ is the average power of the l-th tap related with the j-th transmitter antenna; $\sigma_{j,l}^2 = 0$ if the channel response at this tap is zero. Assuming that $\boldsymbol{\Sigma}_h$ is known (or measured with the aid of pilot symbols), $\boldsymbol{\Sigma}_h^{\dagger} \triangleq \mathrm{diag}[\gamma_{1,1}, \ldots, \gamma_{1,L_f'}, \gamma_{2,1}, \ldots, \gamma_{2,L_f'}]$ is defined as the pseudo inverse of $\boldsymbol{\Sigma}_h$ as $\gamma_{j,l} = 1/\sigma_{j,l}^2$ if $\sigma_{j,l}^2 \neq 0$ and $\gamma_{j,l} = 0$ otherwise. Using Eq. 7.5 and 7.26, $Q(\boldsymbol{X}|\boldsymbol{X}^{(i)})$ is computed as

$$Q(\boldsymbol{X}|\boldsymbol{X}^{(i)}) = -\mathrm{E}_{\boldsymbol{h}|(\boldsymbol{y},\boldsymbol{X}^{(i)})}\left\{\|\boldsymbol{y}-\boldsymbol{X}\boldsymbol{W}\boldsymbol{h}\|^2\right\} + \text{const.}$$

$$= -\sum_{k=0}^{K-1} \underbrace{\left\{\left[y[k] - \boldsymbol{x}^{\mathrm{H}}[k]\boldsymbol{W}'_f(k)\hat{\boldsymbol{h}}\right]^2 + \left[\boldsymbol{x}^{\mathrm{H}}[k]\hat{\boldsymbol{\Sigma}}_h(k)\boldsymbol{x}[k]\right]\right\}}_{q(\boldsymbol{x}[k])} + \text{const.}\,, \tag{7.27}$$

with $\boldsymbol{x}[k] \triangleq [x_1[k], x_2[k]]^{\mathrm{H}}_{2\times 1}$,

$$\boldsymbol{W}'_f(k) \triangleq \begin{bmatrix} \boldsymbol{w}_f^{\mathrm{H}}(k) & \boldsymbol{0} \\ \boldsymbol{0} & \boldsymbol{w}_f^{\mathrm{H}}(k) \end{bmatrix}_{2\times(2L'_f)} ,$$

$$\hat{\boldsymbol{\Sigma}}_h(k) \triangleq \begin{bmatrix} [\boldsymbol{W}\hat{\boldsymbol{\Sigma}}_h\boldsymbol{W}^{\mathrm{H}}]_{(k+1,k+1)} & [\boldsymbol{W}\hat{\boldsymbol{\Sigma}}_h\boldsymbol{W}^{\mathrm{H}}]_{(K+k+1,k+1)} \\ [\boldsymbol{W}\hat{\boldsymbol{\Sigma}}_h\boldsymbol{W}^{\mathrm{H}}]_{(k+1,K+k+1)} & [\boldsymbol{W}\hat{\boldsymbol{\Sigma}}_h\boldsymbol{W}^{\mathrm{H}}]_{(K+k+1,K+k+1)} \end{bmatrix}_{2\times 2} ,$$

where $[\boldsymbol{W}\hat{\boldsymbol{\Sigma}}_h\boldsymbol{W}^{\mathrm{H}}]_{(i,j)}$ denotes the (i,j)-th element of the matrix $[\boldsymbol{W}\hat{\boldsymbol{\Sigma}}_h\boldsymbol{W}^{\mathrm{H}}]$.

Next, based on Eq. 7.27, the M-step proceeds as follows

$$\begin{aligned} \boldsymbol{X}^{(i+1)} &= \arg\max_{\boldsymbol{X}} \left[Q(\boldsymbol{X}|\boldsymbol{X}^{(i)}) + \log P(\boldsymbol{X})\right] \\ &= \arg\max_{\boldsymbol{X}} \left[-\sum_{k=0}^{K-1} q(\boldsymbol{x}[k]) + \sum_{k=0}^{K-1} \log P(\boldsymbol{x}[k])\right] \qquad (7.28) \\ &= \sum_{k=0}^{K-1} \arg\min_{\boldsymbol{x}[k]} \left[q(\boldsymbol{x}[k]) - \log P(\boldsymbol{x}[k])\right] , \end{aligned}$$

$$\text{or} \quad \boldsymbol{x}^{(i+1)}[k] = \arg\min_{\boldsymbol{x}[k]} \left[q(\boldsymbol{x}[k]) - \log P(\boldsymbol{x}[k])\right] , \qquad k = 0, 1, \ldots, K-1 , \tag{7.29}$$

where Eq. 7.28 follows from the assumption that $\boldsymbol{X}$ contains independent symbols. It is seen from Eq. 7.29 that the M-step can be decoupled into K independent minimization problems, each of which can be solved by enumeration over all possible $\boldsymbol{x} \in \Omega \times \Omega$ (Recall that Ω denotes the set of all modulated symbols). Hence the total complexity of the maximization step is $\mathcal{O}(K|\Omega|^2)$. Note that unlike in [8], here the maximization in the M-step is carried out without taking the LDPC coding constraints into considerations, i.e., the symbols in $\boldsymbol{X}$ are treated as uncoded symbols. The LDPC coding structure is exploited by the turbo iteration as well as the LDPC decoder.

Within each turbo iteration, the above E-step and M-step are iterated for I times. At the end of the I-th EM iteration, the extrinsic *a posteriori* LLR's of the LDPC code bits are computed, and then fed to the soft LDPC decoder. At each OFDM subcarrier, two transmitter antennas transmit two coded symbols, which correspond to $(2\log_2|\Omega|)$ LDPC code bits. Based on Eq. 7.29, after I times EM iterations, the extrinsic *a posteriori* LLR of the j-th $(j = 1, \ldots, 2\log_2|\Omega|)$ LDPC code bit at the k-th subcarrier $d_{k,j}$ is computed at the output of the MAP-EM demodulator as follows,

$$\lambda_1^l(d_{k,j}) = \log \frac{P\left(d_{k,j}=+1|\boldsymbol{y}\right)}{P\left(d_{k,j}=-1|\boldsymbol{y}\right)} - \underbrace{\log \frac{P\left(d_{k,j}=+1\right)}{P\left(d_{k,j}=-1\right)}}_{\lambda_2^{l-1}(d_{k,j})}$$

$$= \log \frac{\sum_{\boldsymbol{x}^+ \in \mathcal{C}_j^+} \exp\Big(-q(\boldsymbol{x}^+) + \log P(\boldsymbol{x}^+)\Big)}{\sum_{\boldsymbol{x}^- \in \mathcal{C}_j^-} \exp\Big(-q(\boldsymbol{x}^-) + \log P(\boldsymbol{x}^-)\Big)} - \lambda_2^{l-1}(d_{k,j}) \quad (7.30)$$

Initialization of MAP-EM Demodulator

The performance of the MAP-EM demodulator (and hence the overall receiver) is closely related to the quality of the initial value of $\boldsymbol{X}^{(0)}[p]$ [cf. Eq. 7.24]. At each turbo iteration, $\boldsymbol{X}^{(0)}[p]$ needs to be specified to initialize the MAP-EM demodulator. Except for the first turbo iteration, $\boldsymbol{X}^{(0)}[p]$ is simply taken as $\boldsymbol{X}^{(I)}[p]$ given by Eq. 7.28 from the previous turbo iteration. We next discuss the procedure for computing $\boldsymbol{X}^{(0)}[p]$ at the first turbo iteration.

The initial estimate of $\boldsymbol{X}^{(0)}[p]$ is based on the method proposed in [21, 23], which makes use of pilot symbols and decision-feedback as well as spatial and temporal filtering for channel estimates. The procedure is listed in Table 7.1. In Table 7.1, Freq-filter denotes either the least-square error estimator (LSE) or the minimum mean-square-error estimator (MMSE) as

$$\text{LSE:} \qquad \textsf{Freq-filter}\Big\{\boldsymbol{y}, \boldsymbol{X}\Big\} \triangleq (\boldsymbol{W}^{\mathrm{H}}\boldsymbol{X}^{\mathrm{H}}\boldsymbol{X}\boldsymbol{W})^{-1}\boldsymbol{W}^{\mathrm{H}}\boldsymbol{X}^{\mathrm{H}}\boldsymbol{y}\,,$$

$$\text{MMSE:} \qquad \textsf{Freq-filter}\Big\{\boldsymbol{y}, \boldsymbol{X}\Big\} \triangleq (\boldsymbol{W}^{\mathrm{H}}\boldsymbol{X}^{\mathrm{H}}\boldsymbol{X}\boldsymbol{W} + \boldsymbol{\Sigma}_h^{\dagger})^{-1}\boldsymbol{W}^{\mathrm{H}}\boldsymbol{X}^{\mathrm{H}}\boldsymbol{y}\,,$$

where $\boldsymbol{X}$ represents either the pilot symbols or $\boldsymbol{X}^{(I)}$ provided by the MAP-EM demodulator. Comparing these two estimators, the LSE does not need any statistical information of $\boldsymbol{h}$, but the MMSE offers better performance in terms of mean-square-error (MSE). Hence, in the pilot slot, the LSE is used to estimate channels and to measure $\boldsymbol{\Sigma}_h$; and in the rest data slots, the MMSE is used. In Table 7.1, Temp-filter denotes the temporal filter, which is used to further exploit the time-domain correlation of the channel,

$$\textsf{Temp-filter}\Big\{\hat{\boldsymbol{h}}[p-1], \hat{\boldsymbol{h}}[p-2], \ldots, \hat{\boldsymbol{h}}[p-\iota]\Big\} \triangleq \sum_{j=1}^{\iota} a_j \hat{\boldsymbol{h}}[p-j]\,, \quad (7.31)$$

where $\hat{\boldsymbol{h}}[p-j]\ , j = 1, \ldots, \iota$, is computed from $(\star\star)$ [cf. Tab.1]; $\{a_j\}_{j=1}^{\iota}$ denotes the coefficients of an ι-length ($\iota \leq Pq$) temporal filter, which can be obtained by solving the Wiener equation or from the robust design as in [21, 23]. From the above discussions, it is seen that the computation involved in initializing $\boldsymbol{X}^{(0)}[p]$ mainly consists of the ML detection of $\boldsymbol{X}^{(0)}[p]$ in $(\star)$ and the estimation of $\hat{\boldsymbol{h}}[p]$ in $(\star\star)$. In general, for an LDPC-coded MIMO OFDM system with parameters (N_t, N_r, K, L_f'), the total complexity in initializing $\boldsymbol{X}^{(0)}[p]$ is $\mathcal{O}[(K|\Omega|^{N_t}) + N_r(N_t L_f')^3]$.

Table 7.1. Procedure for computing $\boldsymbol{X}^{(0)}[p]$ for the MAP-EM demodulator (at the first turbo iteration).

pilot slot:	$\hat{\boldsymbol{h}}[0] = \text{Freq-filter}\left\{\boldsymbol{y}[0], \boldsymbol{X}[0]\right\}$	
data slots: for	$p = 1, 2, \ldots, Pq$	
	$\tilde{\boldsymbol{h}}[p] = \text{Temp-filter}\left\{\hat{\boldsymbol{h}}[p-1], \hat{\boldsymbol{h}}[p-2], \ldots, \hat{\boldsymbol{h}}[p-\iota]\right\}$	
	$\boldsymbol{X}^{(0)}[p] = \text{arg max}_{\boldsymbol{X}}\left\{\log p\left[\boldsymbol{y}[p] \middle\vert \boldsymbol{X}, \tilde{\boldsymbol{h}}[p]\right]\right\}$	$(\star)$
	$\boldsymbol{X}^{(I)}[p] = \text{MAP-EM}\left\{\boldsymbol{y}[p], \boldsymbol{X}^{(0)}[p]\right\}$ [cf. Eq. 7.24-7.25]	
	$\hat{\boldsymbol{h}}[p] = \text{Freq-filter}\left\{\boldsymbol{y}[p], \boldsymbol{X}^{(I)}[p]\right\}$	$(\star\star)$
end		

7.2.3 Simulation Results

In this section, we provide computer simulation results to illustrate the performance of the proposed LDPC-coded MIMO OFDM system in frequency- and time-selective fading channels. The characteristics of the fading channels are described in Sec. 7.1.1. In the following simulations the available bandwidth is 1 MHz and is divided into 256 subcarriers. These correspond to a subcarrier symbol rate of 3.9 KHz and OFDM word duration of 256μs. In each OFDM word, a guard interval of 40μs is added to combat the effect of inter-symbol interference, hence $T = 296\mu$s. For all simulations, 512 information bits are transmitted from 256 subcarriers at each OFDM slot, therefore the information rate is $2 \times \frac{256}{296} = 1.73$ bits/sec/Hz. Unless otherwise specified, all the LDPC codes used in simulations are regular LDPC codes with column weight $s = 3$ in the parity-check matrices and with appropriate block-lengths and code rates, constructed by the method in [32]. The modulator uses quadrature phase shift keying (QPSK) constellation with Gray mapping. Simulation results are shown in terms of the OFDM frame-error-rate (FER) versus the signal-noise-ratio (SNR) γ.

Performance with Ideal CSI

Fig. 7.8–7.9 show the performance of multiple-antenna (N_t transmitter antennas and one receiver antenna) LDPC-coded MIMO OFDM systems by using turbo detection and decoding with ideal CSI. Performance is compared for systems with different fading profiles and different number of transmitter antennas. Namely, Ch1 denotes a channel with a single tap at 0μs, Ch2a denotes a channel with two equal-power taps at 0μs and 5μs, Ch2b denotes a channel with two equal-power taps at 0μs and 40μs, and Ch6a denotes a channel with six equal-power taps which equally spaced from 0μs to 40μs. Suffix N2 denotes a system with two transmitter antennas ($N_t = 2$), and similarly denotes N3; suffix P1 denotes that each LDPC-based STC code word spans one OFDM slot ($P = 1$), and similarly denote P5 and P10. Unless otherwise specified, all the STC-OFDM systems are assumed to use two transmitter antennas ($N_t = 2$) and each STC code word spans one OFDM slot ($P = 1$).

First, Fig. 7.8 shows the performance of the LDPC-coded MIMO OFDM system in frequency-selective and time-nonselective channels. The dash-dot curves represent the performance after the first turbo iteration; and the solid curves represent the performance after the fifth iteration. It is seen that the receiver performance is significantly improved through turbo iterations. During each turbo iteration, in the LDPC decoder, the maximum number of iterations is 30; and as observed in simulations, the average number of iterations needed in LDPC decoding is less than 10 when FER is less than 10^{-2}. Compared with the conventional trellis-coded MIMO OFDM system (see figures in [1]), the LDPC-coded MIMO OFDM system significantly improves performance, (e.g., there is around 5 dB performance improvement in Ch2a/Ch2b channels and even more improvement in Ch6a channels). Compared with an enhanced 256-state trellis-coded MIMO OFDM system [25, 26], the LDPC-coded MIMO OFDM system has lower decoding complexity but still has about 1-2 dB performance improvement in all these channels. Moreover, due to the inherent interleaving in LDPC encoder, the proposed LDPC-coded MIMO OFDM scheme narrows the performance difference between Ch2a and Ch2b channels (essentially the outage capacity of these two channels are same). As the selective-fading diversity order L increases from Ch1 to Ch6a, LDPC-coded MIMO OFDM can efficiently take advantage of the available diversity resources and hence can significantly improve the system performance. Moreover, in a highly frequency-selective channel Ch6a, the LDPC-coded MIMO OFDM performs only 3.0 dB away from the outage capacity of this channel (at high information rate 1.73 bits/sec/Hz) at FER of 2×10^{-4}.

Next, Fig. 7.9 shows the performance of the LDPC-coded MIMO OFDM system in frequency- and time-selective ($P \geq 1$) fading channels. The maximum Doppler frequency is 200 Hz (i.e., the normalized Doppler frequency is $f_d T = 0.059$). Again, it is seen that the performance of the system improves as the selective-fading diversity order L (including both the frequency-selectivity and time-selectivity) increases.

Finally, Fig. 7.8 also compares the performance of LDPC-coded MIMO OFDM systems with same multipath delay profiles (Ch2a) but with different number of transmitter antennas ($N_t = 2$ or $N_t = 3$). Since Ch2bN3 has larger outage capacity than Ch2bN2, it is seen that at medium to high SNR's Ch2bN3 starts to perform better than Ch2bN2 with a steeper slope, which shows that the LDPC-coded MIMO OFDM can be flexiblely scaled according to different number of transmitter antennas and can still improve the performance by exploiting the increased spatial diversity, especially at low FER, (which is attractive in data communication applications).

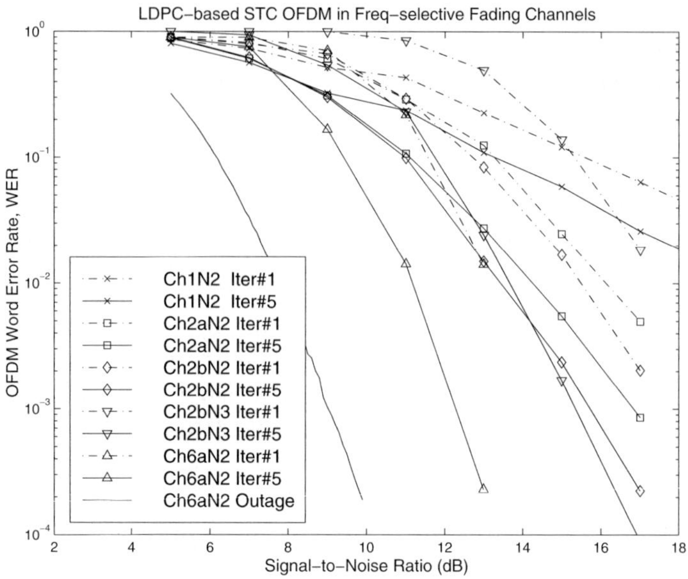

Fig. 7.8. Frame error rate (FER) of an LDPC-coded MIMO OFDM system with multiple antennas ($N_t = \{2, 3\}, N_r = 1$) in frequency-selective and time-nonselective fading channels, with ideal CSI.

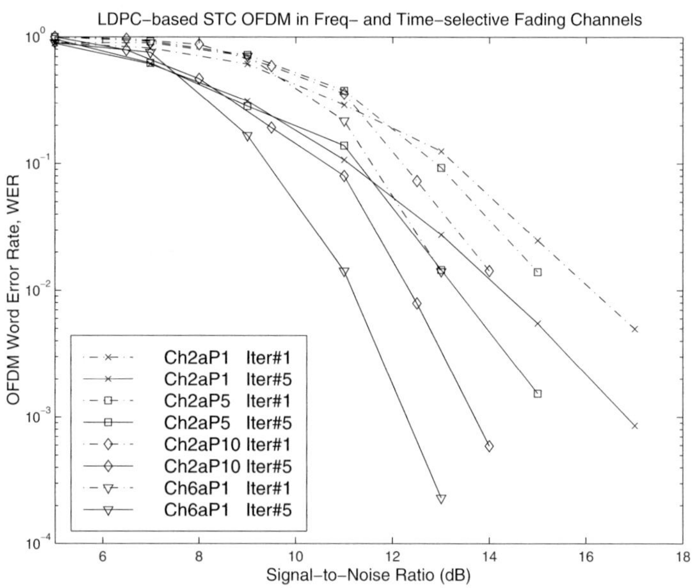

Fig. 7.9. Frame error rate (FER) of an LDPC-coded MIMO OFDM system with multiple antennas ($N_t = 2, N_r = 1$) in frequency-selective and time-selective fading channels, with ideal CSI.

Performance with Unknown CSI

In the following simulations, the receiver performance with unknown CSI is shown. The system transmits in a burst manner as illustrated in Fig. 7.5. Each data burst includes 10 OFDM words ($q = 9, P = 1$), the first OFDM word contains the pilot symbols and the rest 9 OFDM words contain the information data symbols. Simulations are carried out in two-tap (two equal-power taps at 0 μs and 1 μs) frequency- and time-selective fading channels. The maximum Doppler frequency of fading channels is 50 Hz or 150 Hz (with normalized Doppler frequencies 0.015 and 0.044 respectively). Note that in Fig. 7.10–7.11, the energy consumption of transmitting pilot symbols is not taken into account in computing SNR's.

The turbo receiver performance of a *regular* LDPC-coded MIMO OFDM system is shown in Fig. 7.10; whereas that of an *irregular* LDPC-coded MIMO OFDM system is shown in Fig. 7.11 (The average column weight in the parity-check matrix of the irregular LDPC code is 2.30). TurboDD denotes the turbo receiver as simulated in Sec. 7.2.3, except that the perfect CSI is replaced by the pilot/decision-directed channel estimates as proposed in [22]; and TurboEM denotes the turbo receiver with the MAP-EM demodulator as proposed in Sec. 7.2.2. The temporal filter parameters are taken from [21]. The performance of these two receiver structures are compared when using either the regular LDPC codes or the irregular LDPC codes. From the simulations, it is seen that with ideal CSI the receiver performance is close between the regular LDPC-coded MIMO OFDM system and the irregular LDPC-coded MIMO OFDM system. When the CSI is not available, the proposed TurboEM receiver significantly reduces the error floor. Moreover, it is observed that by using the irregular LDPC codes, both the TurboDD receiver and the TurboEM receiver improve their performance, and the TurboEM receiver can even approach the receiver performance with ideal CSI in low to medium SNR's. We believe that the reason for the better performance of irregular LDPC-based STC than regular LDPC-based STC in the presence of non-ideal CSI is due to the better performance of the irregular LDPC codes at low SNR's. In simulations, the turbo receiver takes 3 turbo iterations; and at each turbo iteration, the MAP-EM demodulator takes 3 EM iterations. At the cost of 10% pilot insertion and a modest complexity, the proposed turbo receiver with the MAP-EM demodulator is shown to be a promising receiver technique, especially in fast fading applications.

7.3 Design of LDPC for MIMO OFDM

In this section, we briefly introduce the background of LDPC codes and explain the basic idea of designing and optimizing the LDPC coded MIMO OFDM systems under the framework of turbo signal processing; more details are referred to [29].

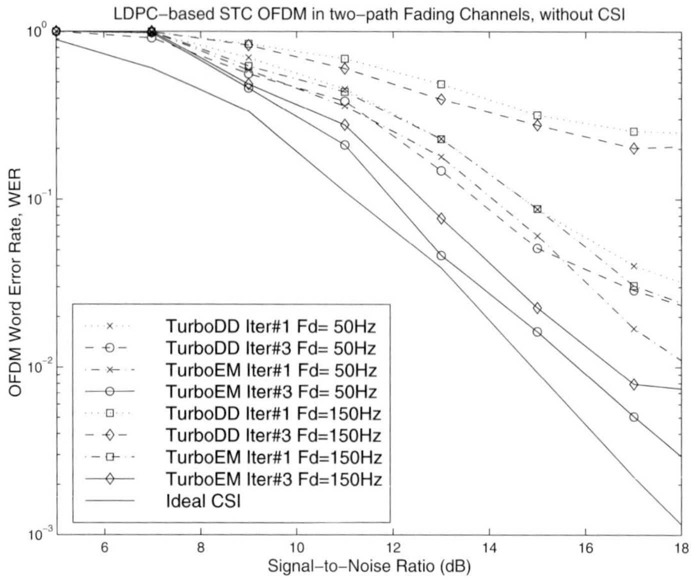

Fig. 7.10. Frame error rate (FER) of a *regular* LDPC-based STC-OFDM system with multiple antennas ($N_t = 2, N_r = 1$) in two-tap ($L = 2$) frequency-selective fading channels, without CSI.

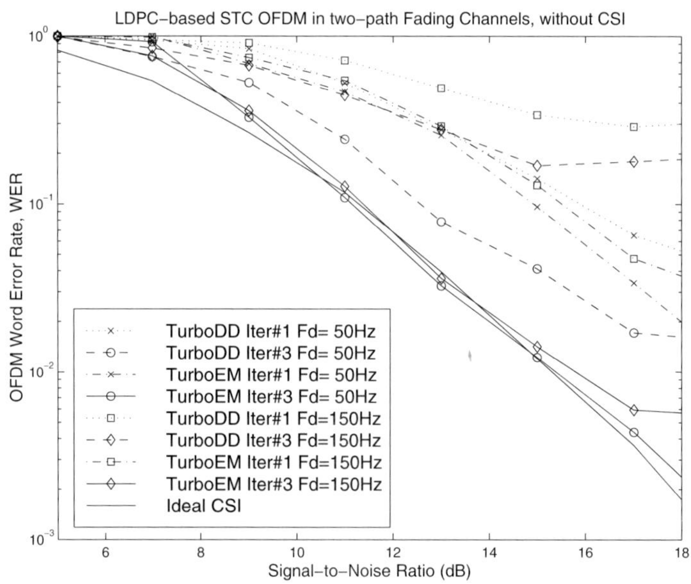

Fig. 7.11. Frame error rate (FER) of an *irregular* LDPC-based STC-OFDM system with multiple antennas ($N_t = 2, N_r = 1$) in two-tap ($L = 2$) frequency-selective fading channels, without CSI.

7.3.1 Low Density Parity Check (LDPC) Codes

A low density parity check (LDPC) code is a linear block code specified by a very sparse parity check matrix. The parity check matrix $\boldsymbol{P}$ of a *regular* (n, k, s, t) LDPC code of rate $r = k/n$ is a $(n-k) \times n$ matrix, which has s ones in each column and $t > s$ ones in each row where $s \ll n$, and the ones are typically placed at random in the parity check matrix. When the number of ones in every column is not the same, the code is known as an *irregular* LDPC code.

The code with parity check matrix $\boldsymbol{P}$ can be represented by a bipartite graph which consists of two types of nodes - variable nodes and check codes. Each code bit is a variable node while each parity check or each row of the parity check matrix represents a check node. An edge in the graph is placed between variable node i and check node j if $P_{j,i} = 1$. That is, each check node is connected to code bits whose sum modulo-2 should be zero. Irregular LDPC codes are specified by two polynomials $\lambda(x) = \sum_{i=1}^{d_{l_{max}}} \lambda_i \, x^{i-1}$ and $\rho(x) = \sum_{i=1}^{d_{r_{max}}} \rho_i \, x^{i-1}$, where λ_i is the fraction of edges in the bipartite graph that are connected to variable nodes of degree i, and ρ_i is the fraction of edges that are connected to check nodes of degree i. Equivalently, the degree profiles can also be specified from the node perspective, i.e., two polynomials $\tilde{\lambda}(x) = \sum_{i=1}^{d_{l_{max}}} \tilde{\lambda}_i \, x^{i-1}$ and $\tilde{\rho}(x) = \sum_{i=1}^{d_{r_{max}}} \tilde{\rho}_i \, x^{i-1}$, where $\tilde{\lambda}_i$ is the fraction of variable nodes of degree i, and $\tilde{\rho}_i$ is the fraction of check nodes of degree i.

7.3.2 Density Evolution Design of LDPC Coded MIMO OFDM

Density evolution is first proposed in [30, 42] as an effective technique to optimize the irregular LDPC codes; later on, it is further extended to analyze and optimize the general communication systems under the framework of turbo iterative signal processing, e.g., [36, 50]. The principal idea of density evolution is to treat the extrinsic information that is passed in the iterative process as random variables. Then, by estimating the pdf of the random variables as a function of SNR γ and iteration number, we can compute the probability of error at every iteration. When the length of the codewords $n \to \infty$, the extrinsic information passed along the edges connected to every check node and variable node can be assumed to be independent variables. This makes it possible to compute the pdf's relatively easily. More specifically, it is known that the extrinsic information passed from soft LDPC decoder to soft MIMO OFDM demodulator can be modeled as mixture symmetric Gaussian distributed [7, 14, 48]; and we propose to approximate the extrinsic information passed from soft MIMO OFDM demodulator to soft LDPC decoder as a mixture of symmetric Gaussian as well [29], therefore we can easily track the pdf's of the extrinsic LLRs within the LDPC code by only evaluating the means of the component Gaussian pdf's rather than the convolution or evaluation of exact pdf's. This significantly reduces the complexity in LDPC code optimization without sacrificing much performance. The optimized LDPC degree profiles $\tilde{\lambda}^*(x)$ and $\tilde{\rho}^*(x)$ and the minimum operational $\mathrm{SNR}_{min.op}$ are such that the means of the extrinsic information in turbo iterative processing tend to $+\infty$. A non-linear optimization procedure called differential evolution [38, 41] can be employed to numerically find the optimal solutions to the above problem.

We remark that besides density evolution, an alternative way to design LDPC code based systems is by extrinsic information transfer (EXIT) charts [49]. Overall, density evolution yields better design results at the cost of relatively higher design complexity, as compared to EXIT charts.

7.3.3 Numerical Results

In this section, we present numerical results for the design and optimization of LDPC coded MIMO OFDM systems. For each transmit-receive-antenna pair, DoCoMo's physical fading channel model, exponentially distributed frequency-selective fading with 88.8ns maximum delay spread, is adopted. OFDM modulation is used with subcarrier spacing 131.836 kHz and cyclic prefix interval of 1.54 μs; as a parameter to be discussed, the number of subcarriers K is specified next. It is clear that the total bandwidth is approximately K times of the subcarrier spacing, and the multipath resolution of the frequency-selective fading channel is the inverse of total bandwidth. For instance, with $K = 1024$ there are 12 resolvable paths in DoCoMo's channel model, but with $K = 512$ the number of resolvable paths is reduced to 6. The modulator uses the QPSK constellation with Gray mapping; for the considered MIMO systems with large number of antennas, the capacity (both ergodic and outage) difference between QPSK signaling and Gaussian signaling is small (e.g., $\sim$ 0.2dB at 4 bits/Hz/sec when $N_t = N_r = 4$). All the LDPC codes designed and optimized below have rate 1/2 and appropriate code lengths. (For clarity, the rate loss due to cyclic prefix is not counted here.)

All the regular LDPC codes are $(s = 3, t = 6)$ codes constructed by the method in [32]. All the irregular LDPC codes are obtained from the proposed design procedure of density evolution with mixture Gaussian approximation. For example, the optimized degree profile for the spatially uncorrelated 2×2 MIMO OFDM systems employing the MAP demodulator is $\lambda(x) = 0.269052x + 0.135031x^2 + 0.024564x^4 + 0.028685x^5 + 0.075819x^6 + 0.033661x^7 + 0.024360x^8 + 0.020951x^9 + 0.018975x^{10} + 0.014373x^{12} + 0.035585x^{13} + 0.015569x^{14} + 0.013611x^{16} + 0.289765x^{19}$ and $\rho(x) = 0.307710x^7 + 0.692290x^8$, and that for the spatially uncorrelated 2×2 MIMO OFDM systems employing the LMMSE-SIC demodulator is $\lambda(x) = 0.294388x + 0.100255x^2 +$

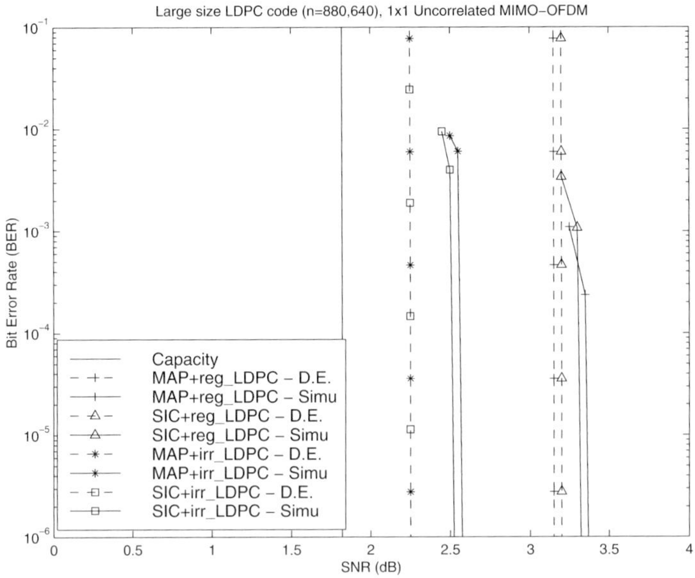

Fig. 7.12. Performance computed by density evolution analysis and computer simulations for ergodic 1×1 MIMO OFDM channels with no spatial correlation.

$0.056131x^3+0.042069x^4+0.032675x^5+0.065028x^7+0.030813x^8+0.027357x^9+0.025533x_{10}+0.029996x_{11}+0.014911x_{15}+0.020255x_{17}+0.013650x_{18}+0.246939x_{19}$ and $\rho(x) = 0.738497x^7 + 0.261503x^8$.

In Sec. 7.3.3–7.3.3, the performance of the LDPC codes in ergodic MIMO OFDM channels is demonstrated by bit-error-rate (BER) versus SNR (see Eq. 7.5); in Sec. 7.3.3, the performance in non-ergodic MIMO OFDM channels is demonstrated by frame-error-rate (FER) versus SNR.

Different Number of Antennas

If only single-transmit-receive-antenna is used, a cellular system designed for 100 Mbps peak rate downlink transmission requires very broad spectrum, as well as broadband transceiver circuitry; either of which could be costly for commercial applications. MIMO techniques provide a promising means to ameliorate this issue. For example, to achieve a fixed data rate of 100 Mbps, traditional single-antenna system requires 100 MHz bandwidth (assume QPSK modulation and coding rate 1/2), whereas a 4-transmit-4-receive-antenna system could potentially transmit the same 100 Mbps data rate using only 25MHz bandwidth. We note that the information rate in the single-antenna system is 1 bit/Hz/sec, whereas in the 4×4 MIMO system it is increased to 4 bits/Hz/sec (higher information rate implies a more efficient use of spectral resource).

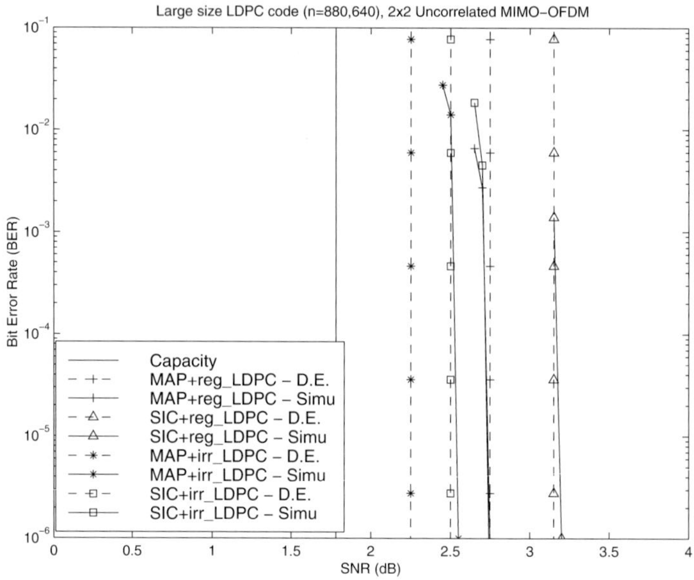

Fig. 7.13. Performance computed by density evolution analysis and computer simulations for ergodic 2×2 MIMO OFDM channels with no spatial correlation.

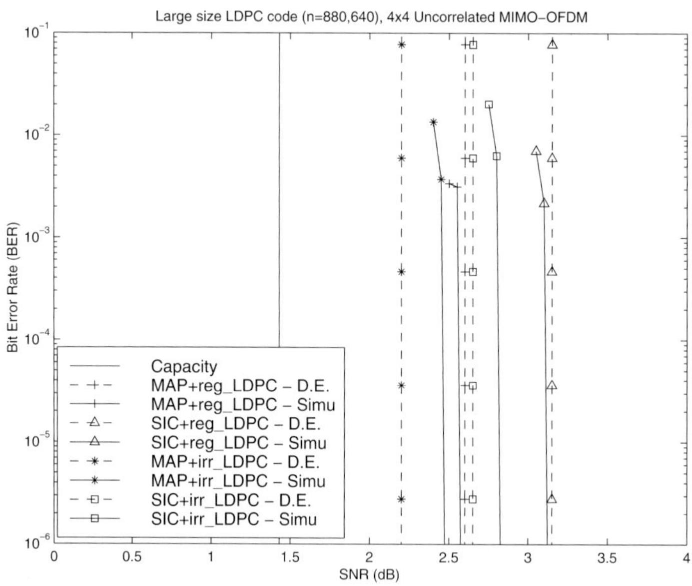

Fig. 7.14. Performance computed by density evolution analysis and computer simulations for ergodic 4×4 MIMO OFDM channels with no spatial correlation.

In our study, it is assumed that the number of receive antennas is the same as the number of transmit antennas, i.e., $N_t = N_r$. We consider 1×1, 2×2 and 4×4 MIMO OFDM systems. The design and optimization results are shown in Fig. 7.12–7.14. In these figures, the ergodic channel capacity computed from Eq. 7.11 and 7.13 is denoted by "Capacity". First, we focus on the performance of iterative receiver employing soft MAP demodulator, i.e., the curves denoted by "MAP+regLDPC –D.E.", "MAP+regLDPC –Simu", "MAP+irrLDPC –D.E." and "MAP+irrLDPC –Simu", where suffix "D.E." denotes the results from density evolution analysis and "Simu" denotes that from computer simulations. In order to achieve ergodic channel capacity, large block-size LDPC codes ($n = 880640$) are used to capture large number of fading channel realizations ($P = 430$). It is seen that by applying MIMO techniques, the information rate is increased to N_t bits/Hz/sec, while the ergodic capacity (the "Capacity" curve) is also slightly improved. Moreover, by employing the optimized irregular LDPC codes and the turbo iterative receiver employing the MAP demodulator, the operational $\text{SNR}_{min.op}$ of LDPC coded MIMO OFDM systems is within 1 dB from the information theoretic ergodic capacity. It is also seen that the performance calculated by density evolution analysis (the "D.E." curves) is in match with that obtained from simulations (the "Simu" curves). At last, we observe that the performance gap between the regular and the irregular LDPC codes tends to be smaller for systems with larger number of antennas.

Different Demodulation Schemes

The performance when employing sub-optimal LMMSE-SIC demodulator is demonstrated in Fig. 7.12–7.14, by the curves "SIC+regLDPC –D.E.", "SIC+regLDPC –Simu", "SIC+irrLDPC –D.E." and "SIC+irrLDPC –Simu". Compared to the MAP demodulator based performance (as in Sec. 7.3.3), the use of the LMMSE-SIC demodulator brings less than 1dB performance loss for 1×1, 2×2 and 4×4 systems. Therefore, in spatially uncorrelated ergodic MIMO OFDM channels, LMMSE-SIC demodulator appears to be a promising choice in practical implementation, for its good performance and relatively low-complexity.

Spatial Correlation

In this subsection, we discuss the performance of MIMO OFDM systems with spatial (antenna) correlation. Following [4], we assume uniform linear antenna placement at both the transmitter and the receiver. The channel model in Eq. 7.5 is modified as

$$\begin{aligned} \boldsymbol{y}[p,k] &= \sqrt{\frac{\gamma}{N_t}}\boldsymbol{R}^{1/2}\boldsymbol{H}[p,k]\boldsymbol{S}^{1/2}\boldsymbol{x}[p,k] + \boldsymbol{n}[p,k]\ , \\ & k = 0,\ldots,K-1, \quad p = 0,\ldots,P-1, \end{aligned} \tag{7.32}$$

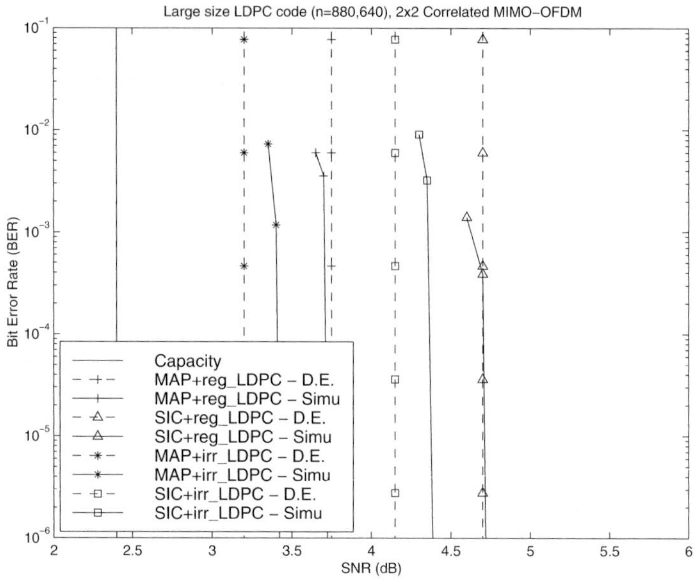

Fig. 7.15. Performance computed by density evolution analysis and computer simulations for ergodic 2×2 MIMO OFDM systems with spatial correlation.

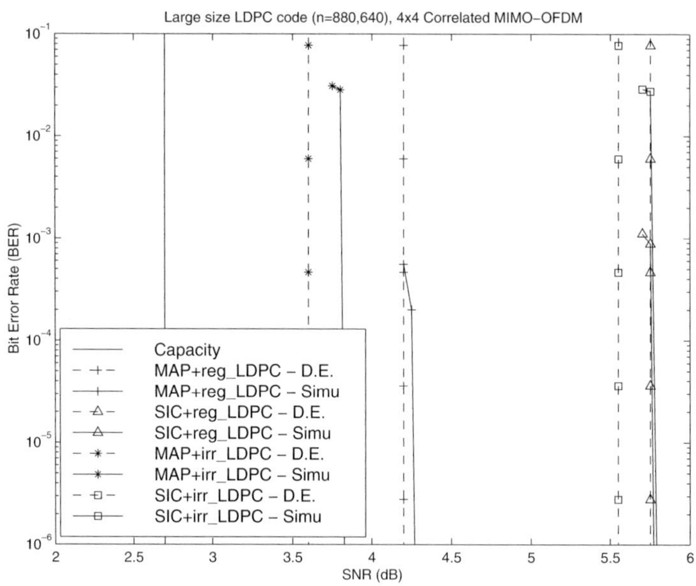

Fig. 7.16. Performance computed by density evolution analysis and computer simulations for ergodic 4×4 MIMO OFDM systems with spatial correlation.

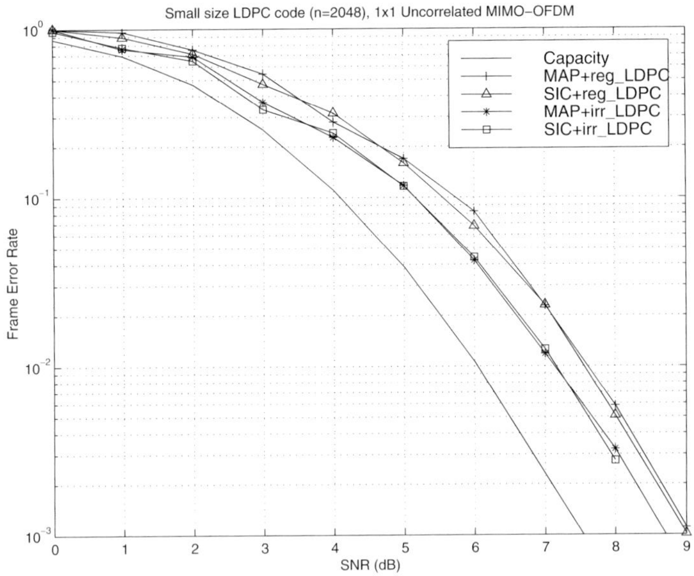

Fig. 7.17. Performance for non-ergodic 1×1 MIMO OFDM channels with no spatial correlation.

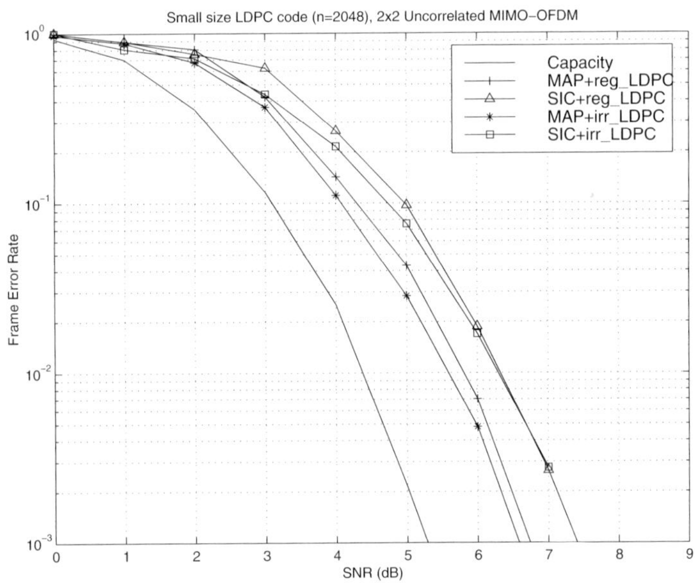

Fig. 7.18. Performance for non-ergodic 2×2 MIMO OFDM channels with no spatial correlation.

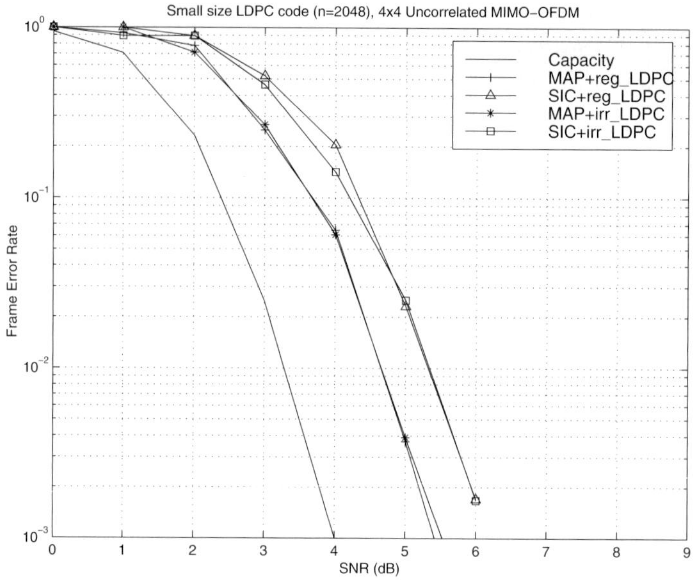

Fig. 7.19. Performance for non-ergodic 4×4 MIMO OFDM channels with no spatial correlation.

where $\boldsymbol{R} = \boldsymbol{R}^{1/2}\boldsymbol{R}^{1/2}$ and $\boldsymbol{S} = \boldsymbol{S}^{1/2}\boldsymbol{S}^{1/2}$ represent the receive and transmit spatial-correlation matrices, and are given by

$$[\boldsymbol{R}]_{(m,n)} = \exp\left(-j2\pi(n-m)d_t\cos(\bar{\theta}_t) - (2\pi(n-m)d_t\sin(\bar{\theta}_t)\sigma_{\theta_t})^2/2\right), \tag{7.33}$$

$$[\boldsymbol{S}]_{(m,n)} = \exp\left(-j2\pi(n-m)d_r\cos(\bar{\theta}_r) - (2\pi(n-m)d_r\sin(\bar{\theta}_r)\sigma_{\theta_r})^2/2\right), \tag{7.34}$$

where d_t denotes the transmitter antenna spacing normalized by carrier wavelength; $\bar{\theta}_t$ denotes the mean angle of departure for each scatterer cluster at the transmitter; $\sigma^2_{\theta_t}$ denotes the root-mean-square (RMS) of angle of departure at the transmitter; d_r, $\bar{\theta}_R$ and $\sigma^2_{\theta_r}$ denote the corresponding variables at the receiver side. In our experiments, we consider an urban micro-cell scenario [45] and for simplicity assume that all L paths follow the same spatial parameters as $\bar{\theta}_t = 53$, $\bar{\theta}_r = 18$, $\sigma_{\theta_t} = 8$, $\sigma_{\theta_r} = 2$; we also let $d_t = 4.0$ and $d_r = 0.5$ to reflect the situations that the antennas at base station are easier to be sparsely placed than the antennas at mobile devices. It is worth to note that some parameters (e.g., $\sigma_{\theta_r} = 2$) here are intentionally set to be worse than typical scenarios in order to highlight the effect of spatial correlation. Going through the same design and optimization procedure, we obtain the analysis and design results in Fig. 7.15–7.16. (The issue of antenna correlation does not exist for 1×1 systems.) Compared to spatially uncorrelated systems, antenna correlation causes channel capacity loss for the systems considered here. Nevertheless, the optimized irregular LDPC codes along with the MAP demodulator based iterative receiver can yield a performance within 1dB from the capacity of correlated channels. This demonstrates again the generality and efficacy of the methods of density evolution with mixture Gaussian approximation in optimizing LDPC-coded OFDM MIMO systems. However, compared to the corresponding result in spatially uncorrelated channels (4×4 systems in particular) the performance of the LMMSE-SIC based receivers is degraded, this is because the correlation matrices $\boldsymbol{R}$ and $\boldsymbol{S}$ lead to a larger conditional number of matrix $\boldsymbol{R}^{1/2}\boldsymbol{H}[p,k]\boldsymbol{S}^{1/2}$ and therefore the matrix operations in LMMSE-SIC (e.g., matrix inverse) are more subject to numerical instability.

Small Block-Size LDPC Coded MIMO OFDM

So far, we have focused on the design and optimization of the LDPC-coded MIMO OFDM systems aiming to achieve the ergodic capacity. In doing so, large block-size LDPC codes were employed for the following reasons. (1) In order to achieve the ergodic channel capacity, the LDPC code word must be long enough to experience a very large number of fading channel realizations. (2) The results of the density evolution analysis are based on the assumption that extrinsic messages connected to each check node and variable node are independent, which holds valid when LDPC code block-size is very large. (3) In the procedure of the density evolution analysis and design, optimized degree profiles, $\lambda(x)$ and $\rho(x)$, are first obtained, from which irregular LDPC codes are then randomly constructed; according to the theorem of concentration around ensemble average and the theorem of convergence to cycle-free ensemble average [41], such randomly constructed LDPC codes are guaranteed to have vanishing probability of error above the $\mathrm{SNR}_{min.op}$ threshold (correspond to the optimized $\lambda(x)$ and $\rho(x)$), when its code block-size is very large. In reality, however, the price paid for achieving the ergodic channel capacity (or error-free communications) by employing very large block-size codes is large decoding delay. Usually, if small amount of fading outage is tolerable, it is a more common practice to employ a small block-size LDPC code, which spans a small number of fading channel states. The sensible performance measure accordingly is outage capacity [cf. Eq. 7.12]. Next, a heuristic design approach which claims no theoretical optimality is proposed. The design begins with the degree profiles that have been optimized above for the ergodic channels (i.e., $P \to \infty$). Based on these degree profiles, a small block-size LDPC code is randomly constructed with proper girth conditioning, e.g., the bit-filling algorithm in [6].

The heuristically constructed small block-size LDPC codes ($n = 2048$) are simulated in non-ergodic MIMO OFDM channels ($P = 1$). In Fig. 7.17–7.19, the performance of regular and irregular LDPC codes when employed in systems with different number of antennas and different types of demodula-

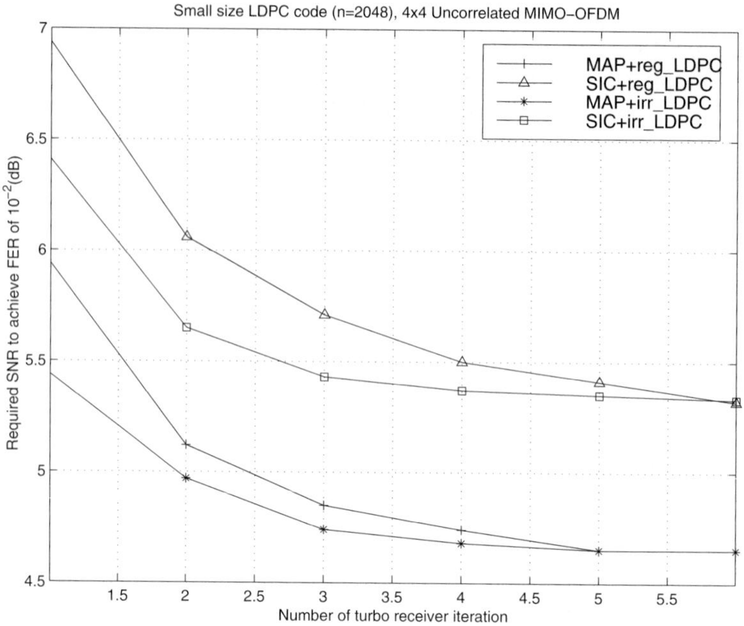

Fig. 7.20. For short block-size LDPC codes in 4×4 MIMO OFDM systems with no spatial correlation, the performance is plotted as the required SNR (dB) to achieve the FER of 10^{-2} versus the number of turbo receiver iteration. Note that flatter curves indicate faster receiver convergence.

tor is presented. Similar to the conclusions we drew in ergodic channels, the proposed LDPC coded MIMO OFDM systems can achieve both information rate increase and performance improvement when using multiple antennas; the MAP demodulator based iterative receiver can perform 1.5dB from the outage capacity; and the low-complexity LMMSE-SIC demodulator based receiver incurs additional small performance loss ($<$ 1dB). Furthermore, in order to demonstrate the process of receiver convergence, the results shown in Fig. 7.17–7.19 are presented in another form in Fig. 7.20, namely the required SNR (dB) to achieve a FER of 10^{-2} versus the number of turbo receiver iteration. In spatially uncorrelated 4×4 MIMO OFDM systems, for both MAP and LMMSE-SIC receivers, we see that although the performance difference between regular and irregular LDPC codes after receiver convergence (the curve "Iter #6") is negligible, the irregular LDPC codes based designs have faster receiver convergence speed. Around 0.5dB gain is achieved after the first receiver iteration for both MAP and LMMSE-SIC receivers. This observation suggests another benefit of optimizing LDPC codes, i.e., to help reduce the number of receiver iterations and consequently the receiver complexity in non-ergodic MIMO OFDM channels.

Table 7.2. A mismatch study to demonstrate the reward of channel-matching irregular LDPC design. The performance is obtained through simulations in the spatial-uncorrelated MIMO OFDM channels. *LDPC.I* denotes the irregular LDPC codes optimized for the corresponding MIMO OFDM channels; *LDPC.II* denotes the irregular LDPC codes originally designed for AWGN channels. The performance (SNR) of the short block LDPC coded MIMO OFDM is measured at FER of 10^{-2}.

	Large Block Irregular LDPC			Small Block Irregular LDPC		
SNR (dB)	LDPC.I	LDPC.II	Channel-specific Design Gain	LDPC.I	LDPC.II	Channel-specific Design Gain
MAP (1×1)	2.57	2.57	0.00	7.08	7.08	0.00
MAP (2×2)	2.56	2.61	0.05	5.57	5.72	0.15
MAP (4×4)	2.46	2.65	0.19	4.48	4.81	0.33
SIC (1×1)	2.52	2.52	0.00	7.06	7.06	0.00
SIC (2×2)	2.75	2.92	0.17	6.32	6.44	0.12
SIC (4×4)	2.82	3.17	0.35	5.33	5.70	0.37

A Mismatch Study

In the above, the performance of the optimized LDPC coded MIMO OFDM is demonstrated, with LDPC codes being optimized for specific MIMO channels. The reward of the channel-specific LDPC code design is exhibited in Table 7.2, by comparing the performance of both MIMO-channel-optimized irregular LDPC codes and AWGN-channel-optimized irregular LDPC codes in MIMO OFDM channels. In general, the channel-specific design gain increases for systems with larger number of antennas. In addition, in non-ergodic channels, a good AWGN-optimized irregular LDPC code also exhibits the faster convergence of turbo iterative receiver than the non-optimized regular LDPC codes.

7.4 Conclusion

In this chapter, we have presented the applications of turbo principle in receiver design and code optimization of an LDPC coded MIMO OFDM system, which achieves high spectral efficiency and capacity-approaching performance in wireless channels. In order to bring the proposed LDPC-based MIMO OFDM system into practice, we considered the receiver design when channel state information (CSI) is not directly available and only a low percentage of known pilot symbols is transmitted, and developed a novel turbo receiver which employs an MAP-EM demodulator and a soft LDPC decoder. Simulations show that in fast fading channels the proposed turbo receiver can approach the receiver performance with ideal CSI in low SNRs and significantly reduce error floors which exist in conventional non-iterative receivers. In order to further improve the system performance, we considered the LDPC code optimization by density evolution, and optimized the degree profiles of irregular LDPC codes for various demodulation schemes and spatial channel characteristics. Simulations show that the optimized LDPC MIMO OFDM systems can perform less than 1.0 dB from ergodic capacity and less than 1.5 dB from outage capacity of MIMO OFDM channels

References

1. D. Agrawal, V. Tarokh, A. Naguib, and N. Seshadri. "Space-time coded OFDM for high data-rate wireless communication over wideband channels." In *IEEE Vehicular Technology Conference, 1998.*, Ottawa, Canada, May 1998.
2. C. Berrou and A. Glavieux. "Near optimum error correction coding and decoding: Turbo-codes." *IEEE Trans. Commun.*, 44:1261–1271, Oct. 1996.
3. C. Berrou, A. Glavieux, and P. Thitimajshima. "Near shannon limit error-correction coding and decoding: Turbo codes." In *Proc. 1993 International Conference on Communications. ICC'93*, Geneva, Switzerland, May 1993.
4. H. Bölcskei, D. Gesbert, and A. J. Paulraj. "On the capacity of OFDM-based spatial multiplexing systems." *IEEE Trans. Commun.*, pages 225–234, Feb. 2002.
5. G. Caire and G. Colavolpe. "On low-complexity space-time coding for quasi-static channels." *IEEE Trans. Inform. Theory*, 49:1400–1416, June 2003.
6. J. Campello, D. S. Modha, and S. Rajagopalan. "Designing LDPC codes using bit-filling." In *IEEE International Conference on Communications (ICC) 2001*, pages 55–59, Oct. 2001.
7. S.-Y. Chung, G. D. Forney, T. J. Richardson, and R. Urbanke. "On the design of low-density parity-check codes within 0.0045 dB of the shannon limit." *IEEE Commun. Lett.*, 52:58–60, Feb. 2001.
8. C. Cozzo and B. L. Hughes. "Joint detection and estimation in space-time coding and modulation." In *Thirty-Third Asilomar Conference on Signals, Systems & Computers*, pages 613–617, Asilomar, CA, Oct. 1999.
9. M.O. Damen, H. El Gamal, and N.C. Beaulieu. "Linear threaded algebraic space-time constellations." *IEEE Trans. Inform. Theory*, 49:2372–2388, Oct. 2003.
10. C. Douillard et al. "Iterative correction of intersymbol interference: Turbo equalization." *European Trans. on Telecommun.*, 6:507–511, Sept.-Oct. 1995.
11. G. J. Foschini and M. J. Gans. "On limits of wireless communications in a fading environment when using multiple antennas." *Wireless Personal Communications*, 6:311–335, Mar. 1998.
12. R. G. Gallager. "Low density parity check codes." *IRE Trans. Inform. Theory*, 8:21–28, 1962.
13. H. El Gamal, G. Caire, and M. O. Damen. "Lattice coding and decoding achieve the optimal diversity-vs-multiplexing tradeoff of MIMO channels." *IEEE Trans. Inform. Theory*, 2004.
14. H. El Gamal and A. R. Hammons. "Analyzing the turbo decoder using the Gaussian assumption." *IEEE Trans. Inform. Theory*, pages 671–686, Feb. 2001.
15. C. N. Georghiades and J. C. Han. "Sequence estimation in the presence of random parameters via the EM algorithm." *IEEE Trans. Commun.*, 45:300–308, Mar. 1997.

16. J. Hagenauer. "The Turbo principle: Tutorial introduction and state of the art." In *Proc. International Symposium on Turbo Codes and Related Topics*, Brest, France, Sept. 1997.
17. B. Hassibi and B. Hochwald. "High-rate codes that are linear in space and time." *IEEE Trans. Inform. Theory*, 48:1804–1824, 2002.
18. H. Imai and S. Hirakawa. "A new multilevel coding method using error-correcting codes." *IEEE Trans. Inform. Theory*, 23:371–377, May 1977.
19. L.H.-J. Lampe, R. Schober, and R.F.H. Fischer. "Multilevel coding for multiple-antenna transmission." *IEEE Trans. Wireless Commun.*, 3:203–208, 2004.
20. H. J. Landau and H. O. Pollack. "PSWFs-III: The dimension of the space of essentially time- and band-limited signals." *Bell Syst. Tech. J.*, pages 1295–1320, July 1962.
21. Y. Li, L. J. Cimini, and N. R. Sollenberger. "Robust channel estimation for OFDM systems with rapid dispersive fading channels." *IEEE Trans. Commun.*, 46:902–915, July 1998.
22. Y. Li, N. Seshadri, and S. Ariyavisitakul. "Channel estimation for OFDM systems with transmitter diversity in mobile wireless channels." *IEEE J. Select. Areas Commun.*, 17:461–471, Mar. 1999.
23. Y. Li and N. R. Sollenberger. "Adaptive antenna arrays for OFDM systems with cochannel interference." *IEEE Trans. Commun.*, 47:217–229, Feb. 1999.
24. Y. Liu, M.P. Fitz, and O.Y. Takeshita. "Full rate space-time turbo codes." *IEEE J. Select. Areas in Commun.*, 19:969–980, 2001.
25. B. Lu and X. Wang. "Space-time code design in OFDM systems." In *IEEE Globecom Conference, 2000*, San Francisco, CA, Nov. 2000.
26. B. Lu and X. Wang. "A space-time trellis code design method for OFDM systems." *Wireless Personal Commun.*, 24(3), pp.403-418, Feb. 2003. 2000.
27. B. Lu, X. Wang, and K. R. Narayanan. "LDPC-based space-time coded OFDM systems over correlated fading channels." *IEEE Trans. Commun.*, 50:74–88, Jan. 2002.
28. B. Lu, X. Wang, and Y.G. Li. "Iterative receivers for space-time block coded OFDM systems in fading channels." *IEEE Trans. Wireless Commun.*, 1(2), pp.213-225, Apr. 2002.
29. B. Lu, G. Yue, and X. Wang. "Performance analysis and design optimization of LDPC coded MIMO OFDM systems." *IEEE Trans. Signal Processing*, 52:348–361, Feb. 2004.
30. M. G. Luby, M. Mitzenmacher, M. A. Shokrollahi, and D. A. Spielman. "Analysis of low density codes and improved designs using irregular graphs." In *Proceeding of 30th ACM Symp. on Theory of Computing*, 1998.
31. D. J. C. MacKay. "Good error-correcting codes based on very sparse matrices." *IEEE Trans. Inform. Theory*, 45:399–431, Mar. 1999.
32. D. J. C. MacKay and R. M. Neal. "Near shannon limit performance of low density parity check codes." *Electronic Letters*, 33:457–458, Mar. 1997.
33. İ. E. Telatar. "Capacity of multi-antenna Gaussian channels." *European Trans. Telecommun.*, 10:585–595, Nov.-Dec. 1999.
34. G. J. McLachlan and T. Krishnan. *The EM Algorithm and Extensions.* John Wiley & Sons, Inc., New York, USA, 1997.
35. M. Moher. "An iterative multiuser decoder for near-capacity communications." *IEEE Trans. Commun.*, 46:870–880, July 1998.

36. K. R. Narayanan, X. Wang, and G. Yue. "Design of low density parity check codes for SIC/MMSE turbo equalization." *Submitted to IEEE Trans. Wireless Commun.*, Feb. 2002.
37. H. V. Poor. "Turbo multiuser detection: A primer." *J. Communictions and Networks*, 3:196–201, Sep. 2001.
38. K. Price and R. Storn. "Differential evolution - a simple and efficient heuristic for global optimization over continuous spaces." *Journal of Global Optimization*, 11:341–369, 1997.
39. John G. Proakis. *Digital Communications.* McGraw-Hill, Inc., New York, USA, 3rd edition, 1995.
40. D. Reynolds and X. Wang. "Low-complexity turbo equalization for diversity channels." *Signal Processing*, 81:989–995, May. 2001.
41. T. J. Richardson, M. A. Shokrollahi, and R. L. Urbanke. "Design of capacity-approaching irregular low-density parity-check codes." *IEEE Trans. Inform. Theory*, 47:619–637, Feb. 2001.
42. T. J. Richardson and R. L. Urbanke. "Capacity of low density parity check codes under message passing decoding." *IEEE Trans. Inform. Theory*, 47:599–618, Feb. 2001.
43. M. Sellathurai and G.J. Foschini. "Stratified diagonal layered space-time architectures: signal processing and information theoretic aspects." *IEEE Trans. Sig. Proc.*, 51:2943–2954, Nov. 2003.
44. P.H. Siegel, D. Dvisalar, E. Eleftheriou, J. Hagenauer, and D. Eowitch (Eds.). "Special issue on the turbo principle: From theory to practice i & ii." *IEEE Journal on Selected Areas in Communications*, 19, May & Sept. 2001.
45. M. Stege, M. Bronzel, and F. Fettweis. "MIMO-cpacities for COST 259 scenarios." In *Proc. Inter. Zurich Seminar on Broadband Commun.*, Feb. 2002.
46. V. Tarokh, H. Jafarkhani, and A. R. Calderbank. "Space-time block codes from orthogonal designs." *IEEE Trans. Inform. Theory*, 45:1456–1467, July 1999.
47. V. Tarokh, N. Seshadri, and A. R. Calderbank. "Space-time codes for high data rate wireless communication: Performance criterion and code construction." *IEEE Trans. Inform. Theory*, 44:744–765, Mar. 1998.
48. S. ten Brink. "Convergence of iterative decoding." *Electronics Letters*, pages 1117–1118, June 1999.
49. S. ten Brink, G. Kramer, and A. Ashikhmin. "Design of low-density parity-check codes for modulation and detection." *IEEE Trans. Commun.*, 52:670–678, Apr. 2004.
50. M. Tuchler, R. Koetter, and A.C. Singer. "Turbo equalization: principles and new results." *IEEE Trans. Commun.*, 50:754–767, May 2002.
51. G. Ungerboeck. "Channel coding with multilevel/phase signals." *IEEE Trans. Inform. Theory*, 28:55–67, Jan. 1982.
52. J.-J. van de Beek, O. Edfors, M. Sandell, S. K. Wilson, and P. O. Börjesson. "On channel estimation in OFDM systems." In *IEEE Vehicular Technology Conference, 1995. VTC'95*, Chicago, IL, July 1995.
53. X. Wang and R. Chen. "Adaptive Bayesian multiuser detection for synchronous CDMA with Gaussian and impulsive noise." *IEEE Trans. Signal Processing*, 48:2013–2028, July 2000.
54. X. Wang and H. V. Poor. "Iterative (Turbo) soft interference cancelation and decoding for coded CDMA." *IEEE Trans. Commun.*, 47:1046–1061, July 1999.

55. Z. Yang, B. Lu, and X. Wang. "Bayesian Monte Carlo multiuser receivers for space-time coded multicarrier CDMA systems." *IEEE J. Select. Areas Commun.*, 19:1625–1637, Aug. 2001.
56. E. Zehavi. "8-PSK trellis codes for a Rayleigh channel." *IEEE Trans. Commun.*, 40:873–884, May 1992.

Chapter 8

Space-Time Turbo Coded Modulation for Future Wireless Communication Systems

Djordje Tujkovic

University of Oulu, Finland

Inspiring the recent renaissance in spatial and temporal signal processing, the latest results in information theory [1, 2] supported by massive field measurement campaigns [3, 4] have demonstrated the potentially large capacity increase for certain multiple-input multiple-output (MIMO) wireless channels [5, 6]. As an eventual evolution of space-time (ST) processing techniques [7], ST coding [8, 9] is a signalling method intended to approach the MIMO channel capacity by simultaneously encorrelating signals across spatial and temporal domains.

Owing to its integrated trellis coding and modulation design with a redundancy expansion in both the signal and antenna space, space-time trellis codes (STTrCs) [10, 11] can be characterized as a generalization of classical trellis coded modulation (TCM) [12, 13] to multi-antenna systems. The high computation complexity compelled by the lack of systematic procedure for reducing the cardinality of the searched class of codes, restrained true STTrC optimization mostly to low constraint-length, up to 64-state trellis codes. Being primarily designed to achieve the maximum transmit diversity gain, such STTrCs commonly settled with rather modest coding gains. As another disadvantage of limited freedom, their optimization was almost exclusively bound for either slow or fast fading channels. Therefore, in practical applications characterized by extremely variable Doppler frequencies, STTrCs from the literature fail to demonstrate much needed robustness.

In an analogy to *Robertson et al.* single antenna turbo coded modulation from [14], a design method for constituent recursive STTrCs (Rec-STTrCs) and parallel concatenated space-time turbo coded modulation (ST-TTCM) was introduced recently by author in his Ph.D. thesis [15]. The new ST coding framework, applicable to any STTrC from the literature, originally integrated code concatenation into a random-like ST coding approach [16]. Similar to single antenna turbo codes [17], the large equivalent constraint-lengths and randomness were jointly provided by an information interleaver, a building block that due to iterative decoding does not add considerably to decoding complexity.

K. Sripmanwat (ed.), Turbo Code Applications: a journey from a paper to realization, 193–222.

Paper [16] considered a punctured recursive constituent code (CC) design based on 4-, 8- and 16-state *Tarokh et al.* STTrCs (Tarokh-STTrCs) from [10], handcrafted for a spectral efficiency of 2 bps/Hz with two transmit antennas and QPSK modulation. The approach was further extended in [18, 19] to 3 and 4 bps/Hz employing 8PSK and 16QAM modulation, respectively. Similar schemes were studied in parallel in [20, 21], [22] and their accompanied papers [23–25], with distinct CCs and information interleaving structural differences that will become apparent after examining the system model in the next section.

The promising results were obtained when considering the application of ST-TTCM to different multi-antenna air-interfaces. This included orthogonal frequency division multiplexing (OFDM) systems in [26], wide-band code division multiple access (WCDMA) systems in [27, 28] and multi-carrier CDMA systems in [29]. The unified approach to single- and multi-antenna TTCM design, performance analysis and constituent TCM optimization was further introduced in [30]. ST-TTCM was analyzed in terms of the two-dimensional distance spectrum (DS), the truncated union bound (UB) and the iterative decoding convergence. The derived bounds appeared to be tight over fast fading channels but rather optimistic over quasi-static fading channels. Nevertheless, the rank deficient part of the DS has been well enumerated, which turned out to be crucial for the accurate design of the full transmit diversity ST-TTCM in the second half of [30].

Modified design criteria for space-time (ST) codes over quasi-static and fast fading channels were further introduced to capture the joint effects of error coefficients and multiplicities in the code's DS. A recursive systematic form with a primitive equivalent feed-backward polynomial was assumed for constituent codes (CCs) to assure good convergence of iterative decoding. The DS optimization produced new sets of CCs for ST-TTCM resulting in a robust ST coding scheme with improved performance over both quasi-static and fast fading channels. Full transmit diversity was achieved with no constrains on the implemented pseudo-random information interleaving. The results of code optimization for the special case of two-transmit antenna ST-TTCM over quasi-static and fast fading channels were further reported in [31] and [32], respectively.

The unified approach was further applied to single-antenna TTCM over AWGN and fading channels in [33]. The upper bounds appeared as rather tight over both AWGN and fast fading channels. Owing to accurate DS interpretation, the new set of constituent TCMs resulted in robust TTCM in terms of both FER and BER. Compared to *Robertson et al.* [14], the newly designed set of constituent TCMs resulted in robust TTCM in terms of both FER and BER. Considerable gains were achieved in the error floor region over AWGN and fast fading channels.

In this book chapter, the recent findings on single- and multi-antenna turbo coded modulations for future wireless communication systems are summarized. The ST-TTCM, as a new paradigm for MIMO signal transmission which combines space-time coding, turbo coding and higher order modulation under the same framework, will be put into the broader perspective of prior art in channel coding theory.

8.1 System Model

This section introduces the system model and main notations used throughout the chapter.

8.1.1 Encoder

Consider a general ST-TTCM system with N transmit and M receive antennas depicted in Fig. 8.1, which for $N = M = 1$ includes the special case of *Robertson et al.* TTCM from [14]. As in [16, 19], the transmitter employs a parallel concatenation of two CCs followed by puncturing, multiplexing and channel interleaving. The input information frame $\boldsymbol{d} = [\boldsymbol{d}(1), \ \boldsymbol{d}(2), \ldots, \boldsymbol{d}(L)]$ consisting at each time instant $l, l = 1, 2, \ldots, L$, of a sequence of K bits $\boldsymbol{d}(l) = [d_1(l), \ d_2(l), \ldots, d_K(l)]$ is first encoded by the first CC to produce a coded frame $\boldsymbol{x}^1 = \left[\boldsymbol{x}^1(1), \ \boldsymbol{x}^1(2), \ldots, \boldsymbol{x}^1(L)\right]$. At each time instant l, a coded frame consists of N complex symbols $\boldsymbol{x}^1(l) = \left[x_1^1(l), x_2^1(l), \ldots, x_N^1(l)\right]$ each belonging to a 2^Z complex constellation with unit average energy. The interleaved version of the input information sequence is then encoded by the second CC to produce a coded frame $\boldsymbol{x}^2 = \left[\boldsymbol{x}^2(1), \boldsymbol{x}^2(2), \ldots, \boldsymbol{x}^2(L)\right]$, $\boldsymbol{x}^2(l) = \left[x_1^2(l), x_2^2(l), \ldots, x_N^2(l)\right]$. To preserve the overall rate, the two coded frames are punctured and multiplexed so that at one time instant, only one of the component encoders has access to N transmit antennas. Unlike [16, 19], only the trellis of the first CC was terminated at the end of the transmission frame while the trellis of the second CC was left open. The resulting ST-TTCM coded frame is denoted by $\boldsymbol{x} = [\boldsymbol{x}(1), \boldsymbol{x}(2), \ldots, \boldsymbol{x}(L)]$, $\boldsymbol{x}(l) = [x_1(l), x_2(l), \ldots, x_N(l)]$ with $\boldsymbol{x}(l) = \boldsymbol{x}^1(l)$ for l odd and $\boldsymbol{x}(l) = \boldsymbol{x}^2(l)$ for l even. Symbol $\sqrt{E_s}x_n(l)$, $n = 1, 2, \ldots, N$ is assigned to transmit antenna n during time instant l. Prior to transmission, symbol streams on transmit antennas are interleaved with the same symbol-wise channel interleaver π_c.

8.1.2 Information Interleaver

Apart from distinct CCs, the information interleaver structure is the main feature that distinguishes different ST-TTCMs proposed in parallel in [16, 20–22]. In [20] a single pseudo-random bit-wise interleaver of length LK was applied and no puncturing was assumed resulting in ST-TTCM with a reduced rate as compared to its constituent STTrCs. In [21] a single symbol-wise pseudo-random interleaver of length L was utilized. Due to additional symbol deinterleaving employed after the second CC, the input information bits were

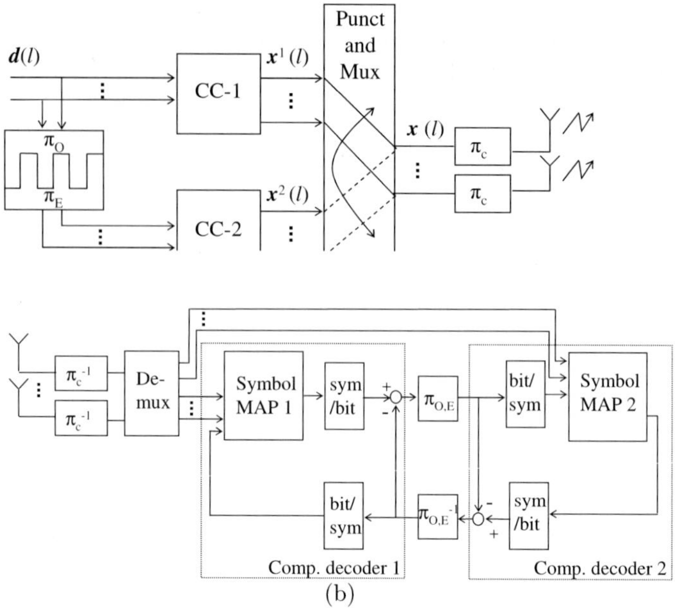

Fig. 8.1. System model. (a) Parallel concatenated encoder. (b) Iterative decoder.

equally protected. However with no constrains imposed on the information interlever, the second CC was non-uniformly punctured. In [22] the same symbol-wise odd–even information interleaver from [14] was adopted. Like the one implemented in [21], such an interleaver assures that the positions of the input bits in the input symbol $\boldsymbol{d}(l) = [d_1(l), d_2(l), \ldots, d_K(l)]$ remain unchanged. The interleaver utilized in [16, 19] and in this paper consists of two pseudo-random bit-wise interleavers, π_{O} for odd l and π_{E} for even l input positions, i.e., it enables the mapping of any input bit from a given even (odd) input symbol to any bit position of any even (odd) symbol.

8.1.3 Decoder

The block diagram of the ST-TTCM decoder is depicted in Fig. 1(b). Prior to iterative decoding, channel de-interleaving and received signal de-multiplexing is undertaken. Due to quite severe puncturing, constituent TCMs for $N = 1$ easily become catastrophic. Therefore, to trigger the decoding process, *a priori* inputs of the first constituent TCM are sub-optimally doped by the outputs of an additional soft-output demodulator (not showed in Fig. 1(b)) in the first iteration [34]. Notice that such an additional complexity is avoided in $N > 1$ case due to the increased cardinality of the set of output TCM symbols [15]. The component codes are decoded with symbol-by-symbol MAP decoders. Due to the assumed bit-wise information interleaving, component decoders include additional symbol-to-bit and bit-to-symbol reliability transforms. Hence, the exchange of extrinsic information between the two component decoders during iterative decoding is done on the bit level. When symbol-wise information interleaving is employed, the exchange of extrinsic information is done directly on the symbol level, i.e., the schemes proposed in [21, 22] operate on the symbol level. The superior performance of the bit-level ST-TTCM from [16] over the symbol-level ST-TTCM from [21] was demonstrated in [19].

8.2 Performance Analysis

The performance of single-antenna and multi-antenna TTCM is analyzed in this section. After the short introduction of the pair-wise error probability (PEP) in Sec. 8.2.1, the generalized two-dimensional DS, as a key ingredient for UB analysis is introduced in Sec. 8.2.2. Analytical UB and simulation results for single-antenna and multi-antenna TTCM are presented in Sec. 8.2.3 and 8.2.4, respectively. Finally, dynamics of sub-optimal iterative decoding algorithm are analyzed in Sec. 8.2.5.

8.2.1 Upper Bounds over AWGN and Fading Channels

Assuming maximum likelihood (ML) decoding with a perfect knowledge of channel state information (CSI) at the receiver, the UB on the FER and BER for a given code over fading channels with AWGN is given by

$$P_{\mathrm{UB}} = \sum_{\mathcal{D}} H\left(\Delta_{\mathrm{H}}, \Delta_{\mathrm{P}}\right) P_{\mathrm{PEP}}\left(\Delta_{\mathrm{H}}, \Delta_{\mathrm{P}}\right) \tag{8.1}$$

where $\Delta_{\mathrm{H}} = \Delta_{\mathrm{H}}\left(\boldsymbol{x}, \boldsymbol{e}\right)$ and $\Delta_{\mathrm{P}} = \Delta_{\mathrm{P}}\left(\boldsymbol{x}, \boldsymbol{e}\right)$ denote the effective Hamming distance (EHD) and the effective product distance (EPD), respectively, between the transmitted sequence $\boldsymbol{x}$ and its erroneously decoded counterpart $\boldsymbol{e}$. The set $\mathcal{D}$ includes all distinct $\langle \Delta_{\mathrm{H}}, \Delta_{\mathrm{P}} \rangle$ pairs in a two-dimensional code's DS. $H\left(\Delta_{\mathrm{H}}, \Delta_{\mathrm{P}}\right)$s in Eq. 8.1 are multiplicities in a two dimensional code's DS, i.e., the average number of error events $\{\boldsymbol{x} \rightarrow \boldsymbol{e}\}$ with a given EHD and EPD. The second term in the product from Eq. 8.1 is the PEP as a function of error coefficients and signal-to-noise ratio (SNR=$N E_{\mathrm{s}}/N_0$) [35]

$$P_{\mathrm{PEP}}\left(\Delta_{\mathrm{H}}, \Delta_{\mathrm{P}}\right) \leq \binom{2\Delta_{\mathrm{H}} M - 1}{\Delta_{\mathrm{H}} M - 1} \Delta_{\mathrm{P}}^{-\Delta_{\mathrm{H}} M} \left(\frac{\mathrm{SNR}}{N}\right)^{-\Delta_{\mathrm{H}} M} \tag{8.2}$$

On quasi-static fading channels, $\Delta_{\mathrm{P}}\left(\boldsymbol{x}, \boldsymbol{e}\right)$ is defined as the geometric mean of non-zero eigenvalues of the $N \times N$ epoch-by-epoch calculable squared codeword difference matrix (SCDM)

$$\boldsymbol{A}(\boldsymbol{x},\boldsymbol{e}) = \sum_{l=1}^{L} \left[\Delta_1^{ex}(l),\ldots,\Delta_N^{ex}(l)\right]^{\mathrm{H}} \left[\Delta_1^{ex}(l),\ldots,\Delta_N^{ex}(l)\right] \tag{8.3}$$

with $\Delta_n^{ex}(l) = e_n(l) - x_n(l)$. $\Delta_{\mathrm{H}}(\boldsymbol{x},\boldsymbol{e})$ is given as a rank of $\boldsymbol{A}(\boldsymbol{x},\boldsymbol{e})$. Therefore on quasi-static fading channels, the EHD is bounded by $1 \leq \Delta_{\mathrm{H}} \leq N$. Notice that when $\Delta_{\mathrm{H}}(\boldsymbol{x},\boldsymbol{e}) = N$, the EPD is equal to a determinant of the SCDM, i.e., $\Delta_{\mathrm{P}}(\boldsymbol{x},\boldsymbol{e}) = \det(\boldsymbol{A}(\boldsymbol{x},\boldsymbol{e}))$. On fast (perfectly interleaved) fading channels, $\Delta_{\mathrm{P}}(\boldsymbol{x},\boldsymbol{e})$ is given as the geometric mean of non-zero epoch-by-epoch calculable terms

$$\lambda_l(\boldsymbol{x},\boldsymbol{e}) = \left[\Delta_1^{ex}(l),\ldots,\Delta_N^{ex}(l)\right]\left[\Delta_1^{ex}(l),\ldots,\Delta_N^{ex}(l)\right]^{\mathrm{H}} \tag{8.4}$$

and $\Delta_{\mathrm{H}}(\boldsymbol{x},\boldsymbol{e})$ denotes the cardinality of a set of time instances $\{l\}$ in which terms in Eq. 8.4 are non-zero. On AWGN channels, tight PEP bound can be obtained from

$$P_{\mathrm{PEP}} = \frac{1}{2}\operatorname{erfc}\sqrt{\frac{E_s \Delta_{\mathrm{P}}}{4N_0}} \tag{8.5}$$

where the EPD represents the squared Euclidean distance between $\boldsymbol{x}$ and $\boldsymbol{e}$

$$\Delta_{\mathrm{P}}(\boldsymbol{x},\boldsymbol{e}) = \sum_{l=1}^{L}\sum_{n=1}^{N} \left|\Delta_n^{ex}(l)\right|^2 \tag{8.6}$$

8.2.2 Distance Spectrum Interpretation

Consider the two pair-wise error events from $\{\boldsymbol{x}^1 \to \boldsymbol{e}^1\}$ and $\{\boldsymbol{x}^2 \to \boldsymbol{e}^2\}$ in the first and the second CC, respectively. Due to its additive property, the SCDM from Eq. 8.3 can be decomposed as

$$\boldsymbol{A}(\boldsymbol{c},\boldsymbol{e}) = \boldsymbol{A}\left(\boldsymbol{c}^1,\boldsymbol{e}^1\right) + \boldsymbol{A}\left(\boldsymbol{c}^2,\boldsymbol{e}^2\right) \tag{8.7}$$

with separated terms contributed from error events in the two CCs. For fast fading channels, the EHD and EPD can be directly decomposed from Eq. 8.4

$$\Delta_{\mathrm{P}}(\boldsymbol{x},\boldsymbol{e}) = \left(\Delta_{\mathrm{P}}\left(\boldsymbol{x}^1,\boldsymbol{e}^1\right)\right)^{\frac{\Delta_{\mathrm{H}}(\boldsymbol{x}^1,\boldsymbol{e}^1)}{\Delta_{\mathrm{H}}(\boldsymbol{x},\boldsymbol{e})}} \left(\Delta_{\mathrm{P}}\left(\boldsymbol{x}^2,\boldsymbol{e}^2\right)\right)^{\frac{\Delta_{\mathrm{H}}(\boldsymbol{x}^2,\boldsymbol{e}^2)}{\Delta_{\mathrm{H}}(\boldsymbol{x},\boldsymbol{e})}} \tag{8.8}$$

$$\Delta_{\mathrm{H}}(\boldsymbol{x},\boldsymbol{e}) = \Delta_{\mathrm{H}}\left(\boldsymbol{x}^1,\boldsymbol{e}^1\right) + \Delta_{\mathrm{H}}\left(\boldsymbol{x}^2,\boldsymbol{e}^2\right) \tag{8.9}$$

On AWGN channels, the lack of diversity gain reduces the spectral interpretation to one-dimensional DS wrt the EPD. From Eq. 8.6 it directly follows

$$\Delta_{\mathrm{P}}(\boldsymbol{x},\boldsymbol{e}) = \Delta_{\mathrm{P}}\left(\boldsymbol{x}^1,\boldsymbol{e}^1\right) + \Delta_{\mathrm{P}}\left(\boldsymbol{x}^2,\boldsymbol{e}^2\right). \tag{8.10}$$

For an error event $\{\boldsymbol{x} \to \boldsymbol{e}\}$, let $\mathcal{M} = \mathcal{M}(\boldsymbol{x},\boldsymbol{e})$ denote the subset of relevant PEP measures for a given type of a channel. For quasi-static fading and AWGN channels, $\mathcal{M}$ is equal to the SCDF and EPD, respectively. Likewise, for fast fading channels $\mathcal{M}$ includes both the EPD and EHD. Also, let the binary operator "$\uplus$" denote the subset of appropriate operations from Eq. 8.7–8.10 for a given type of channel and related performance measure. That is, on quasi-static fading and AWGN channels, the corresponding measures are summed according to Eq. 8.7 and 8.10, respectively. Likewise, on fast fading channels, "$\uplus$" processes the EPDs and the EHDs of its operands according to Eq. 8.8 and 8.9, respectively. Let $\mathcal{M}^0$ denote the identity element from $\{\mathcal{M}\}$, i.e., $\mathcal{M} = \mathcal{M} \uplus \mathcal{M}^0$. Further, let $\boldsymbol{d}$ and $\hat{\boldsymbol{d}}$ denote the correctly and erroneously decoded information sequences related to $\boldsymbol{x}^1$ and $\boldsymbol{e}^1$, respectively. In addition to set $\mathcal{M}$, each error event in the given CC will be identified by the Hamming distance vector $\boldsymbol{h} = \boldsymbol{h}\left(\boldsymbol{x}^1, \boldsymbol{e}^1\right)$ between $\boldsymbol{d}$ and $\hat{\boldsymbol{d}}$. The two CCs share the same information error event $\{\boldsymbol{d} \rightarrow \hat{\boldsymbol{d}}\}$. That is, correctly and erroneously decoded information sequences for $\{\boldsymbol{x}^2 \rightarrow \boldsymbol{e}^2\}$ are interleaved versions of $\boldsymbol{d}$ and $\hat{\boldsymbol{d}}$, respectively. Therefore, $\boldsymbol{h}\left(\boldsymbol{x}^1, \boldsymbol{e}^1\right) = \boldsymbol{h}\left(\boldsymbol{x}^2, \boldsymbol{e}^2\right)$ holds. The four components of $\boldsymbol{h} = [h_{\mathrm{O}1}, h_{\mathrm{O}2}, h_{\mathrm{E}1}, h_{\mathrm{E}2}]$ enumerate the number of bit positions of type $\{d_k(l), \hat{d}_k(l)\} = \{0, 1\}$ and type $\{d_k(l), \hat{d}_k(l)\} = \{1, 0\}$ in the two information interleavers. The sum over components of $\boldsymbol{h}$, denoted hereafter by h, is equal to the Hamming distance between $\boldsymbol{d}$ and $\hat{\boldsymbol{d}}$. For the rigorous definition of $\boldsymbol{h}$ see [30].

Let $H^j_{1,\psi}\left(\boldsymbol{h}, \mathcal{M}^j\right)$ denote the average number of single error events of length ψ in the CC-j, $j = 1, 2$, with the given measures $\boldsymbol{h}$ and $\mathcal{M}^j$. Multiplicities $H^j_{1,\psi}\left(\boldsymbol{h}, \mathcal{M}^j\right)$ include both the odd- and even-punctured single error events that diverge from an arbitrary state at the time instant $l = 1$. To enumerate the set $H^j_{1,\psi}\left(\boldsymbol{h}, \mathcal{M}^j\right)$ in each of the punctured CCs, the General algorithm presented in [13] for the general class of geometrically non-uniform codes has been modified as in [30] to include the enumeration of both the Hamming distance vector $\boldsymbol{h}$ and the set of performance measures $\mathcal{M}^j$.

The average number of compounded error events with the given $\boldsymbol{h}$ and $\mathcal{M}^j$ that start at an arbitrary time instant $l \geq 1$ and consist of $\Theta \geq 1$ concatenated single error events of total length $\Psi \leq L$ is determined by

$$H^j_{\Theta,\Psi}\left(\boldsymbol{h}, \mathcal{M}^j\right) = \Gamma(\Theta, \Psi) \prod_{\theta=1}^{\Theta} H^j_{1,\psi_\theta}\left(\boldsymbol{h}_\theta, \mathcal{M}^j_\theta\right) \tag{8.11}$$

with $\boldsymbol{h} = \boldsymbol{h}_1 + \ldots + \boldsymbol{h}_\theta + \ldots + \boldsymbol{h}_\Theta$, $\Psi = \psi_1 + \ldots + \psi_\theta + \ldots + \psi_\Theta$, $\mathcal{M}^j = \mathcal{M}^j_1 \uplus \ldots \uplus \mathcal{M}^j_\theta \uplus \ldots \uplus \mathcal{M}^j_\Theta$, where $\boldsymbol{h}_\theta$, ψ_θ and $\mathcal{M}^j_\theta$ correspond to the θth error event in the concatenation. In analogy to [36], the function $\Gamma(\Theta, \Psi)$ is defined as

$$\Gamma(\Theta, \Psi) = \binom{\frac{L}{2} - \lceil \frac{\Psi}{2} \rceil + \Theta}{\Theta} \tag{8.12}$$

The contribution of single error events that diverge at time instant $l > 1$ is also encountered with above. The multiplicities in the two-dimensional DS of the CC-j are calculated as

$$H^j\left(\boldsymbol{h}, \mathcal{M}^j\right) = \sqrt{C(h, L, K)} \sum_{\Theta} \sum_{\Psi} H^j_{\Theta,\Psi}\left(\boldsymbol{h}, \mathcal{M}^j\right). \tag{8.13}$$

For BER evaluation $C(h, L, K) = h/LK$ while for FER evaluation it reduces to $C(h, L, K) = 1$. To enable the analysis tractable, both the odd and even bit-wise information interleavers were assumed to be uniform [36]. The resultant DS of ST-TTCM wrt $\mathcal{M}$ as averaged over all possible $\pi_{\mathrm{O}} - \pi_{\mathrm{E}}$ information interleavers is given by

$$H\left(\mathcal{M} = \mathcal{M}^1 \uplus \mathcal{M}^2\right) = \sum_{\boldsymbol{h}} \frac{H^1\left(\boldsymbol{h}, \mathcal{M}^1\right) H^2\left(\boldsymbol{h}, \mathcal{M}^2\right)}{\binom{LK/2}{h_{\mathrm{O}1}, h_{\mathrm{O}2}} \binom{LK/2}{h_{\mathrm{E}1}, h_{\mathrm{E}2}}}. \tag{8.14}$$

For fast fading and AWGN channels, the DS wrt $\mathcal{M}$ can be directly plugged into Eq. 8.1. Otherwise, for the quasi-static fading case where $\mathcal{M} = \boldsymbol{A}(\boldsymbol{x}, \boldsymbol{e})$, the rank and the geometric mean of the non-zero eigenvalues have to be calculated for each SCDM in Eq. 8.14 to arrive at the two-dimensional DS wrt the EHD and EPD.

8.2.3 Truncated Union Bound for $N = 1$

Let us truncate the UB in Eq. 8.1 by replacing the set $\mathcal{D}$ with its subset $\mathcal{P}$. The subset $\mathcal{P}$ is denoted as the *significant* part of the code's DS if adding any disjoint element from $\mathcal{D}$ to $\mathcal{P}$ has a negligible effect to the truncated sum in Eq. 8.1.

The FER and BER simulation and UB results for *Robertson et al.* TTCM from [34] employing constituent TCM with generator polynomials [11; 02; 04] in octal form over AWGN and fast fading channels are depicted in Fig. 8.2. Notice that extensive computer simulations have been, so far, the only methods for studying the performance of TTCM schemes from [34]. Fig. 8.2 includes results for both bit- and symbol-wise information interleaver. Bit-level TTCMs typically outperform symbol-level ones in the error floor region which comes with the expense of slightly higher convergence thresholds. Such an effect will be more exaggerated for TTCM with the new CCs designed in Sec. 8.3.

The set $\mathcal{P}$ was empirically determined adopting all punctured single error events with a Hamming distance up to $h_{\mathrm{max}} = 4$, a sequence length up to $\psi_{\mathrm{max}} = 8$ and an EPD up to $\Delta_{\mathrm{Pmax}} = 48$ were first enumerated in each of the CCs. Compounded error events consisting of up to $\Theta_{\mathrm{max}} = 3$ concatenated single error events were then formed. The frame size effects of $LK = 260$ and 2048 were comparatively studied. The upper bounds appeared as rather tight over both AWGN and fast fading channels. Similar to binary turbo convolutional codes, the asymptotic FER of TTCM does not depend of the frame size. For BER, considerable performance gains can be achieved increasing the frame size due to the *spectral thinning* effect similar to [37].

8.2.4 Truncated Union Bound for $N = 2$

The choice of parameters that uniquely determine the significant part of the DS is a complex multi-dimensional problem. The main bottleneck of the introduced DS interpretation is not its computational complexity but the memory

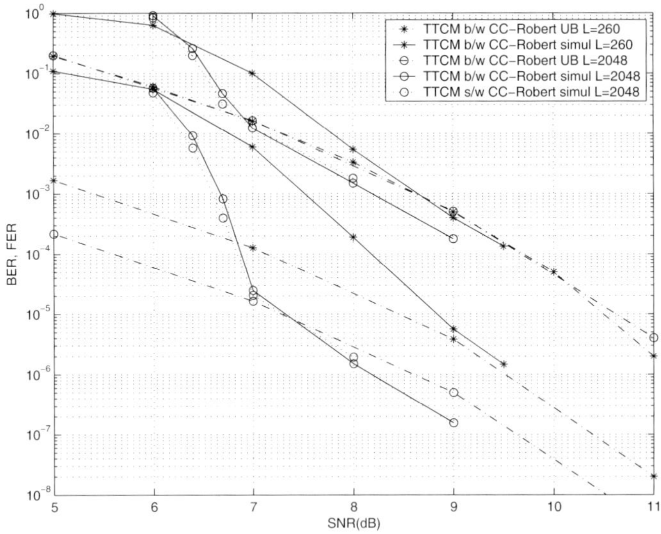

(a)

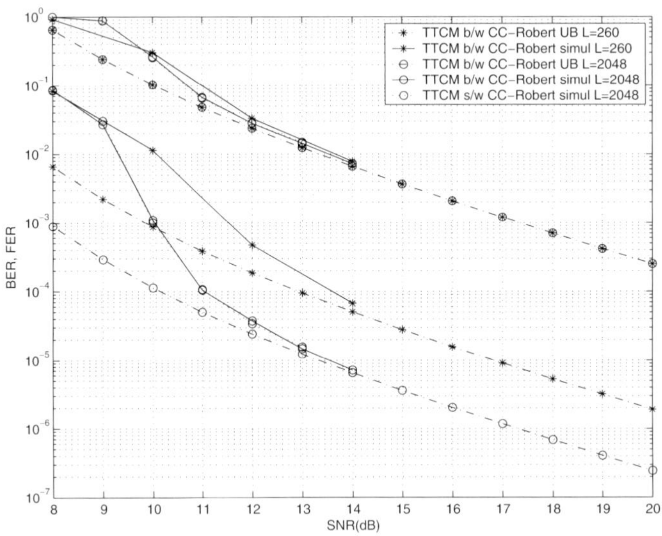

(b)

Fig. 8.2. The simulation and union bound performance of *Robertson et al.* TTCM, $N = M = 1$, 8-state CCs, 10 decoding iterations, $LK = 260$ and 2048 bits. (a) AWGN. (b) Fast Fading.

requirements. The multiplicities of a $4 + N^2$ and 6 dimensional variables are enumerated with the GA over quasi-static and fast fading channels, respectively. With a further error event concatenation inside the CCs and their statistical combing over all possible π_O–π_E realizations, the DS of the ST-TTCM becomes extremely rich.

Given the above memory constrains, for the quasi-static fading case, the following parameters were chosen $h_{\max} = 6$, $\psi_{\max} = 6$, $\Delta_{\mathrm{P}\max} = 16$ and $\Theta_{\max} = 3$. On fast fading channels, the significant part of the DS was identified with $h_{\max} = 6$, $\psi_{\max} = 6$, $\Delta_{\mathrm{P}\max} = 48$ and $\Theta_{\max} = 3$. Despite the large increase in the number of DS entries, a further increase in any of the above parameters for 8-state CC from [16] over fading channels has been empirically found to insignificantly contribute to the sum in Eq. 8.1.

Fig. 8.3 depicts the FER analytical and simulation results of the ST-TTCM over quasi-static and fast fading channels. The two configurations proposed in [16, 19] were comparatively studied. The first one denoted "ST-TTCM 2x8st Rec-STTrC" was built from two 8-state Rec-STTrCs from [16]. In the second configuration denoted "ST-TTCM 2x8st Tarokh&Rec-STTrC", one of the CCs was replaced with a non-recursive Tarokh-STTrC [10]. Such a configuration was shown in [16] to lead to better performance over quasi-static fading channels. Notice that both configurations employ the non-systematic CCs. Both ST-TTCMs achieve a spectral efficiency of 2 bps/Hz applying QPSK modulation ($Z = K = 2$) over $N = 2$ transmit and $M = 1$ receive antennas. The length of the information frame was set to $LK = 260$ bits and 10 decoding iterations were assumed in simulations.

With no additional bounding of the conditional PEP as in [38, 39], upper bounds of the ST-TTCM over quasi-static fading channels in Fig. 3(a) can be characterized as rather optimistic. Nevertheless, the rank deficient error events have been well enumerated, which will appear crucial for the accurate design of the full rank ST-TTCM in Sec. 8.3. The ST-TTCM exhibited the effect of spectral thinning [37], which resulted in a DS with very few codewords with low EHDs and EPDs, and a majority of codewords with average-to-high EHDs and EPDs, similar to the DS expected from random-like ST codes. Evidently, the determinant criterion [10], commonly applied for design of the stand-alone STTrCs may not necessarily be adequate for CC optimization in the ST-TTCM framework.

From Fig. 3(a), it is apparent that without particular CC and/or information interleaving optimization, the punctured parallel concatenation in general does not preserve the full rank of its CCs. That is, the two-dimensional DS of ST-TTCM over quasi-static fading channels has the non-zero entries $H(\Delta_{\mathrm{H}}, \Delta_{\mathrm{P}})$ for $\Delta_{\mathrm{H}} < N$. For the fully recursive configuration, a quite large "code energy" concentrated in the rank deficient part of the DS severely deteriorated the performance. On the other hand, the non-fully recursive configuration experienced a negligible performance deviation away from the second order diversity slope in the range of low-to-medium SNRs typically exhibited in practice.

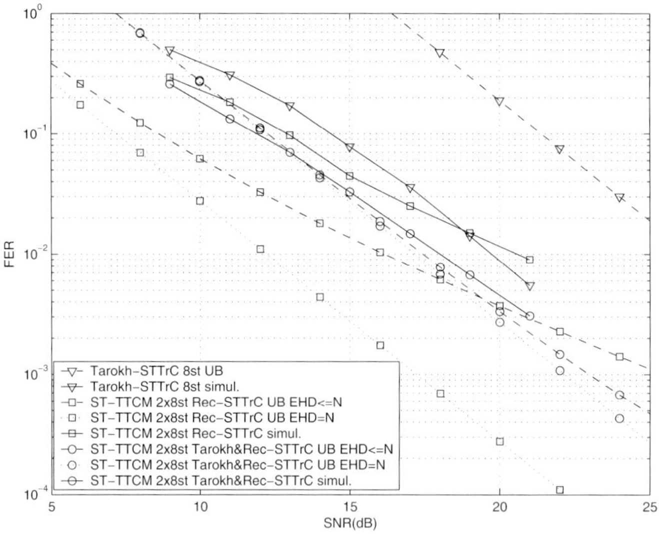

(a)

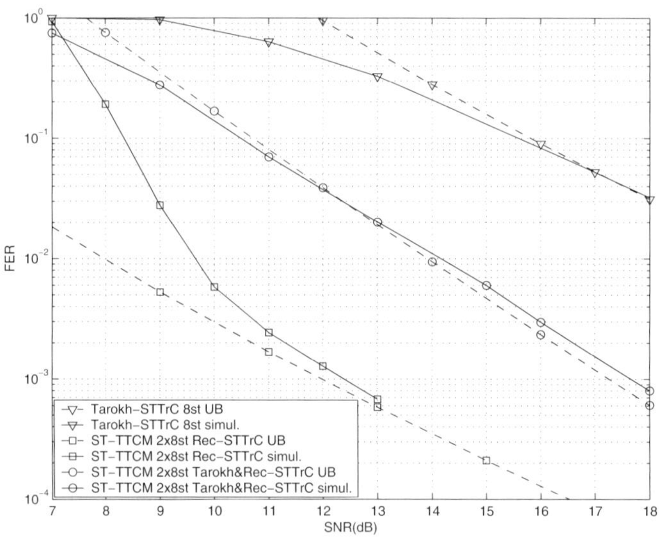

(b)

Fig. 8.3. The simulation and the union bound performance of the fully recursive and the non-fully recursive ST-TTCM, 8-state CCs, 10 decoding iterations, $LK = 260$ bits, $N = 2$, $M = 1$. (a) Quasi-static fading channel. (b) Fast fading channel.

On fast fading channels, UB of ST-TTCM in Fig. 3(b) appeared to be tight. The similar tightness was observed for constituent non-punctured stand-alone STTrCs. The configuration with both Rec-STTrCs exhibited the stronger water-falling effect, i.e., the higher slope in the transition region towards the "error-floor". The error-floor here refers to the asymptotical high SNR behavior with the slope of the FER curve characterized by the diversity gain. The non-fully recursive ST-TTCM experienced no performance degradation in the low SNR region, i.e., the water-fall and error-floor regions were indistinguishable. The reasons for such interesting behavior will become apparent after examining the iterative decoding convergence properties in the next subsection.

8.2.5 Iterative Decoding Convergence

Similar to [40, 41], an i.i.d. symmetric Gaussian approximation of extrinsic information was assumed in this paper to trace the convergence of iterative decoding. Each component decoder from Fig. 1(b) was modelled as a non-linear system with a certain transfer function. The SNR at the input of the jth component decoder, denoted hereafter by $\mathrm{SNR}_{\mathrm{d}j}^{\mathrm{i}}$, was varied and the corresponding output SNR, denoted hereafter by $\mathrm{SNR}_{\mathrm{d}j}^{\mathrm{o}}$, measured to empirically determine the non-linear transfer function G_j. Functions $\mathrm{SNR}_{\mathrm{d1}}^{\mathrm{o}} = G_1\left(\mathrm{SNR}_{\mathrm{d1}}^{\mathrm{i}}\right)$ and $\mathrm{SNR}_{\mathrm{d2}}^{\mathrm{i}} = G_2^{-1}\left(\mathrm{SNR}_{\mathrm{d2}}^{\mathrm{o}}\right)$ were plotted on the same graph. This was done comparatively in Fig. 8.4 for two ST-TTCMs from Sec. 8.2.4 assuming an AWGN channel with SNRs of 5.4 dB and 4.9 dB. The convergence threshold was defined in [40] as the channel SNR value at which two transfer functions just touch. Above this SNR value, the "tunnel" of iterative decoding opens [40] and the $\mathrm{SNR}_{\mathrm{d}j}^{\mathrm{o}}$ increases over iterations without bound. In practise, this means that the probability of error, inversely proportional to such $\mathrm{SNR}_{\mathrm{d}j}^{\mathrm{o}}$ [41], converges towards the error-floor even with finite, large enough value of $\mathrm{SNR}_{\mathrm{d}j}^{\mathrm{o}}$.

From Fig. 4(a), the convergence threshold of the fully-recursive ST-TTCM is 5.4 dB. On the other hand, the tunnel of iterative decoding for the non-fully recursive ST-TTCM is never open. The first CC being non-recursive has a zero asymptotic slope of its extrinsic information transfer function, which conflicts with the asymptotic slope one of the recursive second CC. However, due to the almost complementary outlook of its transfer functions at low $\mathrm{SNR}_{\mathrm{d}i}^{\mathrm{i}}$, the non-fully recursive ST-TTCM enables reasonably good evolution of its extrinsic information within 10 decoding iterations. At the channel SNR=5.4 dB, the non-fully recursive ST-TTCM has a remarkably larger value of $\mathrm{SNR}_{\mathrm{d}j}^{\mathrm{o}}$ than the recursive one whose tunnel is pinched at a very low value of $\mathrm{SNR}_{\mathrm{d}j}^{\mathrm{o}}$. Fig. 4(b) shows that even lower values of channel SNR enable a reasonable evolution of $\mathrm{SNR}_{\mathrm{d}j}^{\mathrm{o}}$ for a non-fully recursive configuration.

The good convergence properties are essential for the satisfactory performance of turbo-like codes over quasi-static fading channels [41]. Adding to it a less deteriorating effect of the rank deficient part of the DS, it is easy

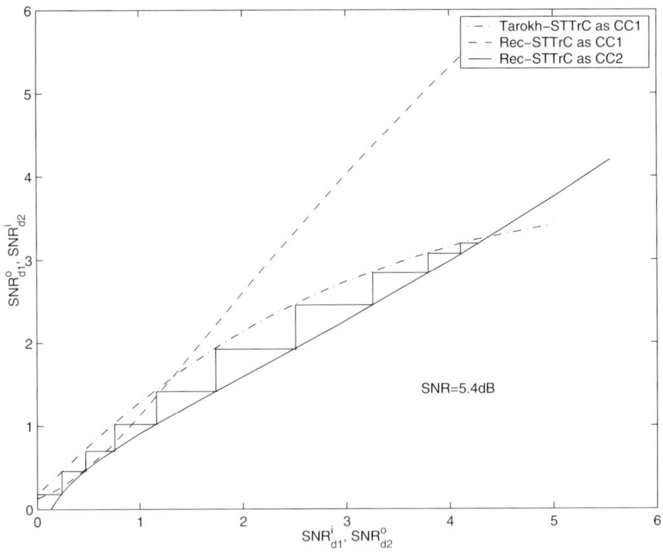

(a)

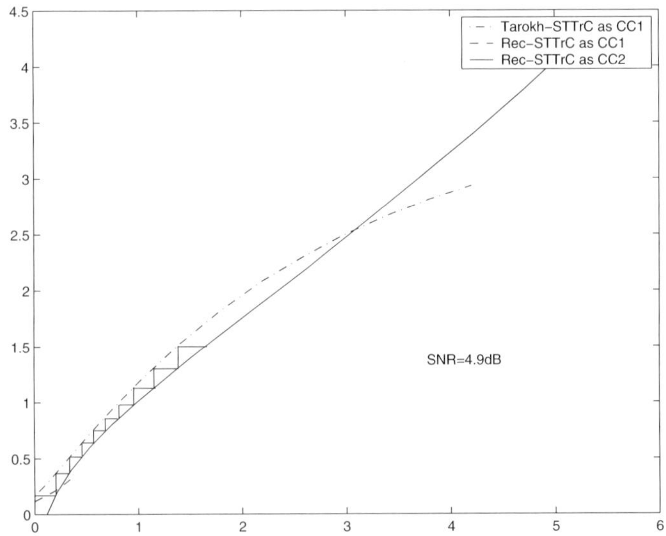

(b)

Fig. 8.4. Iterative decoding convergence for fully recursive and non-fully recursive ST-TTCM on an AWGN channel. (a) SNR=5.4dB. (b) SNR=4.9dB.

to understand the superior performance of the non-fully recursive ST-TTCM over the fully recursive one experienced through simulations over quasi-static fading channels. On fast fading channels where good DS characteristics play a more important role, the fully-recursive ST-TTCM outperformed the non-fully recursive one in spite of its inferior convergence properties.

8.3 Constituent Code Optimization

The design of new CCs for single-antenna and multi-antenna TTCM is addressed in this section. The results of UB and DS optimization of single-antenna TTCM over AWGN and fast fading channels are presented in Sec. 8.3.1. For multi-antenna TTCM, new design criteria are first defined in Sec. 8.3.2. The new system model for constituent multi-antenna TCM is further introduced in Sec. 8.3.3. Finally, the results of DS optimization for multi-antenna TTCM over quasi-static and fast fading channels are presented in Sec. 8.3.4.

8.3.1 Distance Spectrum Optimization for $N = 1$

Authors in [34] optimized TTCM over AWGN channels by extensive TCM code search that maximized the minimal Euclidean distance (also known as *free* distance) of each constituent TCM under assumption that parity bits in every second time instant are randomly picked. Such an optimization did not account for Hamming weight distribution of input information error events. In [42, 43], TTCM was optimized over AWGN channels by maximizing the minimal Euclidean distance in each TCM that corresponds to Hamming weight 2 input information sequences (so called *effective free* distance). For fast fading channel optimization in [42], the output symbol Hamming distance and the product of the squared Euclidean distances along the error events that correspond to Hamming weight 2 input information sequences were maximized in each constituent TCM.

To capture the joint effects of pair-wise error events with error coefficients both equal and greater than the minimum but also the contribution of multiplicities in the DS of TTCM, the search for constituent TCMs that minimize the union bound in Eq. 8.1 is further performed. To reduce the complexity, the sum in Eq. 8.1 was further truncated to $\mathcal{P}$ defined by $h_{\max} = 4$, $\psi_{\max} = 6$, $\Delta_{\mathrm{P}\max} = 48$ and $\Theta_{\max} = 2$. To assure that optimization is performed in the error-floor region, the moderately high SNR of 8dB and 14dB were chosen for AWGN and fast fading channels, respectively. The subset of candidate codes for search was defined by Ungerboeck's single antenna systematic recursive TCM model from [44]. The results of a code search are summarized in Table 8.1. Interestingly enough, the separate DS optimization over AWGN and fast fading channels resulted in the same two best TCMs. The special case of 8-state constituent TCM for $N = 1$, $K = 2$ and $Z = 3$ was considered, i.e.,

Table 8.1. The best $\nu = 3$ 8PSK $N = 1$ constituent TCM for 2 bps/Hz TTCM over AWGN and fast fading channels.

CC	$\boldsymbol{G}$	$P_{\mathrm{BER}}\vert_{\mathcal{P}_{\mathrm{awgn}}}$	$P_{\mathrm{BER}}\vert_{\mathcal{P}_{\mathrm{fastf}}}$
Robert.	$[11; 02; 04]$	1.96×10^{-6}	4.86×10^{-6}
$\mathrm{U_a}$	$[15; 02; 06]$	3.15×10^{-8}	1.48×10^{-7}
$\mathrm{U_b}$	$[13; 04; 06]$	3.15×10^{-8}	1.48×10^{-7}

TTCM for 2 bps/Hz with 8PSK modulation and single transmit antenna. The extension to higher order modulations and bandwidth efficiencies is straightforward but beyond the scope of this book chapter.

8.3.2 Design Criteria $N > 1$

The maximization of the minimum EHD, coined the *rank* criterion on slow fading channels and the *distance* criterion on fast fading channels was identified as the primary importance ST coding design criterion in [10]. The maximization of the minimum EPD, denoted the *determinant* criterion on slow fading channels and *product* criterion on fast fading channels was established as the secondary importance ST coding design criterion therein. In [45] it was further argued that for a moderately large diversity order, i.e., $\Delta_{\mathrm{H}} M \geq 4$, a MIMO fading channel converges reasonably close to an AWGN channel. In such a case the PEP is inversely proportional to the code squared Euclidean distance in Eq. 8.6. The maximization of it, denoted the *trace* criterion, was introduced in [45] as a secondary importance design criterion. The rank and the distance criteria were sustained therein as the primary ones to meet the condition $\Delta_{\mathrm{H}} M \geq 4$ with fewer receive antennas. Similar conclusions on the importance of the code squared Euclidean distance, though as a primary design criterion over quasi-static fading channels for systems with a large number of receive antennas but also those operating in the low SNR region were driven in [46, 47]. In the moderate SNR region, the PEP was further determined in [46] to be inversely proportional to $\det(\boldsymbol{I} + \boldsymbol{A})$ where $\boldsymbol{I}$ denotes the identity matrix. The equal eigenvalue criterion was introduced independently in [48, 49] in a context of the determinant criterion and optimal coding gain, arguing that the maximization of the minimum code squared Euclidean distance renders eigenvalues of the SM to be equal, which maximizes the minimum EPD.

All of the above design criteria concentrated on error coefficients. As a result, different codes with the same minimum EHD and EPD sometimes exhibited different performance in simulations. To capture the joint effects of pair-wise error events, with error coefficients both equal to and greater than the minimum, but also the contribution of multiplicities to the union bound in Eq. 8.1, we introduce the following design criteria for ST codes over quasi-static fading channels:

- *Significant full rank.* To maximize the diversity gain over quasi-static fading channels, the STC must satisfy the full rank criterion in its significant part of the DS, i.e., $H(\Delta_{\mathrm{H}}, \Delta_{\mathrm{P}}) = 0$ for $\forall \langle \Delta_{\mathrm{H}}, \Delta_{\mathrm{P}} \rangle \in \mathcal{P}|_{\Delta_{\mathrm{H}}<N}$
- *Minimum spectral product.* To maximize the coding gain, among all ST codes that satisfy the significant full rank criterion choose the one that minimizes

$$\Delta_{\mathrm{DS}} = \sum_{\langle \Delta_{\mathrm{H}}, \Delta_{\mathrm{P}} \rangle \in \mathcal{P}|_{\Delta_{\mathrm{H}}=N}} H(\Delta_{\mathrm{H}}, \Delta_{\mathrm{P}}) \, \Delta_{\mathrm{P}}{}^{-\Delta_{\mathrm{H}} M}. \tag{8.15}$$

The significant full rank criterion guaranties the full spatial diversity given the large enough cardinality of set $\mathcal{P}$. Notice that Eq. 8.15 follows directly from Eq. 8.1 and 8.2, cancelling the terms independent of the EPD. Once the ST code's DS wrt $\mathcal{M} = \boldsymbol{A}$ in Eq. 8.14 is enumerated, the minimum spectral product optimization can be performed based on either of three EPD definitions, i.e., assuming $\Delta_{\mathrm{P}} = \det(\boldsymbol{A})$, $\Delta_{\mathrm{P}} = \det(\boldsymbol{I} + \boldsymbol{A})$ or $\Delta_{\mathrm{P}} = \mathrm{trace}(\boldsymbol{A})$

The importance of a code's DS, though in the context of stand-alone STTrCs and methods for its optimization over quasi-static fading channels were in parallel also studied in [50]. A recent attempt to apply the similar criteria from Eq. 8.15 to fast fading channels was made in [51]. However, on fast fading channels, it is not possible to accurately remove the SNR term from the DS optimization. The pair-wise error events with the same EPD may in general have different lengths and therefore experience different levels of diversity. Hence, given the SNR point, the DS optimization over fast fading channels in Sec. 8.3.4 will assume the minimization of the truncated UB in Eq. 8.1.

8.3.3 Subset of Candidate Constituent Codes for $N \geq 1$

In contrast to [16, 19] where CCs were defined by trellis diagrams, here we introduce the new multi-antenna transform domain model with binary generator polynomials. To each of K parallel input bits from $\boldsymbol{d}(l) = [d_1(l), d_2(l), \ldots, d_K(l)]$, a binary recursive feedback shift register with memory order ν_k, $k = 1, \ldots, K$, was associated. The sum $\nu = \nu_1 + \nu_2 + \ldots + \nu_K$ denoted the total memory order of the CC and each ν_k satisfied $\lfloor \nu/K \rfloor \leq \nu_k \leq \lceil \nu/K \rceil$. Each $x_n(l)$ was assumed to be the result of mapping Z coded bits $\boldsymbol{c}_n(l) = [c_n^1(l), c_n^2(l), \ldots, c_n^Z(l)]$ to a given 2^Z level amplitude and/or phase modulation. Fig. 8.5 depicts the kth shift register and its relation with the uth register, $u = 1, \ldots, K$. Introducing an indeterminate D to denote a delay operator, the input/output relations can be represented using a binary polynomial arithmetic relation

$$c_n^z(D) = \sum_{k=1}^{K} \left(f_{nzk}^d d_k(D) + f_{nzk}(D) \, a_k(D) \right) \tag{8.16}$$

with $z = 1, \ldots, Z$ and $a_k(D)$ denoting a signal entering the kth shift register

$$a_k(D) = \sum_{k=1}^{K} \left(b_{ku}^d d_u(D) + b_{ku}(D) \, D a_u(D) \right). \tag{8.17}$$

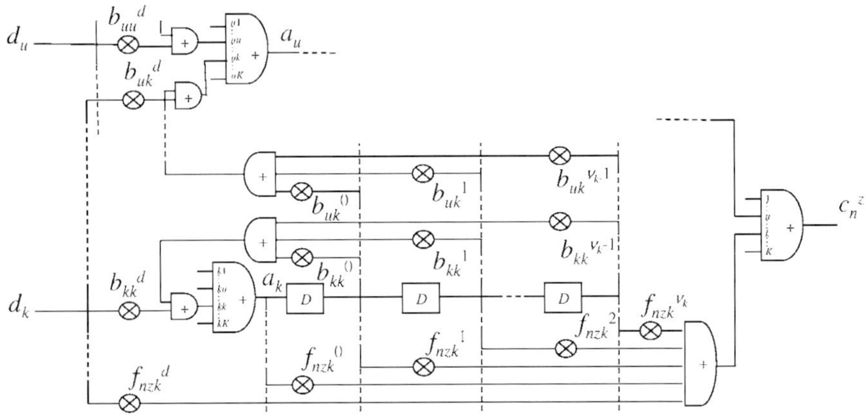

Fig. 8.5. The kth shift register in the constituent code.

f_{nzk}^{d} and b_{ku}^{d} are binary constants, $f_{nzk}(D)$ is a degree ν_k binary feed-forward polynomial from shift register k to the output bit z at the transmit antenna n, $b_{ku}(D)$ is a degree $\nu_u - 1$ binary feed-backward polynomial from shift register u to the input of the shift register k. Assuming that CCs in parallel concatenation are the same, the optimization over the code subset defined with Eq. 8.16 and 8.17 requires a tedious search over $2^{NZ(\nu+2K)+K(\nu+K)}$ different codes. In the following we restrict to the special case of $N = K = Z = 2$, $\nu = 3$ $(\nu_1 = 1, \nu_2 = 2)$ and the natural QPSK mapping, i.e., $x_n(l) = \sqrt{E_s}\exp\{\sqrt{-1}\left(c_n^1(l) + 2c_n^2(l)\right)\pi/2\}$. Therefore, the redundancy expansion is assumed only in the antenna space. The above special case covers a quite large set of 2^{38} 8-state CCs with a spectral efficiency of 2 bps/Hz. Therefore we further constrain the set of candidate codes to:

- Subset of systematic CCs since they were shown in [40] for classical single antenna turbo codes to result in better convergence of iterative decoding. Therefore the following parameters were chosen: $f_{112}^{d} = f_{121}^{d} = 1$, $f_{112}(D) = f_{121}(D) = 0$, $f_{122}^{d} = f_{111}^{d} = 0$, $f_{122}(D) = f_{111}(D) = 0$, $f_{212}^{d} = f_{222}^{d} = f_{211}^{d} = f_{221}^{d} = 0$.
- Fully connected trellises without parallel transitions, i.e., $b_{11}^{d} = b_{22}^{d} = 1$ and $b_{12}^{d} = b_{21}^{d} = 0$ respectively.

Solving Eq. 8.16 and 8.17 with the above assumptions, the input-output relations in the CC can be further simplified as

$$c_1^z(D) = d_{(3-z)}(D) \tag{8.18}$$

$$c_2^z(D) = \sum_{k=1}^{2} \frac{f_{zk}^e(D)}{b^e(D)} d_k(D) \tag{8.19}$$

with $z = 1, 2$ and equivalent feed-forward and feed-backward polynomials given as

$$f_{zk}^e(D) = f_{2zk}(D)\left(1 + Db_{(3-k)(3-k)}(D)\right) + Df_{2z(3-k)}(D)\, b_{(3-k)k}(D) \tag{8.20}$$

$$b^e(D) = 1 + (b_{11}(D) + b_{22}(D))\, D + (b_{11}(D)\, b_{22}(D) + b_{12}(D)\, b_{21}(D))\, D^2 \tag{8.21}$$

Notice that a degree ν binary polynomial $b^e(D)$ is a function of feed-backward polynomials only.

For each k, let the polynomial $h_k(D) = d_k(D) + \hat{d}_k(D)$ denote a sequence of one step Hamming distances between $d_k(D)$ and $\hat{d}_k(D)$. Due to recursive feedback, the summed coefficients of $h_k(D)$ over all k has the minimum value of 2, which occurs in one of the four distinct cases: 1) $h_1(D) = 1 + D^{L_1}$, $h_2(D) = 0$; 2) $h_1(D) = 0$, $h_2(D) = 1 + D^{L_2}$; 3) $h_1(D) = 1$, $h_2(D) = D^{L_3}$; 4) $h_1(D) = D^{L_4}$, $h_2(D) = 1$. Following the intuition from [37], the maximization of $\min(L_1, L_2, L_3, L_4)$ is expected to result in code-word gaining EHD and EPD. Choosing the degree ν primitive polynomial for $b^e(D)$ maximizes the minimum L_1 and L_2 to $L_1 = L_2 = 2^\nu - 1 = 7$. A primitive $b^e(D)$ is also expected to result in better convergence of iterative decoding [40].

Let us define a matrix of polynomial coefficients $\boldsymbol{B}(D) = [b_{11}(D); b_{12}(D); b_{22}(D); b_{21}(D)]$, with $b_{ku}(D) = b_{ku}^0 + b_{ku}^1 D + \ldots + b_{ku}^{\nu_u - 1} D^{\nu_u - 1}$, $k, u = 1, 2$. Letting $b^e(D) = 1 + D + D^3$ results in four solutions for the feed-backward polynomials, i.e., $\boldsymbol{B}_1 = [10; 10; 01; 10]$, $\boldsymbol{B}_2 = [10; 01; 00; 10]$, $\boldsymbol{B}_3 = [00; 11; 11; 10]$ and $\boldsymbol{B}_4 = [00; 01; 10; 10]$. $\boldsymbol{B}_1$ and $\boldsymbol{B}_2$ have $\min(L_3, L_4) = 2$ while $\boldsymbol{B}_3$ and $\boldsymbol{B}_4$ have $\min(L_3, L_4) = 1$. Therefore we restrict further code searches to $\boldsymbol{B}_1$ and $\boldsymbol{B}_2$ cases only. The above optimization of feed-backward polynomials reduced the code search to a subset of $2^{10} = 1024$ candidate CCs for each of $\boldsymbol{B}_1$ and $\boldsymbol{B}_2$. The feed-forward polynomials $f_{2zk}(D)$, $z, k = 1, 2$ were further optimized in the next subsection applying design criteria from Sec. 8.3.2 to the resultant DS of the ST-TTCM enumerated in Sec. 8.2.2.

8.3.4 Distance Spectrum Optimization for $N = 2$

To simplify the DS optimization, both CCs in the parallel concatenation were assumed to be the same. Also, in contrast to [16, 19] the N outputs of the second CC were altered during multiplexing, i.e., $x_n(l) = x_{N-n+1}^2(l)$ for l even. In cases when error events in two CCs are both either odd- or even-punctured, such altering will, due to the systematic nature of CCs, assure that corners of the SCDM on the main diagonal have non-zero entries. This will presumably increase the number of candidate ST-TTCMs that satisfy the significant full rank criterion over quasi-static fading channels with no impact on the performance over fast fading channels.

Table 8.2. The best $\nu = 3$ QPSK $N = 2$ constituent codes for 2 bps/Hz ST-TTCM over quasi-static fading channels.

CC	$\boldsymbol{B}$	$\boldsymbol{F}$	Full rank $\vert_{\mathcal{P}_{\text{quasif}}}$	$\Delta_{\text{DS}}\vert_{\det(\boldsymbol{A})}$	$\Delta_{\text{DS}}\vert_{\det(\boldsymbol{I}+\boldsymbol{A})}$	$\Delta_{\text{DS}}\vert_{\text{trace}(\boldsymbol{A})}$	$P_{\text{FER}}\vert_{\mathcal{P}_{\text{fastf}}}$
Q_a	$[2;2;1;2]$	$[5;6;1;4]$	yes	2.5×10^{-3}	6.2×10^{-4}	5.4×10^{-2}	3.5×10^{-4}
Q_b	$[2;2;1;2]$	$[5;6;4;2]$	yes	2.5×10^{-3}	6.2×10^{-4}	5.4×10^{-2}	3.5×10^{-4}
Q_c	$[2;1;0;2]$	$[6;6;3;4]$	yes	2.5×10^{-3}	6.2×10^{-4}	5.4×10^{-2}	3.5×10^{-4}
Q_d	$[2;1;0;2]$	$[6;6;5;2]$	yes	2.5×10^{-3}	6.2×10^{-4}	5.4×10^{-2}	3.5×10^{-4}

To reduce complexity for the extensive code search over quasi-static fading channels, the cardinality of set $\mathcal{P}$ was reduced compared to the truncated UB analysis from Sec. 8.2.4. To define $\mathcal{P}_{\text{quasi}}$ the following parameters were assumed: $h_{\text{max}} = 3$, $\psi_{\text{max}} = 8$, $\Delta_{\text{Pmax}} = 12$ and $\Theta_{\text{max}} = 3$. Compared to Sec. 8.2.4, h_{max} and Δ_{Pmax} were reduced, which was partly compensated by the slightly increased ψ_{max}. For each of $\boldsymbol{B}_1$ and $\boldsymbol{B}_2$, 30 out of 1024 candidate CCs resulted in ST-TTCM that satisfied the significant full rank criterion. A further search over the minimum spectral product criterion was performed for each of the three EPD definitions from Sec. 8.3.2, i.e., assuming $\Delta_{\text{P}} = \det(\boldsymbol{A})$, $\Delta_{\text{P}} = \det(\boldsymbol{I} + \boldsymbol{A})$ or $\Delta_{\text{P}} = \text{trace}(\boldsymbol{A})$. Interestingly, for each of $\boldsymbol{B}_1$ and $\boldsymbol{B}_2$, the same two CCs satisfied the minimum spectral product criterion for all three EPD definitions. This concedes the equal eigenvalue criterion from [49]. Table 8.2 summarizes the results of a code search over quasi-static fading channels where $\boldsymbol{F} = [f_{212}(D); f_{211}(D); f_{222}(D); f_{221}(D)]$ includes polynomial coefficients of $f_{2zk}(D) = f_{2zk}^0 + f_{2zk}^1 D + \ldots + f_{2zk}^{\nu_k} D^{\nu_k}$. All four codes have the same value of Δ_{DS} for a given EPD definition.

On fast fading channels, the reduced cardinality set $\mathcal{P}_{\text{fast}}$ was determined by $h_{\text{max}} = 4$, $\psi_{\text{max}} = 8$, $\Delta_{\text{Pmax}} = 48$ and $\Theta_{\text{max}} = 2$. Compared to the truncated UB from Sec. 8.2.4, h_{max} and Θ_{max} were decreased, which was partly compensated by increasing ψ_{max}. For each of $\boldsymbol{B}_1$ and $\boldsymbol{B}_2$, the DS optimization produced two new CCs. The moderately high SNR=12dB was assumed to assure that optimization is performed in the error-floor region. Table 8.3 summarizes the results of a code search over fast fading channels. All four codes in Table 8.3 have the same value of the given performance measure.

8.4 Performance Evaluation

The performance of single-antenna and multi-antenna TTCM with newly designed CCs is evaluated in Sec. 8.4.1. Sec. 8.4.2 addresses the main issues related to the choice of bit-level vs. symbol-level information interleaving. Finally, the performance comparison between single-antenna and multi-antenna TTCM is evaluated in Sec. 8.4.3.

Table 8.3. The best $\nu = 3$ QPSK $N = 2$ constituent codes for 2 bps/Hz ST-TTCM over fast fading channels.

CC	$\boldsymbol{B}$	$\boldsymbol{F}$	Full rank $\vert\mathcal{P}_{\text{quasif}}$	Δ_{DS} $\vert\det(\boldsymbol{A})$	Δ_{DS} $\vert\det(\boldsymbol{I}+\boldsymbol{A})$	Δ_{DS} $\vert\text{trace}(\boldsymbol{A})$	P_{FER} $\vert\mathcal{P}_{\text{fastf}}$
F_a	[2; 2; 1; 2]	[3; 4; 4; 6]	no	9.2×10^{-4}	3×10^{-4}	4.3×10^{-2}	2.5×10^{-5}
F_b	[2; 2; 1; 2]	[3; 4; 7; 2]	no	9.2×10^{-4}	3×10^{-4}	4.3×10^{-2}	2.5×10^{-5}
F_c	[2; 1; 0; 2]	[1; 4; 6; 2]	no	9.2×10^{-4}	3×10^{-4}	4.3×10^{-2}	2.5×10^{-5}
F_d	[2; 1; 0; 2]	[1; 4; 7; 6]	no	9.2×10^{-4}	3×10^{-4}	4.3×10^{-2}	2.5×10^{-5}

8.4.1 New versus Old Constituent Codes in TTCM and ST-TTCM

The performance of the TTCM with the new constituent TCM Ua from Table 8.1 over AWGN and fast fading channels is depicted in Fig. 8.6. More than 1dB and 5dB gains are achieved over AWGN and fast fading channels, respectively, in both the asymptotic FER and BER compared to the TTCM from [34].

The DS spectra of the TTCM with the new and the old constituent TCMs are depicted in Fig. 8.7. Despite its much lower *free* distance, the new CCs result in the considerably better overall DS characteristics, i.e., the lower multiplicities for the majority of the spectral lines. This is the key point where turbo-like codes turn the conventional design criteria [34] on its head, making error multiplicities more important than the minimum distance [36].

Fig. 8.8 justifies the assumptions on the importance of the systematic nature of CCs and their primitive equivalent feed-backward polynomial to ST-TTCM decoder convergence. The newly designed CCs result in ST-TTCM with a 0.5 dB lower convergence threshold than with constituent Rec-STTrCs from [16, 19].

Fig. 8.9 depicts the FER performance of the ST-TTCM with the newly introduced CCs from Tables 8.2 and 8.3 over quasi-static and fast fading channels. As expected, different CCs from the same Table resulted in ST-TTCM with exactly the same performance. For brevity, only one curve per Table is plotted. The same parameters were adopted as in Sec. 8.2.4. As seen from UB and simulation results in Fig. 9(a), new CCS from Table 8.2 resulted in full transmit diversity ST-TTCM with improved performance over both schemes studied in [16, 19]. New CCs from Table 8.3 offered additional gain over a fast fading channel and performed exactly the same as CCs from Table 8.2 over a quasi-static fading channel. Although ST-TTCM with CCs Table 8.3 did not satisfied the significant full rank criterion, the noticeable performance degradation was foreseen by UB only for large SNRs, high above those typically exhibited in practise. In summary, ST-TTCM with new CCs achieved more than 1.5 dB gain at FER 10^{-3} over both quasi-static and fast fading channels.

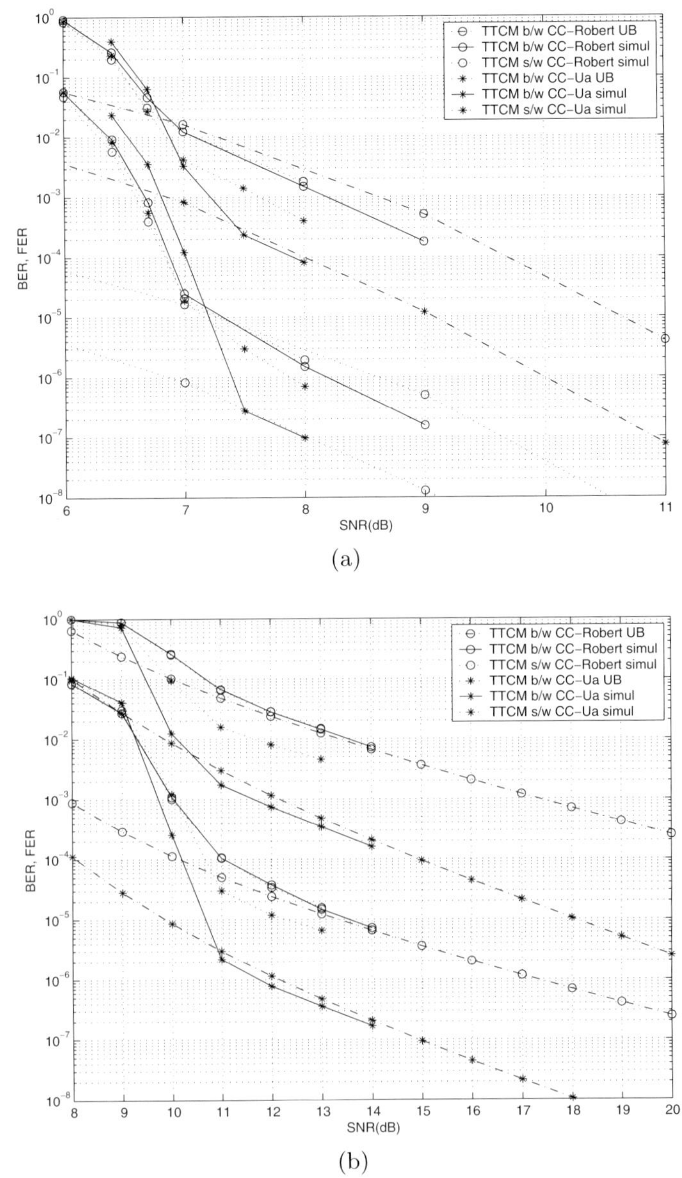

Fig. 8.6. Performance of TTCM with the new constituent TCM, $N = M = 1$, 10 decoding iterations, $LK = 2048$ bits. (a) AWGN. (b) Fast Fading.

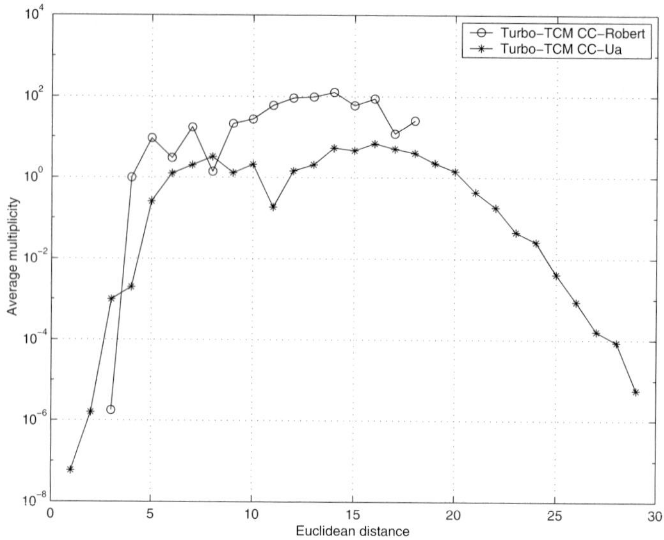

Fig. 8.7. The average number of error events as a function of the Euclidean distance in the FER distance spectrum of TTCM.

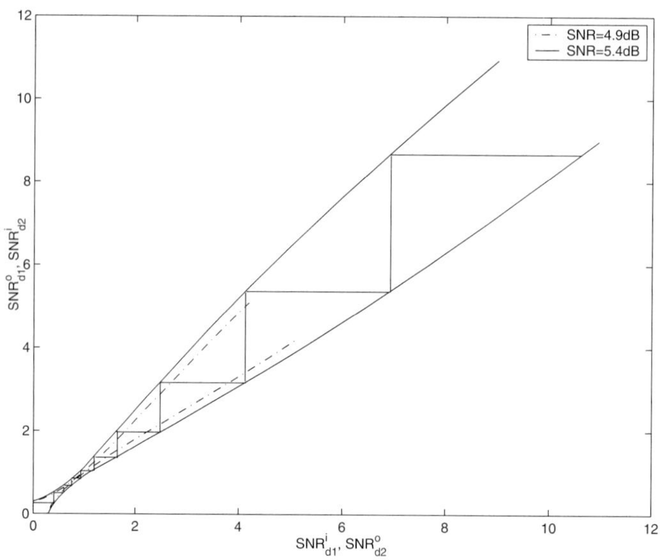

Fig. 8.8. Iterative decoding convergence of the ST-TTCM with the newly designed CC-Q_c, AWGN channel, SNR=4.9dB, SNR=5.4dB.

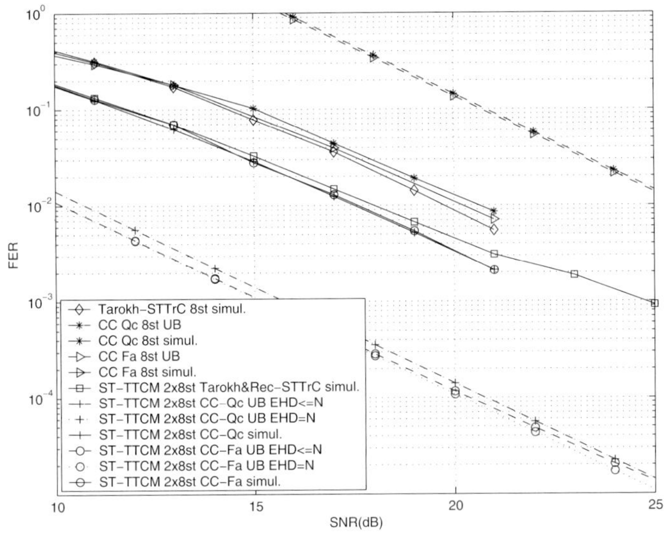

(a)

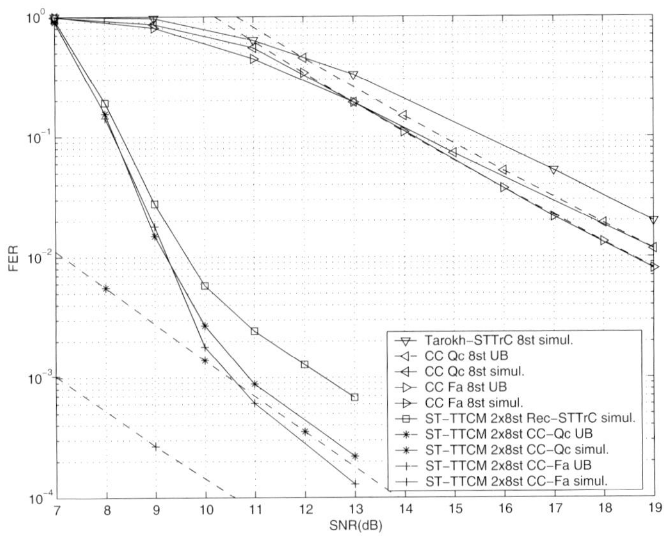

(b)

Fig. 8.9. Performance of ST-TTCM with the new 8-state CCs, 10 decoding iterations, input information frame of $LK = 260$ bits, $N = 2$, $M = 1$. (a) Quasi-static fading channel. (b) Fast fading channel.

8.4.2 Bit versus Symbol Information Interleaving

The new CCs from Table 8.2 were designed to satisfy the significant full rank criterion for an arbitrary realization of the pseudo-random bit-wise information interleaving. This includes also the special case when all the K information interleavers have the same pattern, i.e., for the symbol-wise information interleaving. The comparison between the bit-wise and the symbol-wise interleaved ST-TTCM with the newly designed CC-Q_c over quasi-static and fast fading channels is depicted in Fig. 8.10. The two schemes are denoted as "ST-TTCM b/w 2x8st CC-Q_c" and "ST-TTCM s/w 2x8st CC-Q_c", respectively. Despite the potential loss induced during symbol-to-bit and bit-to-symbol transformations in component decoders [52], no performance degradation of bit- compared to symbol-wise interleaved ST-TTCM was experienced through simulations over quasi-static fading channel. Independent mapping of K input bits from the given input symbol allows spreading the components of one error event to approximately K times more single error events, typically accumulating more diversity [43]. As a result, on a fast fading channel, a significant improvement of the bit compared to symbol level ST-TTCM was exhibited, i.e., more than 3 dB gain at FER 10^{-3}. The similar effect is observed for TTCM in Fig. 8.6 where gains for bit level scheme range from 0.7 dB over AWGN to 3 dB over fast fading channels.

For comparison, Fig. 8.10 also includes the performance of the symbol-level ST-TTCM proposed by Firmanto, Chen, Vucetic and Yuan (FCVY) in [22, 25]. The scheme is denoted as "ST-TTCM s/w 2x8st CC-FCVY" and employs also the 8-state CCs. With no particular optimization other than that regularly applied for CCs as stand-alone STTrCs, the ST-TTCM from [22, 25] failed to achieve the full transmit diversity over quasi-static fading channels. On fast fading channel, a performance loss was induced compared to both the bit-level and the symbol-level ST-TTCMs with the new CCs designed in this paper. Fig. 8.10 also demonstrates that by increasing the frame size, the considerable FER coding gain can be achieved for ST-TTCM when transmission frame experiences more than one independent realization of fading. However, such a coding gain is restricted to convergence region and as such will not effect the asymptotic FER performance.

8.4.3 TTCM versus ST-TTCM in $N = M = 2$ Systems

For multi-antenna transmission and reception ($N = M = 2$), single-antenna TTCM with constituent TCMs from Table 8.1 was further concatenated with inner space-time block code (STBC) [8] and compared to ST-TTCM with CCs from Table 8.2 in Fig. 8.11. With redundancy expansion in spatial rather than higher order modulation domain, ST-TTCM with QPSK modulation offers 1 dB and 1.5 dB gains over quasi-static and fast fading channel, respectively. With single antenna reception ($N = 2, M = 1$), the performance of two

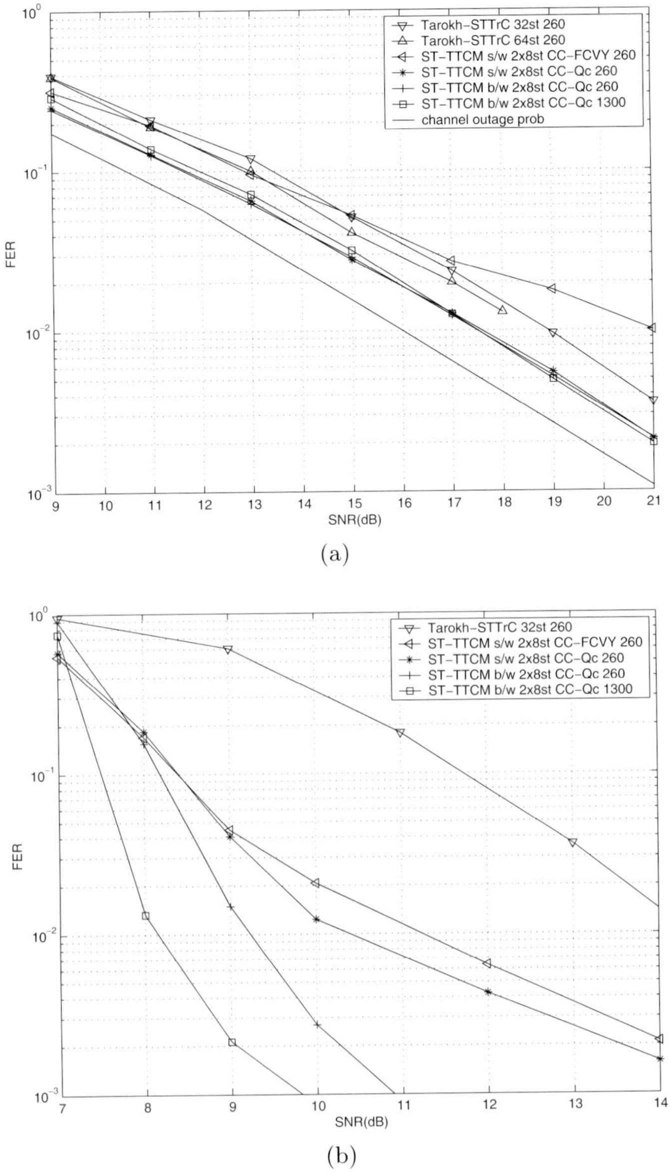

Fig. 8.10. The comparison between ST-TTCMs operating with the bit level and symbol level information interleaver, 8-state CCs, 10 decoding iterations, input information frame of $LK = 260, 1300$ bits, $N = 2$, $M = 1$. (a) Quasi-static fading channel. (b) Fast fading channel.

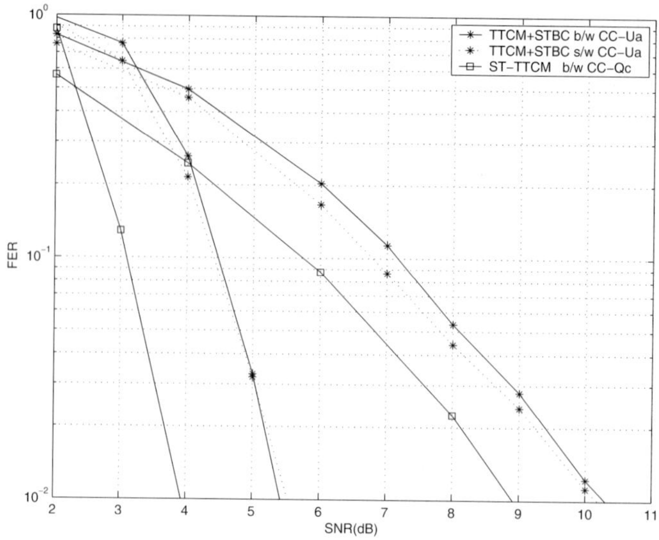

Fig. 8.11. Comparison between 8PSK TTCM with inner STBC and QPSK ST-TTCM, $N = M = 2$, 2bps/Hz, $LK = 260$, quasi-static and fast fading.

schemes is nearly the same (not showed). ST-TTCM is still more advantageous for practical implementation as it achieves the same spectral efficiency with lower order modulation and somewhat reduced decoding complexity, i.e., neither STBC combining nor doping in the first decoding iteration is needed.

8.5 Summary

The recent findings on turbo coded modulation for future wireless communication systems have been summarized in this book chapter. The ST-TTCM, as a new paradigm for MIMO signal transmission which combines space-time coding, turbo coding and higher order modulation has been put into the broader perspective of prior art in channel coding theory. The unified approach to single- and multi-antenna TTCM design, performance analysis and constituent TCM optimization was further introduced. The performance analysis included the two-dimensional DS interpretation, the truncated UB and the iterative decoding convergence analysis. The DS and UB analysis discovered the effect of *spectral thinning* similar to single antenna binary turbo codes. This revealed the importance of multiplicities for the general ST-TTCM design. It was further shown that without particular optimization of CCs and/or information interleaving, the punctured parallel concatenation in general does not preserve full transmit diversity over quasi-static fading channels. The truncated UB appeared to be tight over AWGN and fast fading channels but rather optimistic over quasi-static fading channels. Nevertheless, the rank deficient error events seem to have been well enumerated, which appeared crucial for the accurate design of the full transmit diversity ST-TTCM.

The modified design criteria for TTCM over AWGN, quasi-static and fast fading channels were further introduced that capture the joint effects of error coefficients and multiplicities in the code's DS. To assure the good convergence of iterative decoding, a binary recursive systematic form with a primitive equivalent feed-backward polynomial was assumed to model the subset of candidate CCs. The DS based optimization produced new sets of CCs that resulted in ST-TTCM with more than 1.5 dB gain at FER 10^{-3} over both quasi-static and fast fading channels. The full transmit diversity over quasi-static fading channels was achieved with no constrains on the implemented pseudo-random information interleaving. This leaves plenty of degrees of freedom for the eventual code-matched interleaving design. For single antenna TTCM, more than 1dB and 5dB gains are achieved over AWGN and fast fading channels, respectively, in both the asymptotic FER and BER compared to *Robertson et al.* TTCM. The superior performance of the bit-level compared to the symbol-level ST-TTCM was demonstrated through simulations over fading channels. Finally, for more than one antenna at the receiver, the ST-TTCM was showed advantageous over single-antenna TTCM concatenated with inner STBC.

References

1. E. Telatar, "Capacity of multi-antenna gaussian channels," *European Trans. Telecommun.*, vol. 10, no. 6, pp. 585–595, Nov.-December 1999.
2. G. J. Foschini, "Layered space-time architecture for wireless communications in a fading environment when using multi-element antennas," *The Bell System Technical Journal*, pp. 41–59, 1996.
3. H. Xu, M. Gans, D. Chiznik, J. Ling, P. Wolniansky, and R. Valenzuela, "Spatial and temporal variations of MIMO channels and impacts on capacity," in *Proc. IEEE Int. Conf. Commun.*, vol. 1, 2002, pp. 262–266.
4. A. Molisch, M. Steinbauer, M. Toeltsch, E. Bonek, and R. S. Thom, "Capacity of MIMO systems based on measured wireless channels," *IEEE J. Select. Areas Commun.*, vol. 20, no. 3, pp. 561–569, Apr. 2002.
5. D. Gesbert, H. Bolcskei, D. Gore, and A. Paulraj, "MIMO wireless channels: Capacity and performance prediction," in *Proc. IEEE Global Telecommun. Conf.*, vol. 2, 2000, pp. 1083 –1088.
6. D. Chizhik, G. Foschini, M. Gans, and R. Valenzuela, "Keyholes, correlations, and capacities of multielement transmit and receive antennas," *IEEE Trans. Wireless Commun.*, vol. 1, no. 2, pp. 361–368, Apr. 2002.
7. A. J. Paulraj and C. B. Papadias, "Space-time processing for wireless communications," *IEEE Signal Processing Mag.*, vol. 14, no. 6, pp. 49–83, Nov. 1997.
8. S. Alamouti, "A simple transmit diversity technique for wireless communications," *IEEE J. Select. Areas Commun.*, vol. 16, no. 8, pp. 1451–1458, Oct 1998.

43. C. Fragouli and R. D. Wesel, "Turbo-encoder design for symbol-interleaved parallel concatenated trellis-coded modulation," *IEEE Trans. Commun.*, vol. 49, no. 3, pp. 425–435, Mar. 2001.
44. G. Ungerboeck, "Channel coding with multilevel/phase signals," *IEEE Trans. Inform. Theory*, vol. 28, Jan. 1982.
45. Z. Chen, J. Yuan, and B. Vucetic, "Improved space-time trellis coded modulation scheme on slow Rayleigh fading channels," *Electron. Lett.*, vol. 37, no. 7, pp. 440–441, Mar. 2001.
46. M. Tao and R. S. Cheng, "Improved design criteria and new trellis codes for space-time coded modulation in slow flat fading channels," *IEEE Commun. Lett.*, vol. 5, no. 7, pp. 313–315, July 2001.
47. E. Biglieri, G. Taricco, and A. Tulino, "Performance of space-time codes for a large number of antennas," *IEEE Trans. Inform. Theory*, vol. 48, no. 7, pp. 1794–1803, July 2002.
48. D. M. Ionescu, "New results on space-time code design criteria," in *Proc. IEEE Wireless Commun. and Networking Conf.*, 1999, pp. 684–687.
49. D. M. Ionescu, K. Mukkavilli, Z. Yan, and J. Lilleberg, "Improved 8- and 16-state space-time codes for 4PSK with two transmit antennas," *IEEE Commun. Lett.*, vol. 5, no. 7, pp. 301–303, July 2001.
50. Y. S. Jung and J. H. Lee, "New measure of coding gain for space-time trellis codes," in *Proc. IEEE Int. Symp. Inform. Theory*, Washington DC, USA, June 24–29 2001, p. 198.
51. Y. Sasazaki and T. Ohtsuki, "Improved design criteria and new trellis codes on space-time trellis coded modulation in fast fading channels," in *Proc. IEEE Global Telecommun. Conf.*, 2002.
52. B. Scanavino, G. Montorsi, and S. Benedetto, "Convergence properties of iterative decoders working on bit and symbol level," in *Proc. IEEE Global Telecommun. Conf.*, vol. 2, 2001, pp. 1037–1041.

Chapter 9

Turbo-MIMO for High-Speed Wireless Communications

Mathini Sellathurai and Yvo L.C. de Jong

Communications Research Centre, Canada

Recent research has shown that the performance of multiple antenna systems, often referred to as multiple-input multiple-output (MIMO) systems, can be significantly improved by using the principle of iterative, or "turbo" processing. The resulting class of MIMO systems is referred to as Turbo-MIMO. This chapter discusses a Turbo-MIMO architecture based on space-time bit-interleaved coded modulation (ST-BICM), which is highly effective if used in conjunction with receivers employing iterative detection and decoding. It reviews three recent low-complexity detection schemes that can be used in such systems, i.e., minimum mean-squared based soft-interference cancellation soft-interference cancellation (SIC-MMSE), iterative tree search (ITS) detection, and list sphere detection. Simulation results are presented which show that, with the aid of these or similar schemes, the embodiment of the turbo processing principle with MIMO architectures provides a practical solution to the requirement of high data-rate transmission in a reliable manner for future wireless communication.

9.1 Turbo-MIMO

In 1993, Berrou *et al.* developed the revolutionary iterative, or "turbo" receiver for decoding concatenated convolutional codes [1]. Turbo codes are capable of approaching the Shannon limit of channel capacity in a computationally feasible manner. What is even more profound is the fact that the turbo processing principle has been successfully applied not only to channel decoding, but to a wide range of areas in communications, including channel equalization, coded modulation, multi-user detection, and joint source and channel decoding. This chapter describes the application of the turbo principle to multiple-input, multiple-output (MIMO) wireless

communication systems. As was pointed out by Foschini in 1996 and is well-known by now, such systems have the potential of providing enormous

K. Sripmanwat (ed.), Turbo Code Applications: a journey from a paper to realization, 223–241.

capacity increase as well as diversity gain compared to conventional, single-antenna systems [2]. A large body of work has been devoted to the problem of designing space-time signaling and processing schemes capable of capitalizing on this potential (see [3–12] and the references therein).

It is known that the computational complexity of any optimal, joint detection and decoding scheme for MIMO systems grows exponentially with the burst size, among other factors. This property prohibits the use of optimal MIMO processing in virtually all practical MIMO scenarios. Fortunately, as will be shown in this chapter, near-optimal performance can be achieved by means of iterative detection and decoding, in which the detection stage is effectively decoupled from the channel decoding stage, thus making its complexity independent of the burst size. In an iterative receiver, extrinsic information from the output of the detector is fed to the channel decoder and vice versa, which permits the iterative process to take its natural course in response to the received noisy signal and the code constraints. Both the detector and the channel decoder must be soft-input/soft-output (SISO) modules, i.e., both must be capable of accepting and producing soft-decision information (see [13] for SISO decoding). An interesting feature of this approach is that it can benefit from the use of powerful existing channel codes, such as turbo codes or turbo-like codes, provided the code words are appropriately mapped onto the MIMO channel. This MIMO architecture [3–11] will be referred to herein as Turbo-MIMO. The hallmark of this architecture is that its error performance improves with the number of iterations in the detector/decoder loop and, most importantly, exceeds the performance of correspondingly encoded non-iterative MIMO systems such as V-BLAST [15].

Space-time bit-interleaved coded modulation (ST-BICM) is a highly effective MIMO mapping scheme, in which the bits at the output of the channel encoder are scattered individually across the transmit antennas and the burst length, so as to maximize diversity gain. In ST-BICM MIMO systems, the user's information bits are first encoded using a predetermined linear block or trellis channel code. The encoded binary data stream is permuted using a bit interleaver, which can be random or optimized according to some performance criterion. The coded and interleaved bit stream is then serial-to-parallel converted into parallel substreams, each of which is mapped onto a sequence of PSK or QAM symbols that is transmitted from a separate antenna. Processing of the received signal proceeds in an iterative fashion, as discussed above, with the understanding that the extrinsic information exchanged between the detector and the channel decoder must be appropriately de-interleaved and re-interleaved. Several papers have demonstrated the excellent performance of ST-BICM systems in Rayleigh fading environments, showing that this scheme is capable of achieving a significant portion of the MIMO channel capacity at relatively low complexity [5–7].

The complexity of the iterative receiver is determined by the complexities of the detection and decoding stages. Thanks to the availability of powerful, low-complexity channel codes, for example 4-state turbo codes, the complexity of the channel decoder is generally not predominant with respect to the overall receiver complexity. The complexity of the optimal detection scheme for iterative MIMO receivers, the maximum *a posteriori* probability (MAP) detector, is exponential in the number of code bits transmitted in parallel per symbol duration, and often becomes prohibitively high for practical MIMO systems. Among the sub-optimal detection schemes that have recently been proposed as a consequence of this observation are the SIC-MMSE scheme of [5], the iterative tree search (ITS) detector of [7], and the list sphere detector proposed in [6]. Highly promising performance results have been reported for all of these schemes. This chapter, after a more detailed review of Turbo-MIMO, will present and discuss simulation results demonstrating the performance-complexity trade-offs that can be made using the three detection schemes mentioned above.

9.2 Theory

Turbo-MIMO systems based on ST-BICM exploit the ideas of bit-interleaved coded modulation (BICM) [17] and the turbo processing principle in a space-time coding framework. Fig. 9.1 shows a block diagram of such a system with N_t transmit antennas and N_r receive antennas. In the following, it is assumed that $N_r \geq N_t$. Instead of employing dedicated space-time codes, as proposed in [14], the transmitter uses a one-dimensional forward error-correcting block code to encode the user's information bits. The channel encoder is followed by a pseudo-random interleaver, Π, and a space-time mapper, as discussed below. This configuration can be thought of as a serial concatenation of two constituent encoders separated by the interleaver, the inner encoder being the space-time mapper in conjunction with the channel.

9.2.1 ST-BICM

Denoting a block of information bits by the vector $\mathbf{d}$ and the transfer function of the channel code by G, the codeword at the output of the outer encoder can be written as $\tilde{\mathbf{c}} = G\mathbf{d}$, and $\mathbf{c} = \Pi(\tilde{\mathbf{c}})$ represents the interleaved sequence of code bits. The modulation format is limited herein to be identical for all transmit antennas, and the number of bits per constellation point is denoted by M_c. The space-time mapper partitions $\mathbf{c}$ into L subvectors

$$\mathbf{c}^{(l)} = \left[c_{1,1}^{(l)}, \cdots, c_{1,M_c}^{(l)}, c_{2,1}^{(l)}, \cdots, c_{N_t,M_c}^{(l)}\right]^T, \quad l = 1, \cdots, L \tag{9.1}$$

and maps each of them onto a symbol vector

$$\mathbf{x}^{(l)} = \left[x_1^{(l)}, \cdots, x_{N_t}^{(l)}\right]^T \tag{9.2}$$

according to a unique, predetermined bit mapping scheme. To simplify notation, the superscripts $^{(l)}$ will be omitted whenever possible. If $\mathbf{H} \in \mathcal{C}^{N_r \times N_t}$ is

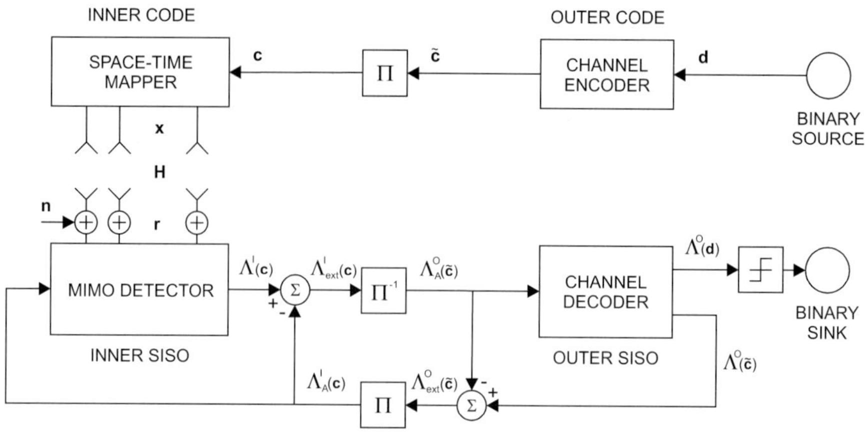

Fig. 9.1. Block diagram of a MIMO system employing ST-BICM and an iterative receiver. Π and Π^{-1} denote interleaving and de-interleaving, respectively. The vectors **c** and **d** refer to coded and uncoded bits, respectively.

used to denote the MIMO channel, which is assumed to be narrowband, the received signal is given by

$$\mathbf{r} = \mathbf{H}\mathbf{x} + \mathbf{n} \tag{9.3}$$

where $\mathbf{n} \in \mathcal{C}^{N_r \times 1}$ is an additive noise vector whose elements are independent, complex-valued Gaussian variables with zero mean and variance σ^2.

In theory, it is possible to model the ST-BICM code, consisting of the two encoders, the interleaver and the channel, as a single Markov process. However, the associated trellis representation is extremely complex and does not lend itself to computationally feasible decoding. The computational complexity of a full search procedure over such a trellis increases exponentially with the product of N_t, M_c, and L. Fortunately, however, the interleaver separating the outer and inner encoders enables near-optimal decoding at reasonable computational complexity, by exploiting the principle of iterative, or turbo processing. This approach, often referred to as iterative detection and decoding (IDD), is similar to the decoding of serially concatenated turbo codes [16].

In general, the outer code of an ST-BICM MIMO system can be any type of error-correcting code that can be decoded with a soft-input soft-output decoder, for example convolutional or turbo codes. Turbo or turbo-like codes are often preferred because of their exceptionally high performance. The space-time mapper provides the flexibility to design the MIMO system such as to achieve the desired trade-off between multiplexing and diversity gain [12]. To achieve the maximum multiplexing gain, each antenna must transmit an independent information stream, i.e., the space-time mapper must be a spatial multiplexer. On the other hand, space-time block codes such as those proposed in [14] can be used to achieve the maximum diversity order.

9.2.2 Iterative Detection and Decoding

Iterative decoding of serially concatenated codes is now a well-established and practical alternative to optimal, joint maximum-likelihood decoding. Fig. 9.1 shows the iterative receiver of an ST-BICM MIMO system, which separates the overall decoding problem into two stages, inner decoding (detection) and outer decoding (channel decoding), and exchanges the information learned from one stage to another iteratively until the receiver converges. The deinterleaver, Π^{-1}, is used to compensate for the interleaving operation used at the transmitter. Furthermore, together with the interleaver, Π, it serves to decorrelate the output from one decoding stage before it is passed to the next. The iterative receiver updates and generally improves the soft decisions on the information bits as well as the code bits at each iteration of the information exchange process. These soft decisions are produced in the form of *a posteriori* log-likelihood ratios (LLRs), defined as

$$\Lambda(c) = \ln \frac{P(c=+1|\cdot)}{P(c=-1|\cdot)} \tag{9.4}$$

In this equation, the probabilities are conditioned on the received signal vector $\mathbf{r}$ or the constraints of the channel code if the LLR is an output of the inner or outer decoder, respectively. Using Bayes' theorem, and assuming statistical independence between bits, which is a reasonable assumption because of the interleaving operation, any *a posteriori* LLR can be written as

$$\underbrace{\ln \frac{P(c=+1|\mathbf{r})}{P(c=-1|\mathbf{r})}}_{\Lambda(c)} = \underbrace{\ln \frac{P(c=+1)}{P(c=-1)}}_{\Lambda_A(c)} + \underbrace{\ln \frac{p(\mathbf{r}|c=+1)}{p(\mathbf{r}|c=-1)}}_{\Lambda_{ext}(c)} \tag{9.5}$$

where $\Lambda_A(c)$ constitutes the *a priori* information on the bit c, and $\Lambda_{ext}(c)$ constitutes the extrinsic information. This extrinsic information is the incremental new information learned from either the received signal vector or the channel code constraints, using the available *a priori* information. The extrinsic information produced by the inner decoding stage is used as *a priori* information by the outer decoder, and vice versa.

With the above definitions, the iterative decoding algorithm, in its optimal implementation, has the following steps:

Step 1: The inner decoder, or detector, generates extrinsic LLRs for each of the $N_t M_c$ code bits mapped onto the symbol vector symbol $\mathbf{x}$. This extrinsic information can be written as

$$\Lambda^I_{ext}(c_{jk}) = \ln \frac{\sum\limits_{\mathbf{c}\in\mathbb{C}_{jk,+1}} \exp \mu(\mathbf{x})}{\sum\limits_{\mathbf{c}\in\mathbb{C}_{jk,-1}} \exp \mu(\mathbf{x})} - \Lambda^I_A(c_{jk}) \tag{9.6}$$

where $\mathbb{C}_{jk,\pm 1}$ are the sets of all possible bit sequences of length $N_t M_c$ for which c_{jk} equals +1 and -1, respectively, i.e.,

$$\mathbb{C}_{jk,\pm 1} = \{\mathbf{c} | c_{jk} = \pm 1\} \tag{9.7}$$

The metric $\mu(\mathbf{x})$ is given by

$$\mu(\mathbf{x}) = -\frac{1}{\sigma^2}\|\mathbf{r} - \mathbf{H}\mathbf{x}\|^2 + \sum_{i=1}^{N_t}\sum_{j\in\mathbb{J}_i} \Lambda_A^I(c_{ij}) \tag{9.8}$$

where

$$\mathbb{J}_i = \{j | j \in \{1, \cdots, M_c\} \text{ and } c_{ij} = +1\} \tag{9.9}$$

During the first iteration, all information bits are assumed equally likely to have been transmitted, i.e., $\Lambda_A^I(c_{ij}) = 0$ for all i, j. The detection algorithm based on Eq. 9.6–9.9 is known as the APP or MAP algorithm. Note that the inner decoder produces sets of $N_t M_c$ extrinsic LLR values for all L symbol vectors transmitted during a block period. The block of extrinsic LLR values associated with the bits in $\mathbf{c}$ is denoted by $\Lambda_{ext}^I(\mathbf{c})$, and becomes available to the outer decoder as *a priori* information after de-interleaving:

$$\Lambda_A^O(\tilde{\mathbf{c}}) = \Pi^{-1}\{\Lambda_{ext}^I(\mathbf{c})\} \tag{9.10}$$

Step 2: The outer decoder, in turn, processes the soft information $\Lambda_A^O(\tilde{\mathbf{c}})$ and computes updated extrinsic information on both the information and code bits, based on the trellis structure of the channel codes. The extrinsic information on the code bits, denoted by $\Lambda_{ext}^O(\tilde{\mathbf{c}})$ is re-interleaved and fed back to the inner decoder:

$$\Lambda_A^I(\mathbf{c}) = \Pi\{\Lambda_{ext}^O(\tilde{\mathbf{c}})\} \tag{9.11}$$

Steps 1 and 2 are repeated until some convergence criterion is achieved. At that point, hard decisions on the information bits are made by taking the signs of the *a posteriori* LLR values $\Lambda^O(\mathbf{d})$.

The complexity of the iterative receiver is determined by the two constituent decoders. If the channel code is a turbo-like code, the outer decoder is usually implemented as a concatenation of soft-input soft-output modules based on the BCJR algorithm [13] or approximations thereof. The complexity of the channel decoder is generally not predominant with respect to the overall receiver complexity. The APP detector, although optimal, is hardly a practical choice for the inner decoder due to its high computational complexity. As can be observed by inspecting Eq. 9.6, this detector performs an exhaustive search over the entire set of length-$N_t M_c$ bit sequences that can be mapped onto a symbol vector $\mathbf{x}$. The associated complexity per bit is therefore asymptotically proportional to $2^{N_t M_c}$. Even for moderate numbers of transmit antennas and commonly used modulation formats, this complexity can be prohibitively high. In practice, therefore, suboptimal, reduced-complexity MIMO detection schemes such as those discussed in the next section are often preferred.

9.3 Suboptimal MIMO Detection

9.3.1 List Sphere Detection

Sphere detection, sometimes also referred to as sphere decoding, is a reduced-complexity approximation to APP detection in which the search space is limited to the set of symbol vectors for which $\gamma(\mathbf{x}) = \|\mathbf{r} - \mathbf{Hx}\|^2 \leq r^2$, i.e., for which $\mathbf{Hx}$ lies within a hypersphere of some predetermined radius r, centered around the noisy received signal. The rationale for this approach is that symbol vectors for which $\|\mathbf{r} - \mathbf{Hx}\|^2$ is large are less likely to contribute significantly to the detector output Eq. 9.6, because their metrics, $\mu(\mathbf{x})$, defined in Eq. 9.8, are more likely to be small. These symbol vectors can therefore be excluded from the search space with limited degradation in performance. Because the reduced search space is, in general, only a small subset of the entire set of possibly transmitted signal points, the complexity of the sphere detector is considerably lower than that of the APP detector.

The procedure employed by the sphere detector to find the set of symbol vectors for which $\gamma(\mathbf{x}) \leq r^2$ avoids an exhaustive search, and is based on the fact that $\gamma(\mathbf{x})$ can be written as

$$\gamma(\mathbf{x}) = \|\mathbf{r} - \mathbf{Hx}\|^2 = (\mathbf{x} - \hat{\mathbf{x}})^H \mathbf{H}^H \mathbf{H} (\mathbf{x} - \hat{\mathbf{x}}) \tag{9.12}$$

where $\hat{\mathbf{x}} = [\hat{x}_1, \cdots, \hat{x}_{N_t}]^T = (\mathbf{H}^H\mathbf{H})^{-1}\mathbf{H}^H\mathbf{r}$ is the unconstrained maximum-likelihood solution. The superscript ^{H} denotes conjugate transpose. Because $\mathbf{H}^H\mathbf{H}$ is Hermitian and positive definite, it has a Cholesky decomposition $\mathbf{H}^H\mathbf{H} = \mathbf{L}^H\mathbf{L}$, in which $\mathbf{L} = [l_{ij}]$ is an $N_t \times N_t$ lower triangular matrix with real, positive diagonal entries. Using this Cholesky decomposition, it is easy to show that Eq. 9.12 can be evaluated in a symbol-by-symbol manner, starting with the first symbol x_1 and proceeding to x_{N_t}, as follows:

$$\begin{aligned} \gamma_1 &= |l_{11}(x_i - \hat{x}_i)|^2 \\ \gamma_i &= \gamma_{i-1} + \left| l_{ii}(x_i - \hat{x}_i) + \sum_{j=1}^{i-1} l_{ij}(x_i - \hat{x}_i) \right|^2, \quad i = 2, \cdots, N_t \\ \gamma(\mathbf{x}) &= \gamma_{N_t} \end{aligned} \tag{9.13}$$

Because the increments in Eq. 9.13 are non-negative, the search criterion $\gamma(\mathbf{x}) \leq r^2$ implies that $\gamma_i \leq r^2$ for $i = 1, \cdots, N_t$. Hence, the sphere detection algorithm starts by selecting the set of symbols x_1 for which $\gamma_1 \leq r^2$. For each of these symbols, the algorithm subsequently selects the symbols x_2 for which $\gamma_2 \leq r^2$, and this procedure continues until $i = N_t$ is reached.

In [9], the authors propose a list version of this scheme, called list sphere detector (LSD). In this scheme, only the M symbol vectors that lie closest to the received signal vector, referred to collectively as the candidate list, $\mathbb{L}$, are used to produce the detector output. Mathematically, the output of the detector output is described by

$$\Lambda^I_{ext}(c_{jk}) = \ln \frac{\sum\limits_{\mathbf{c}\in(\mathbb{L}\cap\mathbb{C}_{jk,+1})} \exp \mu(\mathbf{x})}{\sum\limits_{\mathbf{c}\in(\mathbb{L}\cap\mathbb{C}_{jk,-1})} \exp \mu(\mathbf{x})} - \Lambda^I_A(c_{jk}) \tag{9.14}$$

For $M = 2^{N_t M_c}$ and sufficiently large r, the LSD scheme is identical to the APP detector. Results presented in [18] indicate that its expected worst-case complexity is still exponential in $N_t M_c$, even though it is considerably lower than the APP complexity.

It is noted that, in practice, the log-sum over exponential functions, which is a relatively complex operation, is often approximated by

$$\log \sum_j \exp \mu_j \approx \max_j \mu_j \tag{9.15}$$

with little performance degradation. This is referred to as the max-log approximation. It is further noted that, at each iteration, the quality of the candidate list can be improved by taking into account the *a priori* information computed in the previous iteration. This is utilized in the iterative tree search (ITS) detection scheme proposed in [7], where it is shown that considerably better performance than that of the LSD scheme can be achieved at the same list sizes.

9.3.2 Iterative Tree Search Detection

The ITS detection scheme proposed in [7] has better performance than the list sphere detector, as it produces an improved candidate list at each iteration of the receiver, by taking into account the *a priori* information fed back from the outer decoder. This improved candidate list is generated with the aid of a breadth-first tree search algorithm known as the M-algorithm, as illustrated in Fig. 9.2. The set of all possible symbol vectors can be represented by a tree structure of depth N_t, having a single symbol on each branch and 2^{M_c} branches emanating from each node. Each path within the tree is uniquely associated with a sequence of symbols $s_1, \cdots, s_d$ and a metric μ_d, where $d \leq N_t$ indicates the depth of the path and μ_d can be computed as follows:

$$\mu_1 = |l_{11}(x_i - \hat{x}_i)|^2 + \sum_{j\in\mathbb{J}_1} \Lambda^I_A(c_{1j})$$

$$\mu_i = \mu_{i-1} + \left| l_{ii}(x_i - \hat{x}_i) + \sum_{j=1}^{i-1} l_{ij}(x_i - \hat{x}_i) \right|^2 + \sum_{j\in\mathbb{J}_i} \Lambda^I_A(c_{ij}), \quad i = 2, \cdots, N_t \tag{9.16}$$

Every possible symbol vector corresponds to a path to the maximum depth, and has an associated metric $\mu(\mathbf{x}) = \mu_{N_t}$, as defined in Eq. 9.8. The ITS detector uses the M-algorithm to search for the best paths through the tree,

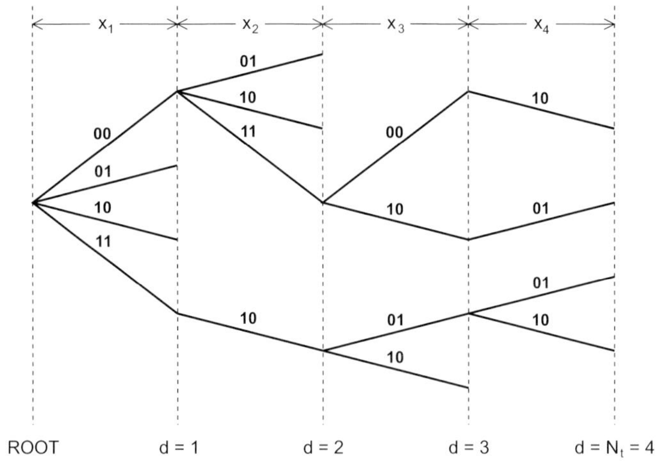

Fig. 9.2. Example of sequential tree search, for $N_t = 4$, $M_c = 2$. At each symbol depth, the best $M = 4$ paths are retained. Deleted paths are not shown.

i.e., those with the largest metrics. At each depth smaller than N_t, the algorithm keeps a candidate list, $\mathbb{L}$, of the best M paths found thus far, and moves forward by extending each of these paths to form $M \cdot 2^{M_c}$ new paths. Path metrics are then updated with the aid of Eq. 9.16, and only the M best new paths are retained in the updated list; the remaining $M(2^{M_c} - 1)$ paths are deleted. When the algorithm reaches the maximum depth, the candidate list $\mathbb{L}$ is used to compute an approximation of the APP detector output, using Eq. 9.14 or the max-log approximation.

Theoretically, the performance of the ITS detector is identical to that of the APP detector only for the maximum possible list size $2^{N_t M_c}$. In practice, however, near-optimum performance is often achieved when M is only a small fraction of the full search space. The complexity per bit of the ITS detector is linearly dependent on N_t and, like the list sphere detector, its dependence on the signal constellation size is of order 2^{M_c}.

9.3.3 Multilevel Mapping ITS Detection

The ITS detector's exponential growth in complexity with increasing M_c is a consequence of the fact that the number of branches departing from each node in the tree structure and, therefore, the number of metric updates at each step of the tree search, is proportional to the alphabet size, 2^{M_c}. Obviously, this rapid complexity increase is a disadvantage if higher-order modulation schemes are employed, which are of special interest if high spectral efficiency is desired. Fortunately, the use of multilevel mapping(MLM) QAM signal constellations can significantly reduce the complexity of the tree search [7].

A QAM constellation with a multilevel bit mapping has the property that it can be partitioned into four equal subsets such that (i) the maximum Euclid-

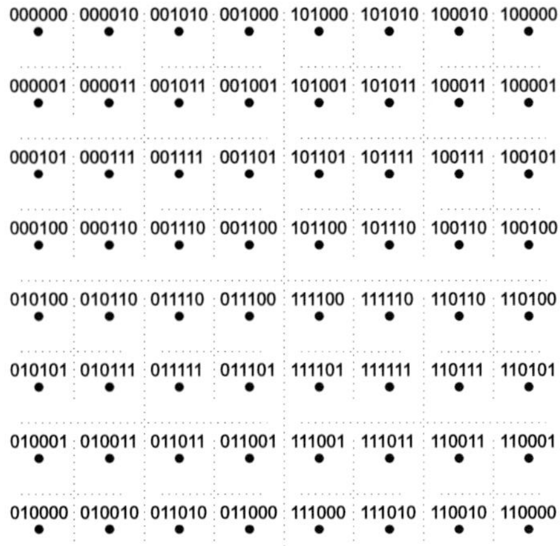

Fig. 9.3. Example of a 64-QAM signal constellation with a multilevel Gray bit mapping.

ean distance between the signal points in each subset is minimized, (ii) each subset is uniquely identified by the first two bits of its signal points, and (iii) the remaining $M_c - 2$ bits of each subset again form a new MLM. An example of a 64-QAM signal constellation with a multilevel Gray bit mapping is shown in Fig. 9.3. As indicated by the dotted lines, this constellation can be successively partitioned into square subsets with minimum mean intra-subset Euclidean distance. At each partitioning level l, the $2l$ most significant bits of a signal point determine in which subset it is located. Because the subsets form dense clusters of signal points, decisions on the first bit pairs in the label can be made without considering the subsequent bits, by selecting the subset whose center of gravity best matches the received signal.

In the MLM-ITS detection scheme, the breadth-first tree search is performed in steps of two bits at a time, even if $M_c > 2$. At any stage, the surviving paths are extended according to the four possible values of the next, say lth, bit pair in the label of the current symbol x_i. The "intermediate" metric associated with each of these extended paths is evaluated from Eq. 9.16, in which x_i is replaced by the center of gravity of the corresponding constellation subset at the lth partitioning level. Furthermore, the *a priori* LLRs $\Lambda_A^I(c_{ij})$ with $j > 2l$ are replaced by zeros in the evaluation of Eq. 9.16. Similar to the ITS detector, after each step, only the M largest intermediate metrics are retained. When the paths departing from the previous symbol depth have been extended by M_c bits, they reach the next symbol depth. At this point, the actual 2^{M_c}-ary QAM signal points can be used in the computation of the path metrics.

Because the number of metric updates per bit in the MLM-ITS detection scheme is $2M$, instead of $M \cdot 2^{M_c}/M_c$ in the basic ITS scheme, its complexity per bit is practically independent of the alphabet size. Moreover, computer simulations have shown that the complexity reduction due to the use of MLM does not lead to any performance degradation relative to the basic scheme.

9.3.4 Soft Interference Cancellation MMSE Detection

The soft interference cancellation minimum mean-squared error (SIC-MMSE) detector proposed in [5] treats each of the symbols in the symbol vector $\mathbf{x}$ as being distorted by the other $N_t - 1$ symbols in $\mathbf{x}$ and noise. It uses the *a priori* information available at its input to estimate and cancel the interference due to the other symbols, and suppresses the residual interference and noise with the aid of the MMSE criterion. This scheme was originally developed for constant-envelope modulation, but can also be extended for QAM modulation.

Let x_j be the desired symbol, then Eq. 9.3 can be rewritten as

$$\mathbf{r} = \underbrace{\mathbf{h}_j x_j}_{\text{desired signal}} + \underbrace{\sum_{i \neq j} \mathbf{h}_i x_i}_{\text{interference}} + \underbrace{\mathbf{n}}_{\text{noise}} \tag{9.17}$$

where $\mathbf{h}_j \in \mathcal{C}^{N_r \times 1}$ denotes the jth column vector of $\mathbf{H}$. The SIC-MMSE scheme produces an estimate of x_j by subtracting the estimated interference due to the other symbols from $\mathbf{r}$, and suppressing the residual interference and noise by means of adaptive linear filtering, as follows:

$$\hat{\hat{x}}_j = \mathbf{w}_j^H (\mathbf{r} - \sum_{i \neq j} \mathbf{h}_i \hat{x}_i) \tag{9.18}$$

Estimates of the interfering symbols are obtained as

$$\hat{x}_i = \mathcal{E}\{x_i\} \tag{9.19}$$

and the corresponding mean-squared estimation errors are denoted by

$$\sigma^2_{\hat{x},i} = \mathcal{E}\{|\hat{x}_i - x_i|^2\} \tag{9.20}$$

In these equations, the expectation $\mathcal{E}\{\cdot\}$ is taken over the 2^{M_c} possible realizations of x_i, weighted by their respective *a priori* probabilities, which are determined from $\Lambda^I_A(\mathbf{c})$. The optimal weighting vector $\mathbf{w}_{j,opt} \in \mathcal{C}^{N_r \times 1}$ minimizes the mean-squared error in the estimate of x_j after soft interference cancellation and linear filtering, i.e.,

$$\mathbf{w}_{j,opt} = \arg \min_{\mathbf{w}_j} \mathcal{E}\{\left|\hat{\hat{x}}_j - x_j\right|^2\} \tag{9.21}$$

The solution to this optimization problem can be obtained by standard minimization techniques, and is given by

$$\mathbf{w}_{j,opt} = \left(\mathbf{h}_j \mathbf{h}_j^H + \sum_{i \neq j} (\sigma^2_{\hat{x},i}/\sigma^2_x) \mathbf{h}_i \mathbf{h}_i^H + (\sigma^2/\sigma^2_x) \mathbf{I}_{N_r}\right)^{-1} \mathbf{h}_j \tag{9.22}$$

where σ_x^2 is the average power used on each of the transmit antennas. In arriving at this solution, use was made of the fact that all symbols are statistically independent of the noise and among themselves.

Based on the assumption that the error $\hat{\hat{x}}_j - x_j$ is a zero-mean complex Gaussian variable, statistically independent of the errors on the other symbols in $\mathbf{x}$, the SIC-MMSE detector approximates the APP detection output Eq. 9.6 as

$$\Lambda^I_{ext}(c_{jk}) = \ln \frac{\sum\limits_{\mathbf{c}_j \in \mathbb{C}_{jk,+1}} \exp \zeta(x_j)}{\sum\limits_{\mathbf{c}_j \in \mathbb{C}_{jk,-1}} \exp \zeta(x_j)} - \Lambda^I_A(c_{jk}) \tag{9.23}$$

in which $\mathbf{c}_j = [c_{j1}, \cdots, c_{jM_c}]^T$ is the vector of bits mapped onto x_j, and

$$\zeta(x_j) = -\frac{|\hat{\hat{x}}_j - x_j|^2}{\sigma_x^2(1 - \mathbf{h}_j^H \mathbf{w}_j)} + \sum_{l \in \mathbb{J}_j} \Lambda^I_A(c_{jl}) \tag{9.24}$$

In Eq. 9.23, $\mathbb{C}_{jk,\pm1}$ has a slightly different definition than in Eq. 9.7, namely

$$\mathbb{C}_{jk,\pm1} = \{\mathbf{c}_j | c_{jk} = \pm 1\} \tag{9.25}$$

The log-sum function in Eq. 9.23 can be approximated using the max-log approximation.

For large N_r, the complexity of the SIC-MMSE detector is dominated by the inversion of the Hermitian positive definite matrix in Eq. 9.22, and is of order N_r^3. For large M_c, on the other hand, the overall complexity is dominated by the evaluation of Eq. 9.19 and 9.20, and the complexity per bit is in the order of 2^{M_c} .

9.4 Simulation Results

In this section, results are provided of computer simulations of an ST-BICM MIMO system employing an iterative receiver. The first three examples illustrate the performance of the SIC-MMSE inner decoder in various scenarios. The last example provides a comparison, in terms of BER performance, of the suboptimal, reduced-complexity inner decoding schemes described in the previous section.

Example 1: Convolutional versus Turbo Codes

The outer codes considered in the first example are 8-state, rate-1/2 turbo and convolutional codes. The turbo codes are composed of two identical 8-state convolutional codes generated by using a recursive feedback generator polynomial [1011] and a feedforward polynomial [1101]. No effort was

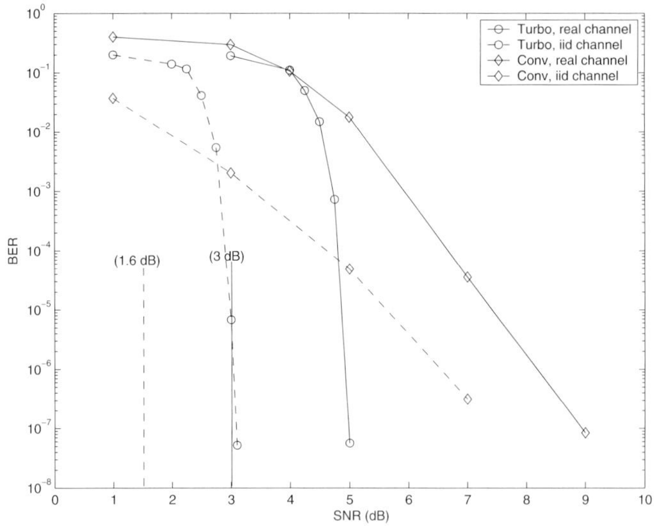

Fig. 9.4. Error performance of a 4×4 ST-BICM MIMO system employing the SIC-MMSE detector, using 8-state, rate-1/2 turbo and convolutional codes as channel codes. Modulation format is QPSK, channel is measured or i.i.d., and known perfectly.

made to optimize either the pseudo-random interleaver used in the turbo encoder and decoder, or the interleaver used to separate the channel encoder from the space-time mapper. The antenna configuration considered here is $N_t = N_r = 4$. The modulation format is QPSK, and the MIMO detection scheme is SIC-MMSE. Both real-life measurements, taken in Manhattan using a (16,16) MIMO testbed [10], and a theoretical model were used to generate the MIMO channel coefficients. The theoretical model generates temporally and spatially uncorrelated complex-Gaussian distributed channel coefficients for each symbol interval. This channel is "ideal" in the sense that it provides maximum diversity, and is referred to as an identically and independently distributed (i.i.d.), or uncorrelated Rayleigh fading channel. In this simulation, the number of outer (detector-decoder) iterations is fixed to 5, and the number of inner (turbo decoding) iterations is 1, 2, 4, 8 and 10 in successive detector-decoder iterations. The transmission is organized in blocks of 3200 information bits. The SNR on the horizontal axis is the average signal-to-noise ratio at each receive antenna. Perfect knowledge of the channel state information (CSI) is assumed.

In Fig. 9.4, the performance results obtained by using real-life channel data are represented by solid lines, and those obtained with the aid of the i.i.d. channel model by dashed lines. The figures show that the use of a turbo channel code results in a gain of more than 4 dB (at a BER of 10^{-6}) over the same system employing the convolutional code. Note that the performance

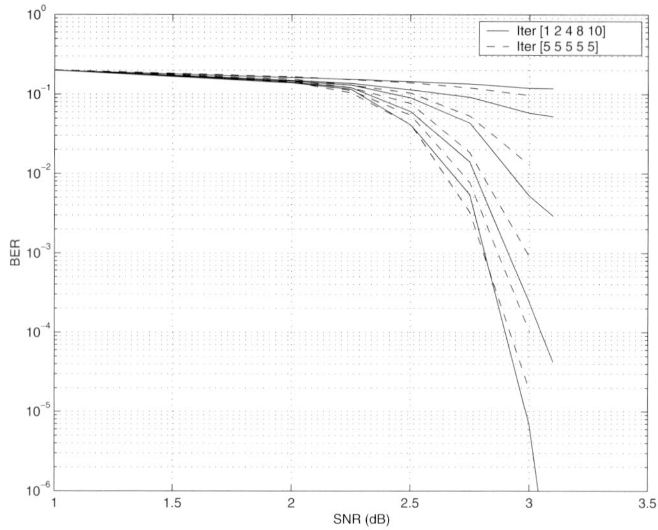

Fig. 9.5. Error performance of a 4 × 4 ST-BICM MIMO system employing the SIC-MMSE detector, using an 8-state, rate-1/2 turbo code. Modulation format is QPSK, channel is i.i.d., and known perfectly.

in the theoretical MIMO channel is 1-2 dB better than the performance for the real-life data. This is because the capacity limit associated with real-life channels, which are typically correlated in space and time, is lower than that of an ideal, uncorrelated MIMO channel. In the figures, the ergodic capacity limits [2] corresponding to both the uncorrelated Rayleigh fading channel and the measured channel are indicated by vertical lines.

Example 2: Performance versus Number of Iterations

Fig. 9.5 shows the error performance of the SIC-MMSE receiver for 1, 2, 3, 4 and 5 detector-decoder iterations. The channel (outer) code is the 8-state turbo code discussed in the previous example, and two different sets of channel decoding iterations are considered. Uncorrelated Rayleigh fading is assumed, the number of information bits per block is 3200, and the channel is assumed to be known perfectly. For both sets of channel decoding iterations, the performance of the receiver improves significantly with an increasing number of detector-decoder iterations. However, due to increasing correlation among the SISO outputs, the incremental gains become smaller after the first few iterations. These gains can be further increased by increasing the interleaver size. With the first set of channel decoding iterations (1, 2, 4, 8 and 10 per detector-decoder iteration), the receiver tends to perform better than with the second set (5 channel decoding iterations in each detector-decoder iteration).

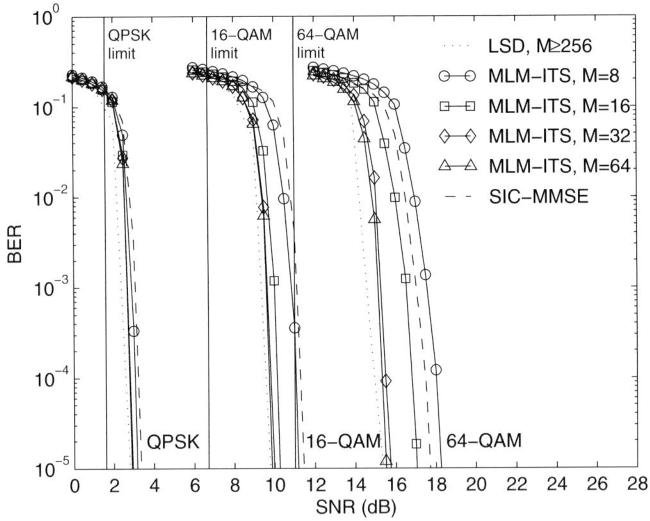

Fig. 9.6. Error performance of a 4×4 ST-BICM MIMO system employing the LSD, MLM-ITS and SIC-MMSE detector. Channel code is a 4-state turbo code of rate 1/2.

Example 3: Comparison Between Suboptimal Detection Schemes

The last example provides a comparison of BER performance for the reduced-complexity MIMO detection schemes discussed in the previous section. Again, the antenna configuration is $N_t = N_r = 4$. The channel code is a turbo code of rate 1/2 and memory 2, whose feedforward and feedback generators are [101] and [111], respectively. The elements of the channel matrix are samples of independent, complex-valued, zero-mean Gaussian processes. The channel is assumed to remain constant over blocks of 192 information bits, and every new channel realization is statistically independent of previous blocks. The block length of the channel code and, therefore, the size of the interleaver Π, is equal to 18432 code bits. The number of iterations in the detector/decoder loop is 4, and that of the turbo decoder is limited to 8 per outer iteration. All interleavers are pseudo-random; no attempt was made to optimize their design. The modulation formats considered are QPSK, 16-QAM and 64-QAM, all with multilevel Gray mapping.

Fig. 9.6 shows performance results for the LSD scheme, which were taken from [6], as well as for MLM-ITS and SIC-MMSE detection. Note that the list size of the LSD scheme is 256 for QPSK, and 512 for 16-QAM and 64-QAM modulation. The performance of the basic ITS detection scheme is identical to that of MLM-ITS, and is not shown in the figure. Ergodic MIMO capacity bounds are shown as a reference. On the basis of the results presented in the figure, the following observations can be made:

- As expected, the performance of MLM-ITS improves if the list size M is increased, in this case from 8 to 64.
- The performance of MLM-ITS can be made equal to that of SIC-MMSE by tuning the list size. Interestingly, simulation results indicate that the complexities of both schemes are comparable at the same performance level. However, an advantage of MLM-ITS is that its performance can be traded off for lower complexity.
- To achieve comparable error performance, the list size of the LSD scheme must be chosen approximately 8 times larger than that of MLM-ITS.
- The ST-BICM system with the best performance shown in Fig. 9.6, which is the LSD scheme, operates at approximately 1, 3 and 4 dB from the capacity limit (at a BER of 10^{-5}) for QPSK, 16-QAM and 64-QAM, respectively.

9.5 Applications

Although MIMO is currently still a relatively young technology, the wireless industry has shown considerable interest, and is making significant progress in the development of MIMO products for wireless data networks and third-generation (3G) mobile wireless.

Theoretically, transmission rates supported by existing wireless systems can be increased proportionally to the number of transmit antennas. Manufacturers of telecommunications devices such as Lucent Technologies and Agere Systems have advocated the use of MIMO in next-generation wireless local area network (LAN) systems based on the IEEE 802.11 standards [21], to achieve data rates over several hundreds of Mbps. MIMO technology is also proposed for metropolitan area networks (MANs), referred to as IEEE 802.16a [21]. Such systems typically operate in the non-line-of-sight channel conditions that are beneficial to MIMO, and support continuously varying traffic levels at speeds approaching those offered by fiber optic networks. Because these systems need to operate at a very high spectral efficiency, techniques such as MIMO will likely be applied in IEEE 802.16a. Furthermore, MIMO is expected to be adopted in a next phase of 3G mobile wireless standardization, in which a technology referred to as high-speed downlink packet access (HSDPA) will enable transmission rates up to 20 Mbps for multimedia services.

The use of multiple-antenna technologies in wireless devices requires additional analog hardware, as well as sophisticated processing capabilities, especially for signal detection and decoding at the receiver. The Turbo-MIMO architecture discussed in this chapter provides superior performance compared to non-iterative MIMO techniques, but is also more computationally complex. Research focused on low-complexity MIMO transceiver design and low-complexity detection, such as the work discussed in this chapter, will be key to the success of Turbo-MIMO in practical systems.

9.6 Summary and Discussion

In this chapter, a number of reduced-complexity detection schemes for space-time bit-interleaved coded modulation (ST-BICM) Turbo-MIMO systems were reviewed. Such systems distinguish themselves from more conventional space-time architectures such as BLAST by the use of iterative processing at the receiver. They generally have considerably better performance than non-iterative MIMO systems, although currently available schemes do not operate as close to the Shannon limit as similarly encoded single-antenna systems. At the same time, their computational complexity is reasonably low. In contrast to the optimal APP detector, the reduced-complexity schemes discussed in this chapter (LSD, ITS and SIC-MMSE detection) do not exhibit the same exponential growth in complexity as the number of transmit antennas is increased. The complexity per bit of the multilevel mapping ITS detector is linear in the number of transmit antennas, and practically independent of the signal constellation size. Like the LSD scheme, it also offers the possibility of trading off performance for lower complexity, by changing the list size parameter of the algorithm. The SIC-MMSE scheme is quite different from the other schemes in that it is based on an adaptive filtering approach, but has comparable performance as MLM-ITS at roughly the same computational cost.

Although the Turbo-MIMO architectures discussed herein are farther away from theoretical capacity limits than similarly encoded single-antenna wireless systems, they can provide considerably larger spectral efficiency. Possible ways to decrease the remaining MIMO capacity gap include

- signal-constellation shaping [6],
- improved turbo codes by using designed interleavers for turbo codes to achieve good minimum distance [20],
- designing turbo-like codes such as low-density parity check codes (LDPC) or repeat accumulate codes that optimize the performance of the ST-BICM MIMO systems [9].
- asynchronous transmission of spatially multiplexed data streams [19], and
- stratified processing uses the fact of multilevel coding principle [22].

Acknowledgments

The authors are grateful to Prof. Simon Haykin, McMaster University, Hamilton, Ontario, Canada, and Lucent Bell Labs, Crawford Hill, New Jersey, USA, for providing access to the MIMO measured channel database.

References

1. C. Berrou, A. Glavieux, and P. Titmajshima, "Near Shannon limit error-correction coding and decoding: turbo codes", in *Proc. International Conference on Communications*, pp. 1064–1090, Geneva, Switzerland, May 1993.
2. J. G. Foschini, "Layered space-time architecture for wireless communication in a fading environment when using multi element antennas", *Bell Labs Technical Journal*, vol. 2, pp. 41–59, Autumn 1996.
3. S. L. Ariyavisitakul, "Turbo space-time processing to improve wireless channel capacity", *IEEE Trans. Communications*, vol. 48, no. 8, pp. 1347–1359, Aug. 1999.
4. H. El Gamal and A. R. Hammons, Jr., "New approach to layered space-time coding and signal processing", *IEEE Trans. Information Theory*, vol. 47, pp. 2321–2334, Sept. 2001.
5. M. Sellathurai and S. Haykin, "Turbo-BLAST for wireless communications: theory and experiments", *IEEE Trans. Signal Processing*, vol. 50, no. 20, pp. 2538–2546, Oct. 2002.
6. B.M. Hochwald and S. ten Brink, "Achieving near-capacity on a multiple-antenna channel", *IEEE Trans. Communications*, Vol. 51, no. 3, March 2003.
7. Y.L.C. de Jong and T.J. Willink, "Iterative tree search detection for MIMO wireless systems", in *Proc. VTC 2002-Fall*, Vancouver, BC, vol. 2, pp. 1041–1045, Sept. 2002.
8. Y.L.C. de Jong and T.J. Willink, "Iterative tree search detection for MIMO wireless systems", submitted to *Trans. Communications*, Sept. 2003.
9. S. ten Brink and G. Kramer, "Design of repeat-accumulate codes for iterative detection and decoding", *IEEE Trans. Signal Processing*, vol. 51, no. 11, pp. 2764–2772, Nov. 2003.
10. S. Haykin and M. Sellathurai, "Turbo-BLAST with multi-loop feedback receiver", in *Proc. 3rd International Symposium on Turbo Codes and Related Topics*, Brest, France, Sept. 2003 (8 pages).
11. H. El Gamal, M. O. Damen, "Universal space-time coding", *IEEE Trans. Information Theory*, vol. 49, no. 5, pp. 1097–1119, May 2003.
12. L. Zheng and D.N.C. Tse, "Diversity and multiplexing: a fundamental trade off in multiple-antenna channels", *IEEE Trans. Information Theory*, vol. 49, no. 5, p. 1073–1096, May 2003.
13. L. Bahl, J. Cocke, F. Jelinek, and J. Raviv, "Optimal decoding of linear codes for minimizing symbol error rate", *IEEE Trans. Information Theory*, vol. 20, no. 2, pp. 284–287, Mar. 1974.
14. V. Tarokh, N. Seshadri, and A. R. Calderbank, "Space-time codes for high data rate wireless communications: performance criterion and code construction", *IEEE Trans. Information Theory*, vol. 44, pp. 744–765, Mar. 1998.
15. G. D. Golden, J. G. Foschini, R. A. Valenzuela, and P. W. Wolniansky, "Detection algorithm and initial laboratory results using V-BLAST space-time communication architecture", *Electronics Letters*, vol. 35, pp. 14–15, Jan. 1999.
16. S. Benedetto, D. Divsalar, G. Montorsi, and F. Pollara, "Serial concatenation of interleaved codes: performance analysis, design and iterative decoding", *JPL TDA Progress Report*, vol. 42–126, Aug. 1996.
17. G. Caire, G. Taricco, and E. Biglieri, "Bit-interleaved coded modulation", *IEEE Trans. Information Theory*, vol. 44, no. 3, pp. 927–946, May 1998.

18. B. Steingrimsson, Z.-Q. Luo, and K.M. Wong, "Soft quasi-maximum-likelihood detection for multiple-antenna wireless channels", *IEEE Trans. Signal Processing*, vol. 51, no. 11, pp. 2710–2719, Nov. 2003.
19. Y.L.C. de Jong and T.J. Willink, "Iterative trellis search detection for asynchronous MIMO systems", in *Proc. VTC 2003-Fall*, Orlando, FL, 2003.
20. S. Crozier and P. Guinand, "High-performance low-memory interleaver banks for turbo-codes", in *Proc. VTC 2001-Fall*, Atlantic City, NJ, USA, Oct. 2001.
21. "Part 11: Wireless LAN Medium Access Control (MAC) and Physical Layer (PHY) Specifications: High speed physical layer in the 5GHz Band, IEEE Standard 802.11a-1999. Local and Metropolitan Area Networks-Part 16," *IEEE Standard 802.16a*.
22. M. Sellathurai and G. Foschini, "A stratified diagonal layered space-time architecture: information theoretic and signal processing aspects", *IEEE Trans. Signal Processing*, vol. 51, no. 11, pp. 2943–2954, Nov. 2003.

Chapter 10

Turbo Codes in Broadband Wireless Access Based on the IEEE 802.16 Standard

Poramate Tarasak[1] and Theeraputh Mekathikom[2]

[1] University of Victoria, Canada
[2] National Electronics and Computer Technology Center-NSTDA, Thailand

Due to the success of wireless communications industry, it is now the time to reconsider whether the traditional wireline Internet access using copper wire or optical fiber is the best technology. Broadband wireless access (BWA) is emerging as an alternative way to achieve connectivity to the core network. In the physical layer, turbo code is once again one of the most crucial part that improves the system performance, capacity and/or coverage. This chapter reviews the application of advance turbo coding schemes that have recently been approved in the IEEE 802.16 standard [1].

10.1 Brief Overview of BWA based on the IEEE802.16 Standard

Traditional broadband Internet access technologies, which are ADSL and cable modem, are now facing a challenge by broadband wireless access (BWA). The inherent advantage of BWA lies on its wireless transmission medium which is a large cost saving factor and allows fast scalable growth. Therefore, as in Fig. 10.1, BWA can complement ADSL and cable modem very well in the area where telephone lines and optical fibers are not worth installed, e.g., a rural area. BWA can also be an alternative Internet access choice for customers in the urban and suburban areas when ADSL and cable modem services are disrupted or almost saturated. It is envisioned that at the end of the day the customer can buy the BWA equipment and perform indoor self-installation to access the Internet at the speed greater than 20 Mbit/s [2, 3]. To reach this goal and ensure interoperability among vendors, a working group IEEE802.16 was established in August 1998. Until now, the working group has already published two main standards for the BWA air interface. The first one, IEEE802.16 published in 2002 under the name WirelessMAN (Wireless Metropolitan Area Network), addresses the air interface whose operating frequency is from 10 to 66 GHz. The second one, IEEE802.16a published in 2003, is an amendment of the IEEE802.16 in order to accommodate the

K. Sripmanwat (ed.), Turbo Code Applications: a journey from a paper to realization, 243–254.

- WirelessMAN-OFDMA: Orthogonal frequency division multiple access (OFDMA) is applied with 2048-point transform with 1702 useful subcarriers. OFDMA divides total bandwidth into groups of contiguous subcarriers. Each subchannel is composed of a subcarrier from each group. Similar to OFDM subcategory, FEC applies a concatenated Reed-Solomon and RCPCC with optional BTC or CTC. OFDMA enjoys the advantages of OFDM and also has better interference rejection and enables single frequency network where all neighbor cells use the same set of frequencies.
- WirelessHUMAN (Wireless High-speed Unlicensed Metropolitan Area Networks): The system in this category follows WirelessMAN-SCa, WirelessMAN-OFDM or WirelessMAN-OFDMA with the additional constraints on the transmitted spectral density. Its advantage is the free frequency resource which is available nationwide.

10.2 Turbo Codes in the IEEE802.16 Standard

To ensure high transmission performance in the hostile propagation environment, turbo coding schemes are proposed as optional FEC in IEEE802.16. BTC and CTC are two competing schemes which are mainly contributed by AHA corporations and ICORE coding respectively.

10.2.1 Block Turbo Code

Block turbo code (BTC) or in another name turbo product code (TPC) was originally an optional FEC in WirelessMAN-SC. BTC can be illustrated as in Fig. 10.2. It is composed of a two dimensional array of information bits. Suppose the size of this array is $k_2 \times k_1$. Each row and each column is encoded by a systematic block component code of rate $\frac{k_1}{n_1}$ and $\frac{k_2}{n_2}$ respectively. The parity bits are appended to the information bits at the corresponding rows and columns. Normally the encoding is done row-wise first and then column-wise. Each column of the parity bits in the last $n_1 - k_1$ columns is encoded by the column encoder. The resulting parity bits are appended in the corresponding columns. The codeword of BTC is then an array of size $n_2 \times n_1$. The overall code rate is $\frac{k_1 k_2}{n_1 n_2}$. In general case, more than two dimensional arrays are possibly constructed and the encoder in each dimension can be different.

The codeword is read out starting from the first bit in the first row of the codeword array. This continues row-by-row until the last bit. Note that an interleaver is not required to encode BTC but it can be incorporated as well. A simple block interleaver may be implemented just by reading out bits column-by-column starting from the first bit in the first column.

Iterative decoding of BTC performs soft decoding for each row first and then feeds the soft output to the column decoder. The iterative process is continued until some stopping criterion is reached. The advantage of BTC in this regard is that the block component code usually has a well-known

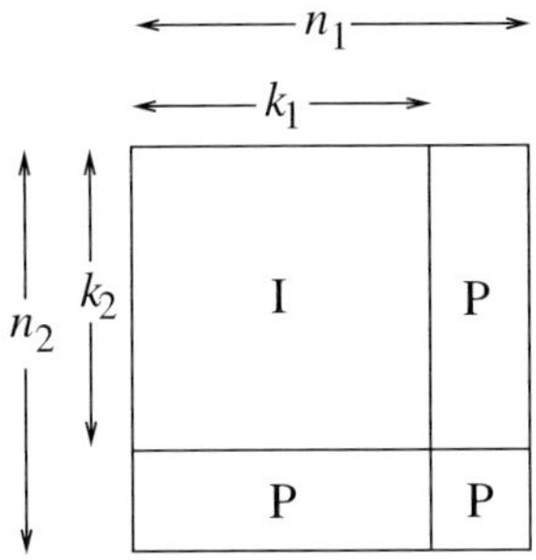

Fig. 10.2. TPC codeword

algebraic structure. This allows soft algebraic decoding which has much lower complexity than soft maximum likelihood decoding. Other advantages of BTC includes [7]

- BTC has a large minimum Hamming distance. The minimum Hamming distance of the BTC is the product of minimum Hamming distances of the row and column component codes. The minimum Hamming distance affects the code performance at the error floor region, i.e. at the high SNR or very low BER. Hence, BTC does not suffer from the error floor as much as the CTC does.
- Matching the codeword to the required frame size is easy. This is usually done by shortening first few rows and columns which is equivalent to filling those rows and columns with zeros and not transmitting them. Since zero input bits do change the state of the encoder, the encoder needs to know only how many bits is to be encoded in each row and column. Puncturing parity bits of the corresponding information bits is not needed. (Puncturing parity bits may result in poor performance compared to the mother code as in the case of CTC.)
- Performance analysis is easier than CTC since BTC has a well-known algebraic structure. This can be useful when selecting an appropriate code matching to the system requirement.
- BTC has less latency than CTC. This means that for a given decoding time BTC might exploit higher number of iterations which results in an improved performance.
- BTC chip is widely available since its implementation less complex than CTC.

WirelessMAN-SC recommends an extended Hamming code as a constituent code of the two dimensional TPC. The extended Hamming codes is derived from a Hamming code appended by a single parity check (SPC) bit. With the SPC code, the minimum Hamming distance of BTC increases significantly with very small additional complexity. For example, WirelessMAN-SC defines (32,26) extended Hamming code which is derived from a (31,26) Ham-

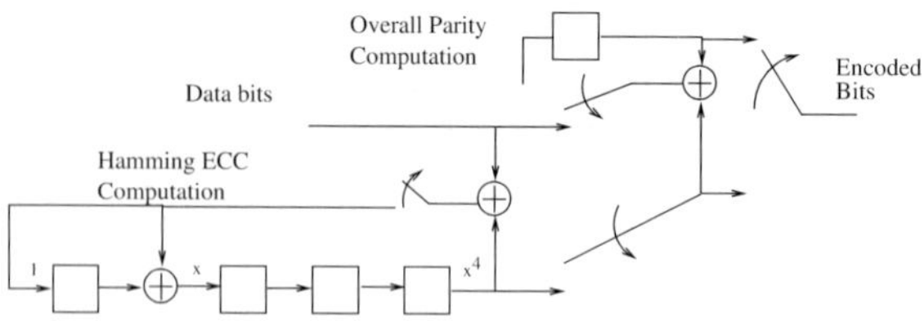

Fig. 10.3. (16,11) BTC encoder.

ming code and the SPC code. The minimum distance of the (31,26) Hamming code is three. With one additional single parity check bit, the minimum distance of the extended Hamming code becomes four. Hence, the minimum Hamming distance of the BTC is 16 which is largely increases from the constituent Hamming code alone.

An example of (16,11) BTC encoder with the generator x^4+x+1 is shown in Fig. 10.3. The (32,26) and (64,57) BTCs which were defined in IEEE802.16 can be designed with similar concept.

10.2.2 Convolutional Turbo Code

Recent advance in convolutional turbo code (CTC) attracts much interest in its applications. Conventional CTC suffers from high error floor due to its relative small minimum Hamming distance and suffers from performance degradation due to puncturing. Nonbinary CTC has recently emerged and it seems to solve many flaws of classical binary CTC. In addition, the concept of tail-biting convolutional code has been applied to CTC. The concept of tail-biting code improves the spectral efficiency of CTC since it solves the problem of tail bits used to terminate the state of the encoder.

Tail-Biting and Circular Convolutional Codes

Conventional encoding of a nonrecursive convolutional code usually add a number of zero bits to ensure that the terminating state is zero. The number of required zero bits is equal to m, the number of encoder memory. This results in loss of spectral efficiency since these zero bits carry no information.

Tail-biting code was proposed to terminate the known state without loss of spectral efficiency. Tail-biting code takes the last m information bits as the initial state of the encoder. This results in the code whose initial state and terminating state are the same. Although the decoder does not know the initial or terminating state, it can assumes each of the possible state as initial and terminating states and find the ML sequence. However, this approach of tail-biting code works only on the nonrecursive code. For the recursive code, the terminating state depends on the whole information sequence. Therefore, taking only the last m information bits as the initial state will not ensure the same terminating state.

Circular convolutional code is a variant of a tail-biting code but accommodates a recursive code. Circular encoding ensure that after encoding operation the terminating state ends up with the initial state. Such a state is called circular state which exists only when N, the length of the information sequence, is not a multiple of the generator. The way to find the initial state is to pre-encode the information sequence with the zero initial state. The resulting terminating state, S_N^0, is used to compute the initial state, S_c from $S_c = (I + G^N)^{-1} S_N^0$, where G is the generator matrix. In practice, S_c can be found from a precomputed table lookup. Since CTC composes of two constituent codes, pre-encoding has to be performed two times to find the initial state for each encoder.

Nonbinary Turbo Codes

Nonbinary turbo codes take more than one bit input at a time. Other structures follow classical binary turbo codes which are parallel concatenated of recursive systematic convolutional codes composing of two constituent encoders, an interleaver and an optional puncturer. The interleaver has a unique structure in which it introduces 'local disorder' in the input bits. For example, if two-bit input enters the encoder at a time (double-binary code), the interleaver permutes its location pairwise and also introduces local disorder by swapping the two-bit input at every even interval. The local disorder is needed to enhance the randomness of the interleaved sequence since pair interleaving has some regular structure to handle low weight error. Therefore, nonbinary turbo code requires two-step interleaving.

Recent CTC combines two concepts above and it enjoys many advantages [8].

- CTC has better convergence than BTC and classical turbo code. The structure of CTC reduces the correlation between outputs from two constituent encoders. It seems that CTC has significant performance gain in the waterfall region.
- CTC has larger minimum distance than classical turbo code due to two-level interleaving. Therefore, the error floor of CTC is much less visible as well.
- Although puncturing for CTC is needed to match the frame size, it turns out that CTC is quite robust to it. Thanks to the nonbinary structure, CTC requires less puncturing of parity bits. Performance of high rate code is therefore improved.
- Since more than one bit at a time enters the encoder, latency is reduced.
- It was observed that CTC is robust to suboptimal decoding such as Max-Log-MAP. This might be because the block size is smaller for nonbinary turbo codes. This leads to less complex receiver.
- Compared to BTC, CTC requires less messaging in the MAC layer.

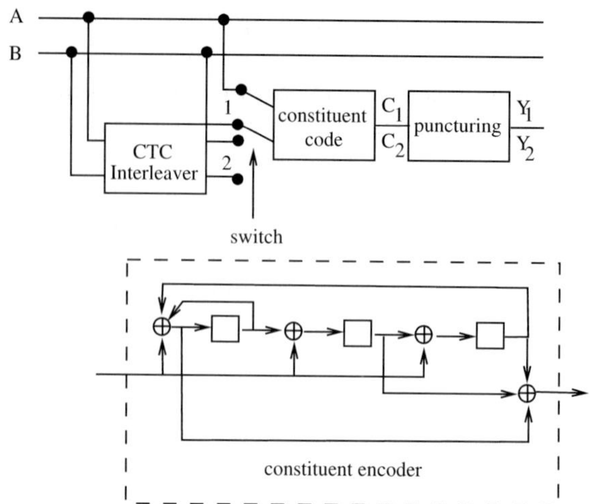

Fig. 10.4. CTC encoder.

IEEE802.16 defines CTC as an optional code in 2-11GHz frequency range. It uses a double binary circular recursive systematic convolutional code. The CTC encoder is illustrated in Fig. 10.4. The generator polynomial for the parity bit Y is $1 + D^2 + D^3$ and the generator polynomial for the feedback branch is $1 + D + D^3$. Interleaving is done in two steps as follows.

1. The couple bits are switched $(A, B) = (B, A)$ in every even encoding interval.
2. The interleaved address i of the couple j is computed from
   - if j mod $4 = 0$ or $1, i = (P_0 \cdot j + 1)_{\text{ mod } N}$
   - if j mod $4 = 2$ or $3, i = (P_0 \cdot j + 1 + N/4)_{\text{ mod } N}$

Noted that the CTC encoder structure in Fig. 10.4 has also been adopted in HIPERMAN and DVB-RCS standards [9]. HIPERMAN is another broadband wireless access standard. DVB-RCS stands for digital video broadcasting-return channel via satellite. Both standards origanated from Europe.

It is important to note that a fair comparison between CTC and BTC is a difficult task. Since both have their own advantages and drawbacks, IEEE802.16 includes both BTC and CTC as optional coding schemes to cover most operating conditions and to acquire more flexibility.

10.3 Performance Analysis of BTC

Since BTC is a linear code, a classical union bound approach can be used to evaluate the performance and it can be assumed that all zero sequence is transmitted. The union bound approach needs to know the weight enumerating function (WEF) of the code. To compute the WEF of BTC, one may apply the knowledge of WEF of the constituent row and column codes. However, the complete WEF of BTC is still an open problem. It is generally convenient to consider those codewords with the minimum weight.

Let us consider BTC codewords with the minimum weight. A minimum nonzero weight codeword of BTC occurs as follows [6]. Suppose we choose one row in the BTC as a minimum nonzero weight row codeword. On this row, we choose one column with bit '1' and then fill this column with same or another minimum nonzero weight codeword. To make it a BTC codeword, we still have to fill other rows and columns with the minimum nonzero weight codewords. However, after we already chose one row and one column, this leaves us no other choice but to have other row and column codewords are identical row-wise and column-wise, respectively, and other rows and columns are all zero codeword. Therefore, the number of minimum weight codewords of BTC, $A_{\tilde{p}}$, is

$$A_{\tilde{p}} = A_{\tilde{r}} \times A_{\tilde{c}} \tag{10.1}$$

where $A_{\tilde{r}}, A_{\tilde{c}}$ are the number of minimum weight codewords of row- and column- constituent codes, respectively. $A_{\tilde{r}}, A_{\tilde{c}}$ can be found from the coefficients of the terms with minimum nonzero order in the WEF of the constituent codes. The WEF of an (n, k) Hamming code is computed from

$$A(z) = \frac{1}{n+1}\left[(1+z)^n + n(1-z)(1-z^2)^{\frac{n-1}{2}}\right] \tag{10.2}$$

For an extended code whose WEF of the original code is $A(z)$, it can be readily found that the WEF is computed as

$$A_e(z) = \frac{1}{2}\left[A(z)(1+z) + A(-z)(1-z)\right] \tag{10.3}$$

Therefore, the WEF of an $(n+1, k)$ extended Hamming code is

$$A_e(z) = \frac{1}{n+1}\left[\frac{(1+z)^{n+1}}{2} + \frac{(1-z)^{n+1}}{2} + n(1-z^2)^{\frac{n-1}{2}+1}\right] \tag{10.4}$$

Suppose the channel is AWGN, the union bound on the bit error performance of an (n, k) linear code is [5]

$$P_b \leq \sum_{d=d_{min}}^{n} A_d \frac{\omega_d}{k} Q\left(\sqrt{\frac{2dRE_b}{N_0}}\right) \tag{10.5}$$

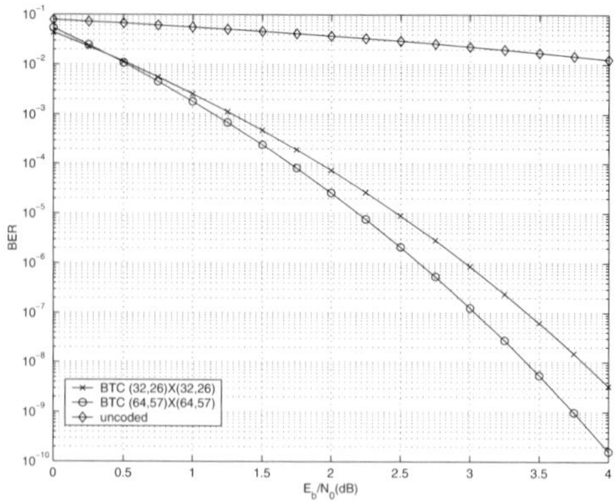

Fig. 10.5. Performance of (32,26)x(32,26) and (64,57)x(64,57) BTCs in AWGN channels.

where A_d is the number of weight-d codewords, ω_d is the average weight of the input sequence yielding a weight-d codeword, R is the code rate. If we consider only the minimum weight codewords, for the BTC composing of extended Hamming codes, the bit error performance is asymptotically given by

$$P_b \approx A_{\tilde{p}} \frac{\omega_{16}^p}{k^2} Q\left(\sqrt{\frac{2(16)R_p E_b}{N_0}}\right) \tag{10.6}$$

where R_p is the code rate of the BTC and ω_{16}^p is computed from $\omega_{16}^p = \omega_4^r \times \omega_4^c$ and ω_4^r, ω_4^r is found by computer simulation.

Fig. 10.5 shows the asymptotic performance of BTC on AWGN channels. Both BTCs used in IEEE802.16 have huge performance gains over uncoded transmission.

Performance analysis of CTC is still an active research problem. Some preliminary results can be found in [10].

10.4 Implementation

Communication Industry has acknowledged the potential impact and huge revenue from BWA business. WiMAX Forum was established in 2003 by industry leaders as a mean to ensure compatibility and inter-operability of BWA equipment.

Although the performance of turbo code is better than those of the required channel coding schemes in the standard, it becomes an optional scheme due to its higher complexity and latency in the decoding. Nevertheless, for a non-real-time application that can afford high complexity, turbo code is an excellent choice to be deployed. Implemenation of turbo code in practice is still an early stage. Few practical implementation will be mentioned as follows.

- Comtech AHA Corporation has produced AHA4525, a standalone IEEE802.16 compliant BTC chip. [17] It has 60Mbit/s channel rate and 50Mbit/s payload data rate for $(64,57) \times (64,57)$ with 3 iterations. The chip integrates independent encoder and decoder functions and can be configured for full or half duplex operation. Its code rate can be flexibly chosen from 0.25 to 0.97 while its supported block size is from 64 bits to 4 Kbits.
- Xilinx has produced separated chips for BTC encoding and decoding functions based on Comtech AHA Corporation TPC Galaxy core [18, 19]. The chips were designed with structural VHDL and can support block size from 64 bits to 4 Kbits. A data rate of 45Mbit/s can be achieved for $(64,57) \times (64,57)$ with 5 iterations while a data rate of 155Mbit/s can be achieved with four SISO option.
- TurboConcept has produced TC3401 and TC3404 for low and high transmission rates, respectively [20, 21]. TC3401 provides 37Mbit/s decoding rate for $(64,57) \times (64,57)$ with 5 iterations while TC3404 provides 140Mbit/s decoding rate. The supported block size is up to 65 Kbits which is considerably larger than the other companies.

Note that only BTC for IEEE802.16 were implemented until now. However, since CTC for IEEE802.16 has similar structure to CTC for DVB-RCS, it is expected that the companies that produce chips for DVB-RCS will soon produce CTC chips for IEEE802.16. TurboConcept and iCoding are two such companies that have CTC products in hand.

Recent implementation effort is to produce a complete 802.16 standard chip (system-on-a-chip WiMAX). This results in a fierce competition among big companies such as Intel and Fujitsu. Intel claimed to ship its samples of the first WiMAX chip, called Rosedale, to key system makers in September 2004. However, not much details were given and the news seems to be quiet since then. Wavesat, a Montreal-based company, claimed to ship the first chip which fully comply with the IEEE802.16 standard [22]. The chip is now available to the market. However, the DM256 chip from Wavesat doesn't include turbo coding feature. Sequans communications, a French company, is another new competitor to produce a WiMAX chip.

10.5 Conclusions

IEEE802.16 standard has finally been finalized in October 2004. Turbo code is an option to be incorporated in the devices and has a significant role to make BWA a practical alternative. There is no doubt that turbo code and its derivative will be a crucial component in the emerging wireless access application while its full potential is yet to be proved.

References

1. IEEE 802.16-2004, "IEEE standard for local and metropolitan area networks – Part 16: Air interface for fixed broadband wireless access systems," Oct. 2004.
2. C. Eklund, R.B. Marks, K.L. Stanwood and S. Wang, "IEEE standard 802.16: A technical overview of the WirelessMAN^{TM} air interface for broadband wireless access," *IEEE Commun. Magazine,* vol. 40, no. 6, pp. 98–107, June 2002.
3. I. Koffman and V. Roman, "Broadband wireless access solutions based on OFDM access in IEEE802.16," *IEEE Commun. Magazine,* vol. 40, no. 4, pp. 96–103, Apr. 2002.
4. Z. Wang, X. Ma and G.B. Giannakis, "OFDM or single–carrier block transmissions?," *IEEE Trans. Commun.,* vol. 52, no. 3, pp. 380–394, Mar. 2004.
5. S. Lin and D.J. Costello Jr., *Error Control Coding: Fundamentals and Applications,* Prentice-Hall, 1983.
6. N.Y. Yu, Y. Kim and P.J. Lee, "Iterative decoding of product codes composed of extended Hamming codes," in *Proc. IEEE Symposium on Computer and Communications,* pp. 732–737, July 2000.
7. R. Pyndiah, "Near-optimum decoding of product codes: block turbo codes," *IEEE Trans. Commun.,* vol. 46, no. 8, pp. 1003–1010, Aug. 1998.
8. C. Berrou, M. Jezequel, C. Douillard and S. Kerouedan, "The advantages of non-binary turbo codes," in *Information Theory Workshop,* pp. 61-63, Sept. 2001.
9. C. Douillard et al., "The turbo code standard for DVB-RCS," in *Proc. 2nd Int. Symposium on Turbo Code,* Sept. 2000.
10. Y. O.-C.-Mouhamedou, S. Crozier and P. Kabal, "Distance measurement method for double binary turbo codes and a new interleaver design for DVB-RCS," in *Proc. IEEE Globecom,* vol. 1, pp. 172-178, Dec. 2004.
11. IEEE 802.16.3c-01/14, "Turbo code comparison (TCC v TPC)," 17 Jan. 2001.
12. IEEE C802.16.3a-02/80, "Method for using concatenated convolutional turbo codes in IEEE 802.16a," 4 July 2002.
13. IEEE 802.16a-02/73r1, "Block turbo codes for 802.16 optional coding," 9 July 2002.
14. IEEE C802.16a-02/81, "TPC/TCC complexity comparison (latency, data rate, size," 9 July 2002.
15. IEEE C802.16a-02/76, "Convolutional turbo codes for 802.16," 2 July 2002.
16. D. Williams, "Turbo product code tutorial," tutorial slides, 1 May 2000.
17. Comtech AHA Corporation, "Product specification AHA4525 IEEE802.16a compliant turbo product code encoder/decoder."
18. Xilinx Logicore, "IEEE802.16 compatible turbo product code encoder v1.0 product specfication."
19. Xilinx Logicore, "IEEE802.16 compatible turbo product code decoder v1.1 product specfication."
20. TurboConcept, "Low complexity turbo product code decoder TC3401."
21. TurboConcept, "Very high-speed turbo product code decoder TC3404."
22. Wavesat, "DM256-Product description IEEE802.16d OFDM PHY."

Part IV

Satellite and Space Communications

Chapter 11

Turbo Codes on Satellite Communications

Sorin Adrian Barbulescu

University of South Australia, Australia

11.1 A New Turbo World

The first modem designed for a satellite application of turbo codes [1] was tested only four years after the publication of the first paper which introduced the concept of turbo codes [2–5]. Such a short span between the invention of a radical new concept and its application is also a first in the brief history of this field. Satellite communications took off in early 70's, with Intelsat being the first international body that coordinated the activities in this sector and acted as an enabler through standards definition process. For example, the IESS308/309/310 standards allowed different satellite modem manufacturers to build Viterbi and Reed-Solomon codecs that were compatible with one another.

In the late 90's, with the move towards privatization, Intelsat's role changed. With the explosion of different turbo-like codes and no standards available, Intelsat issued generic standards (IESS 315/316) which do not specify the details of the coding scheme but only its performance requirements. Therefore, one can see today a wide range of proprietary solutions using turbo-like codes. Performance-wise, there are very small differences between the latest contenders, e.g., in the order of fractions of a dB. Additional features like code rate flexibility, variable delay, performance in non-linear channels (Sec. 11.5.10) or increased security (Sec. 11.5.1) became the differentiator factors, on top of the cost issue.

Another trend encouraged by the iterative decoding techniques introduced with the invention of turbo codes is the joint source-channel decoding, an example being briefly described in Sec. 11.5.2. In conjunction with the latest MPEG4 standard, this could improve significantly the link budget for image/video transmissions.

The chapter is organized in the following sections:

- Sec. 11.2 looks at the latest turbo-like technology that is used or has the potential to be used in satellite applications.

K. Sripmanwat (ed.), Turbo Code Applications: a journey from a paper to realization, 257–299.

- Sec. 11.3 covers the current manufacturers of satellite modems based on turbo-like products.
- Sec. 11.4 presents a survey of current satellite services based on turbo-like codes. It doesn't include a description of the DVB-S2 standard.
- Sec. 11.5 presents the latest developments which can impact future satellite services.

There are also a wide range of turbo-like code proposals for different other services, e.g., ADSL and 802.16, which might also find their application in satellite communications, but are not covered in this section.

Please note that the prices for various equipments or services presented in this Chapter are indicative only and are based on a web survey in late 2004.

11.2 Turbo-like Coding Technology Used in Satellite Services

Convolutional and block turbo codes together with LDPC codes represent different flavors of the turbo coding technology for the new generation of satellite services. This section is focused on the implementation of these turbo-like codes and their performance.

11.2.1 Convolutional Turbo Codes

A survey of some of the most popular satellite communication technologies based on parallel and serial concatenation of convolutional codes is attempted below. Apart from the original turbo codes, most of the codecs are proprietary solutions and the reader is directed to the relative web site for more information. The largest field programable gate array (FPGA) manufacturers like Xilinx (`www.xilinx.com`) or Altera (`www.altera.com`) also offer cores for turbo codecs, but are not reviewed in this section.

DVB-RCS

The new European Digital Video Broadcasting (DVB) standard [6–9] has adopted turbo codes for the return channel over satellite application (DVB-RCS). This standard allows a large number of small terminals to send return signals to a central gateway while they receive IP data (DVB/MPEG2) on the forward link. The users send small packets using multi-frequency time division multiple access (MF-TDMA) on demand assigned multiple access (DAMA). After the link is established, it is permanently available for the IP connection making efficient use of the satellite resources.

The turbo-coding scheme described in [10] allows transmission of data bit rates from 144 kbit/s up to 2 Mbit/s and uses double-binary circular recursive systematic convolutional (CRSC) component codes. It can handle ATM (53 bytes) or MPEG (188 bytes) packets, the actual block sizes being variable from 12 to 216 bytes. The coding rates can be selected from 1/3, 2/5, 1/2, 2/3, 3/4, 4/5 and 6/7. Given the fact that the DVB-RCS is a published standard, there are a few codec manufacturers that provide ASIC or core FPGA solutions such as:

- TC1000 turbo encoder/decoder, from (`www.turboconcept.com`).
- S2000 from (`www.icoding.com`).
- ECC3110 from (`www.eccincorp.com`).

S – tec™

Iterative Connections (`www.iterativeconnections.com`) launched in late 2003 the **S – tec™** family of serial concatenated convolutional codes. These codes allow satellite communications at less than 1 dB from channel capacity for a BER of less than 1E-10. In addition to the power efficiency, they offer the user the option to switch to a secure communication mode. This is achieved by a unique codec configuration between the transmitter and the receiver. This mode, described in more detail in Sec. 11.5.1 can be used in principle by any serial concatenated scheme where the information is not sent in clear. The **S – tec™** family was integrated in the Datum Systems line of modems as a plug-in option (`www.datumsystems.com`) and it is also offered as a new line of satellite modems, Premier 5/20/45, by Iterative Connections.

FlexiCodes

TrellisWare Technologies (`www.trellisware.com`) launched in early 2004 a new family of turbo like codes called FlexiCodes. There are basically three elementary encoders combined in a serial concatenated mode: a 4-state outer rate 1/2 convolutional code, followed by a single parity check (SPC) code and an inner rate 1 convolutional code. Both outer and inner codes have a time varying trellis. The SPC input is organized in groups of J bits, while a single output bit representing the modulo-2 sum of the input is produced. The overall code rate is J/(J+2). This is the way in which various coding rates can be generated. The decoder architecture is a serial concatenation of three decoders. The soft output demapper of the I and Q received channel inputs provides estimates for the outer and inner decoders. The outer decoder produces estimates which, after interleaving, are fed to the SPC decoder; in turn, the SPC decoder output is used by the inner decoder. The feedback loop starts from the inner decoder which passes extrinsic information to the SPC decoder; after deinterleaving, the bit estimates produces by the SPC decoder are input to the outer decoder. This family of codes performs very close to capacity for a wide range of coding rates and modulations.

STMicroelectronics

One of the largest suppliers of set-top boxes, (www.st.com), developed the STV0499 8PSK turbo codec. Live satellite tests in collaboration with French satellite pay-TV operator TPS and Eutelsat were performed over the EUROBIRD 2 at 33 ° East. The chip allows a 50% increased throughput over today's capability or a reduction by more than 33% in dish size. The chip is also fully compliant with DVB-S and DIRECTV specifications. It can use QPSK, 8PSK and 16QAM modulations. In addition to the turbo codec it supports the DiSEqC 2.x technology for bi-directional data communication with the Outdoor equipment. Sec. 11.4.6 gives more details related to chips compatible with the DVB-S2 standard.

Broadcom

The BCM4500 from (www.broadcom.com) is an integrated digital receiver that supports BPSK, QPSK and 8PSK in conjunction with turbo codes for up to 30 Mbaud. It includes a 7-bit A/D converter, demodulator, turbo decoder, interface to the BCM7020 HD graphics and video subsystem (MPEG2 or DIRECTV format), controller for two way communication with LNB. Paired with the BCM3440, a direct conversion satellite tuner, for USD20 in large volumes, it is the cheapest solution for direct broadcast satellite market available today. Another turbo code chip is the TurboQAM BCM3348 single chip cable modem that is compatible with the DOCSIS 2.0 specification which uses Advanced TDMA and synchronous CDMA. It has built in a 200 MHz MIPS32 Communication Processor, 16 bit, 100 MHz SDRAM interface, 10/100 Ethernet MAC and USB interface. It supports 4/16/32/64/128/256/512/1024 QAM FEC decoding. If 1024QAM is unlikely today to be used over the satellite, the capability to use smaller constellations might be also useful for emerging satellite applications compatible with the DOCSIS standard.

iCoding

A few families of high speed iterative decoders are available from (www.icoding.com). The S1000 is a 15 Mbit/s iterative decoder for Xilinx or Altera FPGAs or 45 Mbit/s in an ASIC. The S4000 is a high speed turbo decoder that can run at 200 Mbit/s in Virtex II or 100 Mbit/s in Virtex E.

Small World Communications

High speed MAP decoders are available from (www.sworld.com.au) which can be programmed for different codes and coding rates. They are available as BIT/MCS files for download into Xilinx Virtex or Spartan field programmable gate arrays (FPGA) or EDIF/VHDL core.

11.2.2 Block Turbo Codes

Iterative decoding of two concatenated block codes was introduced in [11]-[14]. Block turbo codes are suitable for high code rates, typically with rates greater than 0.75, for systems that require high spectral efficiencies. Their performance does not depend too much on the interleaver design [11, 15] because the rows and columns of concatenated block codes are independent. Random interleaving of data bits does not produce any significant improvement. Results from [13] and [16] conclude that for spectral efficiencies greater than 4 bits/s/Hz, block turbo coded quadrature amplitude modulation (QAM) systems outperform convolutional coded turbo coded QAM.

The performance improvements of block coded based turbo codes applied to quadrature phase shift keying (QPSK) is not as great as that obtained for 16QAM. Another interesting observation from [16] is that block turbo codes using 16QAM are not degraded in performance when the input data is quantized to only 4 bits, unlike convolutional based turbo codes which exhibit a minimum degradation of 0.5 dB. The product codes used in [13] and [16] were generated using identical Bose-Chaudhuri-Hocquenghem (BCH) codes (C1=C2).

Similarly, [14, 17, 18] report the performance of a sub-optimum iterative decoding algorithm for two dimensional product codes using Reed-Solomon (RS) codes and a Hamming (7, 4) code. The resultant minimum Hamming distance is 147 and the code rate is 0.5. Results indicate that a BER of 1E-6 would be possible at an E_b/N_o of 2.5 dB.

One advantage of using block codes is that error detection can be performed after iterative decoding without using additional check bits for error detection. This solution is attractive for small data block time division multiple access (TDMA) systems with high code rates and where ARQ systems are in place. As an example of the performance that can be achieved with block turbo codes, a $(512, 502, 4)^2$ BCH code achieved a BER less than 1E-5 at 0.8 dB and 0.45 dB from Shannon's limit after four iterations and forty iterations, respectively [11]. The iterative decoding process is similar to turbo codes. A comparison is made in [19] between parallel concatenated block codes (PCBC) using systematic cyclic codes and serial concatenated block codes (SCBC) using Hamming and BCH codes. The SCBCs perform better than PCBCs, but significantly worse than the equivalent structures based on convolutional codes. The performance of RS(63, 56, 8) used in a PCBC structure is remarkable for code rates higher than 0.8 and small data blocks.

There is considerable research effort targeted to developing faster RS decoders. A new technique to decode parallel RS codes based on the Multisequence Shift-register Synthesis algorithm [20] is described in [21]. Another new decoding algorithm that is 2-3 orders of magnitude less complex relative to the Vardy-Be'ery algorithms is described in [22]. This algorithm can be easily modified to produce soft output estimates. It is very likely to see FPGA cores available for block codes based on RS codes in the near future.

Turbo Product Codes (TPC)

A particular implementation of block turbo codes is the Turbo Product Code (TPC) for which very fast ASIC decoders are available from Comtech AHA Corporation (`www.aha.com`). The family of TPCs starts with AHA4501 Astro 36 Mbit/s codec. There is AHA4522 Astro LE 2k and AHA 4524 Astro LE 4k which use 2k or 4k interleavers. The AHA4525 is an IEEE 802.16 compliant codec. AHA4540 Astro OC-3 device is a single-chip TPC encoder / decoder capable of 155 Mbit/s data rates for code rates higher than 0.9. On top of the TPC code there is a CRC computation used to check data integrity at the decoder end. The chip can handle coding rates from 0.25 to 0.98 in fixed block sizes. It includes a soft metric computation module for BPSK, QPSK, 8PSK, 16QAM, 64QAM and 256QAM schemes. The AHA4540B is an improved version of AHA4540A, but mechanically the same size and the same footprint. The B version is approximately 15% faster which allows a wider selection of codes that meet the OC3 standard data rate to be used. AHA4541 is a codec that can run up to 311 Mbit/s (or 360 Mbit/s channel rate). The performance for QPSK modulation is relatively good for BERs down to 1E-8. Some implementations offer a concatenation between TPC and BCH codes in order to lower the error floor. Other manufacturers are (`www.eccincorp.com`), (`www.turboconcept.com`), etc.

11.2.3 LDPC Codes

There are many flavors of LDPC codes developed in the last few years which were targeted to specific emerging standards but could also be used in a closed loop satellite link. This section will not cover any of the LDPC proposals nor the solution selected for DVB-S2 standard. The focus will be on those manufacturers who have already a chip or an FPGA core which could be included in a satellite modem design.

Comtech LDPC/BCH

Comtech EF Data (`www.comtechefdata.com`) has developed in-house a LDPC BCH scheme which achieves very good performance as shown in Fig. 11.1 and 11.2.

V-LDPC

Flarion Technologies (`www.flarion.com`) has developed a Vector-LDPCTM family suitable for satellite applications with a range of coding rates from 1/6 to 5/6. The codes can be designed for either low threshold with high error floor or best BER with higher threshold.

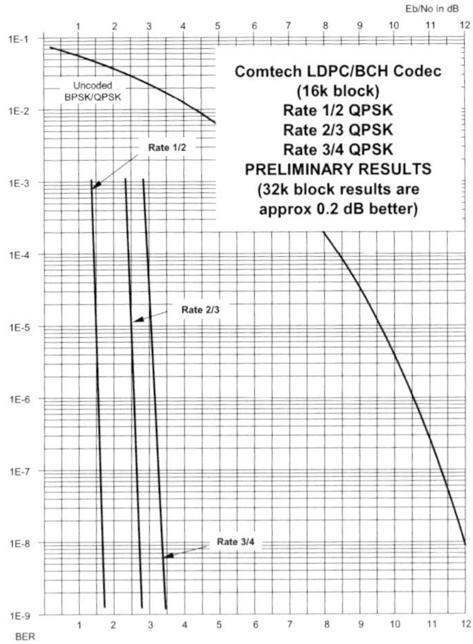

Fig. 11.1. BER Performance of ComtechEFData LDPC/BCH codes and QPSK (Courtesy of Comtech EF Data).

DVB-S2

Some of the manufacturers of set top box silicon tuners for the DVB-S2 standard based on BCH/LDPC codes are mentioned in Sec. 11.4.6.

11.3 Turbo Satellite Modem Manufacturers

Until recently, the only commercially available satellite modems which incorporated turbo-like codes, apart from those compatible with the DVB-RCS standard, used only TPCs. Although TPCs are a significant step forward relative to the Viterbi / Reed-Solomon standard, their performance can be further improved to get closer and closer to capacity. With the new explosion of turbo-like codecs it is likely to see a further breakdown of the market into closed networks which use a specific coding scheme optimized for their particular applications.

11.3.1 Comtech EF Data

Comtech EF Data was the first company to incorporate TPCs in their CDM550T and later on CDM600 series (`www.comtechefdata.com`). From

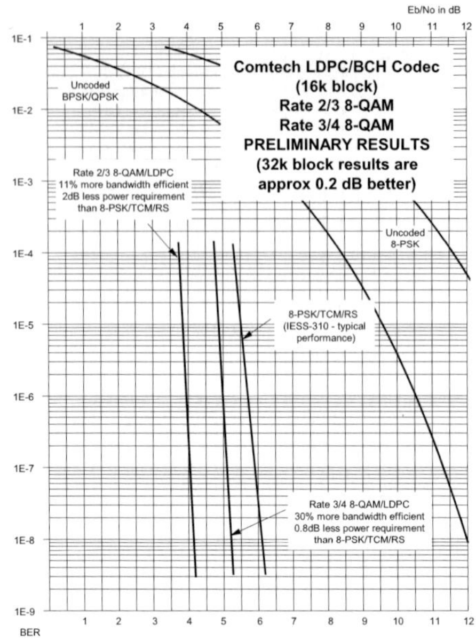

Fig. 11.2. BER Performance of ComtechEFData LDPC/BCH codes and 8QAM (Courtesy of Comtech EF Data).

the first release in August 1999 until early 2003, the company advertises that it sold around 5,000 TPC enabled modems. There is a wide range of turbo satellite modems to choose from depending of maximum data rate, modulation, L-band interface, etc. The company has improved continuously the performance of their TPC and also developed their own LDPC/BCH codec and a new 8QAM modulation technique.

Their latest results are shown in Fig. 11.3.

11.3.2 Radyne

An excellent modem using TPC is the DMD20 satellite modem from Radyne ComStream (`www.radynecomstream.com`). The modem can operate from BPSK to 16QAM, up to 20 Mbit/s data rates, offers L-band interfaces, TPCs or compatibility with IESS 308/309/310/314/315 standards. For more information on this and other products the reader should visit the company web site.

11.3.3 Paradise

Paradise Datacom, (`www.paradisedata.com`), part of the Intelek group, supplies the P300 series of modems which also offer the option of TPC codec.

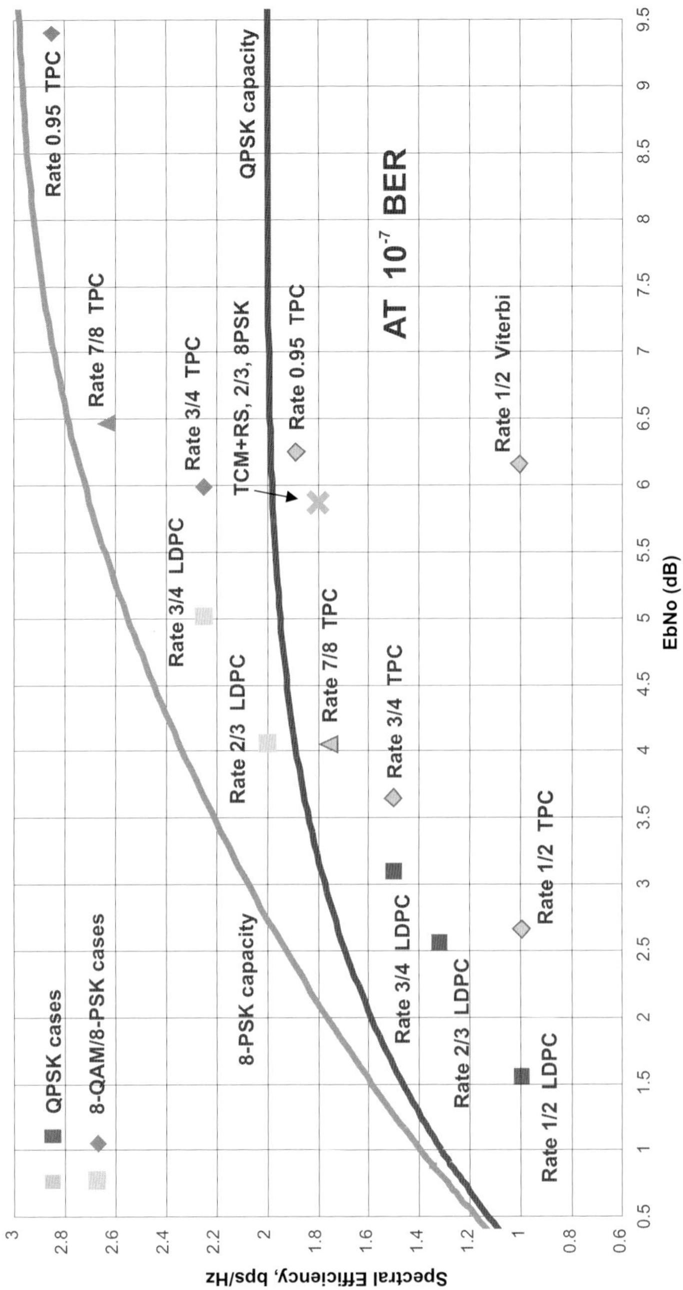

Fig. 11.3. TPC vs LDPC codes (Courtesy of Comtech EF Data).

Fig. 11.4. Premier 5 satellite modem (Courtesy of Iterative Connections).

This family of modems includes the P300VMUX, a combination of a turbo satellite modem and the V100 Multiplexer from Vocality International, targeted to SNG/DSNG operators. It provides up to eight voice/fax ports, IP Bridge/Router for speeds up to 2 Mbit/s.

11.3.4 Advantech

Advanced Microwave Technologies, (`www.advantech.ca`), released in early 2004 their AMT-70 satellite modem with a range from 8 kbit/s to 140 Mbit/s, BPSK to 16QAM, 70/140 MHz or L-band version with an Enhanced TPC option. The SPL-ACT (also an Advantech AMT company), (`www.advantechamt.com`) offers the AMS 8192 series of satellite modems which are compatible with IESS and DVB-S standards, data rates up to 8192 kbit/s with L-band or 70 Mhz interface.

11.3.5 iDirect

The iDirect technology used in their line of satellite modems, (`www.idirect.net`), is based on Turbo Product Codes. The company claims it achieves 25% higher IP throughput than the DVB standard or 62% more data on equivalent satellite bandwidth usage.

11.3.6 ViaSat

ViaSat, (`www.viasat.com`) offers a few products based on turbo-like codes: LinkStar, Surfbeam and WildBlue. LinkStar is a two-way bandwidth-on-demand broadband VSAT system using the DVB-Multi-protocol Encapsulation forward link and the DVB-RCS return link standards. Surfbeam and WildBlue are based on a satellite-enabled version of the Data Over Cable Service Interface Specifications (DOCSIS1.1) standard. This is a mature technology with very significant benefits such as low cost subscriber terminals due to very high volume chip sets, fast time to market, availability of infrastructure products for network control, system management, subscriber management, and billing systems. These terminals will also be used for the latest Canadian Ka-band spot beam satellite, Anik F2. Intelsat and Eutelsat currently allow consumers to choose either LinkStar or SufBeam satellite modems.

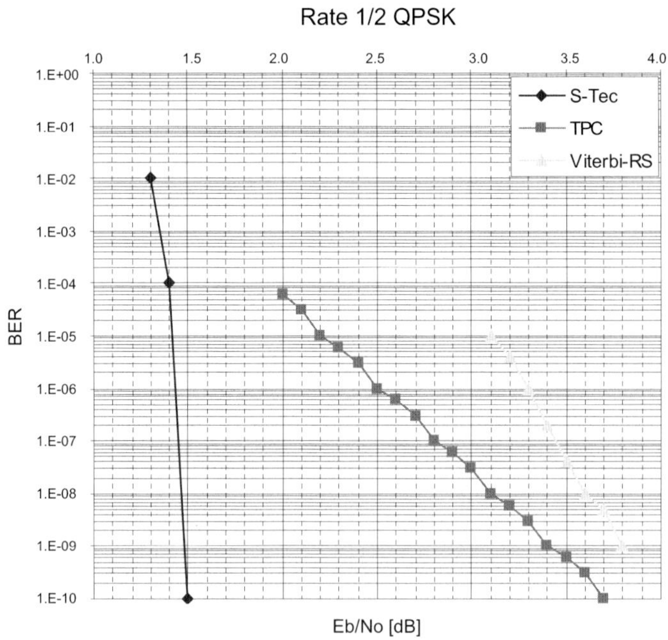

Fig. 11.5. Premier 5 BER curve for rate 1/2 (Courtesy of Iterative Connections).

11.3.7 Iterative Connections

Iterative Connections, (`www.iterativeconnections.com`) has integrated its own new $\mathbf{S}-\mathbf{tec}^{\text{TM}}$ codec technology in an enhanced Datum Systems modem which can achieve synchronization at an E_b/N_o below 1 dB. This new satellite modem, Premier 5, shown in Fig. 11.4 can provide data rates up to 5 Mbit/s in QPSK mode with coding rates 1/2, 3/4 and 7/8.

The performance of Premier 5 is shown in Fig. 11.5 for rate 1/2 and in Fig. 11.6 for rate 3/4.

Premier 5 satellite modem is the most power efficient satellite modem currently available in the market. This is the first in a series that includes the Premier 20 and Premier 45 satellite modems which will provide 8PSK and 16QAM modulations for up to 20 and 45 Mbit/s respectively. Specific to the $\mathbf{S}-\mathbf{tec}^{\text{TM}}$ codec is its capability to work in a secure mode which allows a unique configuration setup for each modem based on a unique interleaver configuration. This makes it extremely hard for an attacker to intercept the communications in a point-to-point link. There are currently up to 45000! combinations for the pseudo-random interleaver to chose from. Even if the attacker finds 90% of the particular interleaver used for a particular frame, the BER is almost 0.3 which makes any decoded data useless as shown in Fig. 11.7 for rate 1/2.

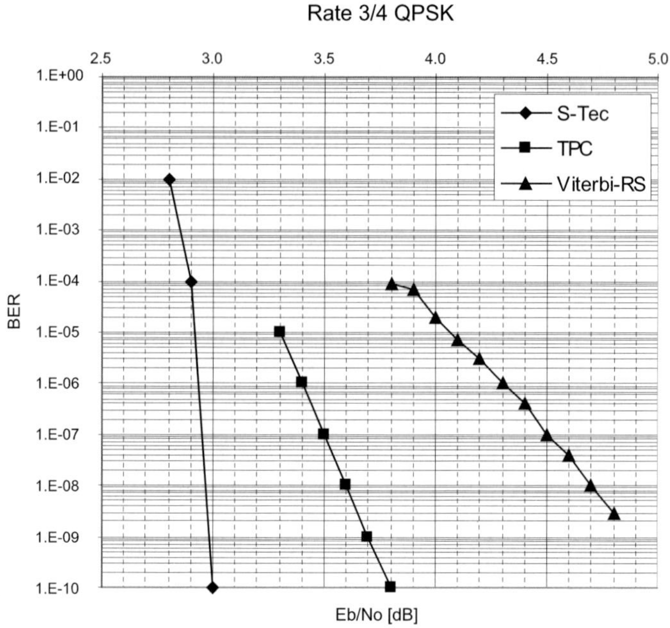

Fig. 11.6. Premier 5 BER curve for rate 3/4 (Courtesy of Iterative Connections).

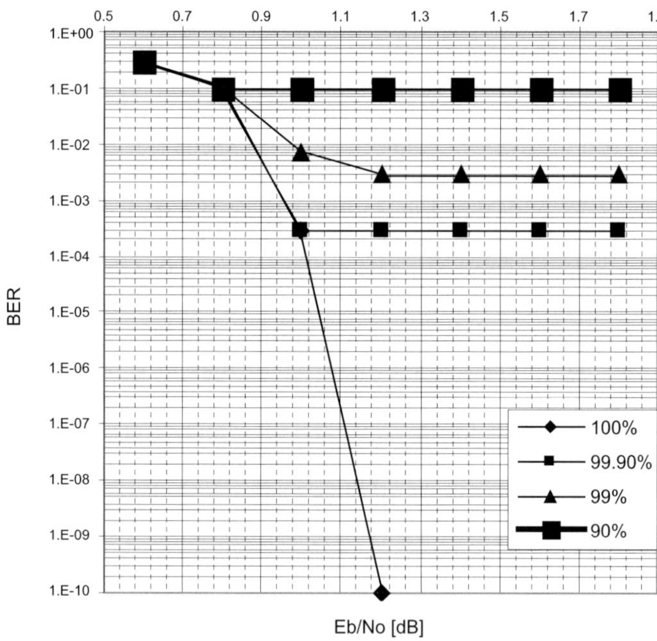

Fig. 11.7. BER sensitivity to interleaver (Courtesy of Iterative Connections).

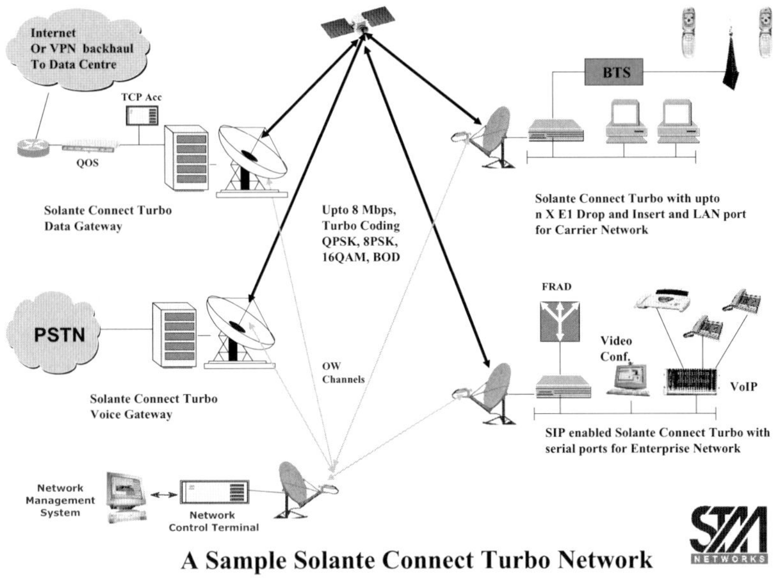

Fig. 11.8. Solante Network (Courtesy of STM Networks).

11.3.8 Datum Systems

Datum Systems satellite modems, (`www.datumsystems.com`), as mentioned in the previous section, do accept plug-in-codec cards that use the new $\mathbf{S-tec^{TM}}$ technology.

11.3.9 STM Networks

The Solante product family, produced by STM Networks, (`www.stmi.com`), is a full mesh interconnected VSAT network system that uses MF-TDMA along with SCPC overlay for on demand or reconfigurable high speed data links. The new Solante Turbo with the $\mathbf{S-tec^{TM}}$ codec can be used for a variety of applications as shown in Fig. 11.8. The product has a built in drop and insert capability, plus an IP port supporting SIP protocol. The modulation and coding rate can be configured on the fly on every on-demand link to meet the throughput requirements using the available satellite capacity.

11.4 Satellite Systems Using Turbo-like Codes

This section describes satellite systems based on turbo codes which are already launched or in field trials.

11.4.1 Inmarsat Broadband Global Area Network

The first commercial service to use of the original turbo codes was Inmarsat's multimedia service [3–5]. This is a new service based on turbo codes and 16QAM that allows the user to communicate with existing Inmarsat-3 spot-beam satellites from a laptop-sized terminal at 64 kbit/s. The Narrowband Technology based on 16QAM and turbo-coding provides significant reduction ($> 50\%$) in the required bandwidth for mobile satellite channels, at the same time improving the satellite power efficiency. An Inmarsat compatible turbo decoder core is offered by (`www.sworld.com.au/products/pcd04i.html`).

Three Inmarsat-4 satellites, built by Astrium will support the new Broadband Global Area Network (BGAN) (`regionalbgan.inmarsat.com`)that will be introduced in 2005. BGAN will be compatible with third-generation (3G) cellular systems. Inmarsat-4 will deliver Internet and intranet solutions, video on demand, videoconferencing, fax, e-mail, phone and LAN access at speeds up to 432 kbit/s, the satellites will be 100 times more powerful than the present generation. The coverage area is Europe, Middle East, India and North Africa.

The BGAN air interface was designed for a land-portable directional antenna. An extension of the service is planned for a mobile environment including maritime and aeronautical applications. This implies support for omnidirectional antenna with the option for low to medium data rates that can be implemented in smaller and low cost terminals. Also the point-to-point S-UMTS service will be extended to a multi-casting architecture that comprises of a multicast server, radio access network and user terminal.

New services targeted are telemedicine, WEB TV, live news broadcast and others. (`www.outfittersatellite.com/rbgan.htm`) advertises terminals for USD1339 or rent at USD120 per week (2004). The Hughes' data satellite terminal can interface through Ethernet, USB or Bluetooth to allow a 144 kbit/s connection to a BGAN user terminal.

11.4.2 SKYPLEX

SkyPlex is the on board processor specified by EUTELSAT and ESA (`www.esa.int`) that allows multiplexing to be performed on board of the satellite. The turbo code scheme is defined by the DVB-RCS standard. Downward-compatible solutions based on hierarchical modulations, with conventional DVB-S and DVB-DSNG reserved for the high-priority stream, could also use turbo coding for low-priority streams or for non real time services such as the DVB Multimedia Home Platform.

This technology eliminates the need for on-ground hub, two terminals being able to communicate with each other enabling LAN to LAN connectivity. A 2 Watt amplifier and DVB PC Card define the Satellite Interactive Terminal (SIT), or the DVB set-top-box. The small dish, between 45 cm and 84 cm allow a hub to SIT data rate of up to 4 Mbit/s and a return link of 16 kbit/s. Also, the estimated price at under €4,000 (2004) makes it attractive for pull services (high speed web browsing where a single user requests a specific item) or push / multicast services (a file or stream is transmitted to many users at the same time).

There are 18 SkyPlex units on four EUTELSAT satellites: W3A (6 units), HotBird 6 (8 units), HotBird 4 (1 unit) and EuroBird 2 (3 units). Each unit can receive uplink signals from different sources, multiplex them on board, and generate a single DVB downlink signal. Some of the services that can be provided are: Intranets, Tele-health, Tele-education, Web TV, Video streaming, IP-multicast, etc.

11.4.3 iPSTAR

iPSTAR (`www.thaicom.net/ipstar.html`) is a spot-beam satellite placed at 120 degree East longitude designed to provide two-way Internet service over satellite, with information capacity up to 40 Gbps, to customers in India, China, East Asia, South East Asia, Australia and New Zealand.

The physical layer uses higher order modulation and turbo product codes (TPC) in a Dynamic Link Assignment (DLA) system designed by Shin Satellite Public Company (Shin Sat). The DLA is able to control both the type of modulation and coding rate function of channel condition, from a maximum of 3.52 bits per symbol using 16QAM modulation with a 0.879 rate TPC to a minimum of 0.65 bits per symbol using QPSK with a 0.325 rate TPC. The iPSTAR satellite provides transponders of 65, 90, 120 and 150 Watts allowing each user a forward link of up to 12 Mbps and a return link of up to 4.1 Mbps. The access for the return link uses Aloha to gain entry to the system, Slotted Aloha for non-real time applications and Multiple Frequency Time Division Multiple Access for high data rate services. The home user has a forward link capacity from 256 kbit/s up to 2 Mbit/s and a return link of 256 kbit/s. The business user has allocated from 384 kbit/s up to 8 Mbit/s in the forward link and 384 kbit/s for the return link. On top of the last mile coverage, iPSATAR offers second mile coverage intended to complement the terrestrial network by providing redistributed ADSL to Cable Heads ends or Wireless Base stations. It can also act as a gateway for ISP's point of presence with capacity of up to 45 Mbit/s.

The dish size can be between 75 cm and 1.8 m with a 1 Watt to 2 Watt block up-converter that can be used with conventional Ku-Band satellites as well as the new multi-beam iPSTAR Satellite. The user terminal can be interfaced with a PC if operated in single-user mode or can be networked via Ethernet interface (`www.ipstar.com.au`).

11.4.4 Satellite IP: Boeing (Connexion)

The Connexion service (`www.connexionbyboeing.com`) is another application of TPCs introduced by Boeing and adopted by airlines like Lufthansa, Japan Airlines, British Airways, Scandinavian Airlines, Korean Airlines and

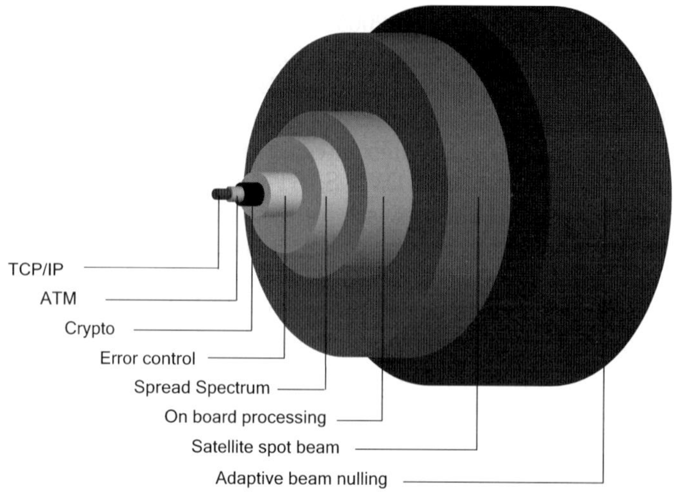

Fig. 11.9. Armored cable model (Courtesy of Jeffrey McCarthy, DSTO).

11.4.8 Australian Federation Satellite

In December 2002 an Australian micro satellite, FedSat, was launched for scientific research and technology demonstrations (`www.crcss.csiro.au/fedsat/default.htm`) The communications payload functions are implemented in a Base-Band Processor (BBP) module with a total power dissipation of less than 5 Watts. A turbo codec, using the original turbo codes, is implemented in a Xilinx FPGA that allows communication with small mobile terminals (e.g., 250 mW transmit power) in a TDMA mode, up to 250 kbit/s. A modified version of the communications payload is on board of the Korean satellite K4.

11.5 New Applications and Technologies

This section covers potential new satellite services and applications based on turbo-like codes. Some of these can be used in military applications, others like joint source and channel coding are general purpose applications.

11.5.1 Improved Security in Satellite Communications

Satellite communications, like any other wireless system, can be very easily intercepted and monitored. There are many layers of protection which are currently used in modern satellite systems. A typical armored cable model, Fig. 11.9, might have all or a combination of the following layers that protect the user application: TCP/IP (IPSEC), ATM, crypto, error control, spread spectrum techniques (direct sequence and/or frequency hopping), ultra high

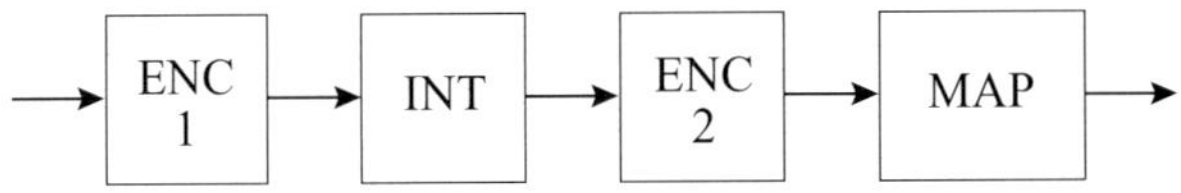

Fig. 11.10. Transmitter block diagram.

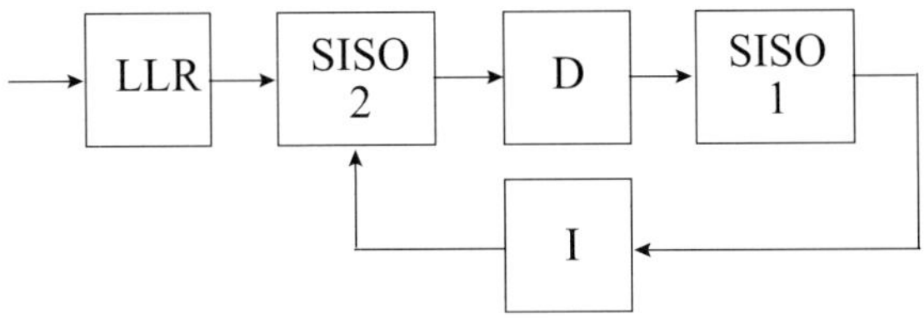

Fig. 11.11. Receiver block diagram.

frequencies (EHF), on board processing (to reduce jammer interference), spot beam discrimination (e.g., steerable dish, phased arrays) and adaptive nulling (e.g., multiple beams).

The error control layer was initially included in this armored cable model due to the use of an interleaver in a serial concatenated scheme that is able to break bursts of errors. The most popular conventional error control scheme used today is a Reed-Solomon outer code concatenated with a inner convolutional code. Enhanced security may be obtained using a serial concatenated turbo-like code in the error control layer. By employing very large interleavers, e.g., up to 1E6 bits, and a secret exchange of encode/decode parameters a very high level of security may be obtained. A typical transmitter sub-system that could be used with this approach is shown in Fig. 11.10.

The binary encrypted data is encoded by a turbo-like scheme made from the serial concatenation of two elementary encoders: ENC1 and ENC2 blocks are two 16 state encoders of rate 2/3 and 3/4 respectively to produce an equivalent rate 1/2 code. Each of them could be implemented from rate half encoders with some puncturing pattern. The INT block represents a large pseudo-random interleaver of size N bits, e.g., N = 1E6. The pseudo-random interleaver table can be automatically generated for a given seed. In theory there are N! combinations. In practice there are slightly less given the particular constraint that the interleaved bits need to be spaced from their original neighbors at a distance greater than s, the separation threshold [27]. The MAP block maps a group of bits to the transmitted constellation symbol. Assume a simple 16QAM constellation, for every 4 bit sequence, there are 24 mapping positions available which can be used. It should be noted that the serial concatenation doesn't send the uncoded data at all as is the case of the original turbo codes (parallel concatenation). Therefore the information bits, even decoded with errors, can't be retrieved without the knowledge of the interleaver table. The receiver sub-system is shown in Fig. 11.11.

The demodulated received symbol is used by the LLR block to calculate the log-likelihood ratio for each bit in the symbol. The same mapping pattern used in the MAP block has to be used by the LLR block. These ratios, soft estimates of each transmitted bit, are then passed to the iterative decoder made by the two soft-in soft-out (SISO) blocks separated by the deinterleaving (D) and interleaving (I) stages. The same pseudo-random interleaver table has to be used by the transmitter INT block and by the receiver D and I blocks. Also the initial states of the two ENC encoders must be the same with the initial state of the two SISO decoders. The particular parameters that need to be established before communication can start are as follows:

- Interleaver size N and the generator seed used to generate the pseudo-random interleaver
- S, the separation threshold used to generate the pseudo-random interleaver
- ENC1 and ENC2 state used to initialize the encoder state at the beginning of each block; the same states are used to initialize both decoder blocks
- Puncturing pattern, only in the case that there is any puncturing that occurs at the output of each encoder block; the same puncturing pattern needs to be used by the decoders.
- Bit to symbol mapping used by the MAP and the LLR blocks.

Some implementations that use unique words (UW) to identify the beginning of a new block restrict the number of parameters that can be used. There are techniques based on codec self-synchronisation, e.g., using probability surface metric, which avoid the use of UW [28]. All the above parameters represent the secret key that has to be exchanged between the transmitter and receiver. The receiver is always configured in a well defined default state, advertised as a public key. This public key contains the following three numbers: p, g and y. The y number was calculated using a secret number a such that $y = g^a \pmod{p}$.

Secret Key Exchange

The secret key could contain the parameters for a sequence of different seeds that can be used on a block by block basis. The secret key exchange is based on Diffie Hellman algorithm and contains the following steps:

- Both the transmitter and receiver are configured in the default mode.
- The transmitter uses a secret number, b, to calculate the secret key, s, that needs to be exchanged with the receiver. This secret key is calculated using the public key y of the receiver: $s = y^b \pmod{p} = (g^a)^b \pmod{p} = g^{ab} \pmod{p}$. More values of b might be tried before a valid seed is achieved.
- Once the transmitter decides that s is a good seed, he sends $z = g^b \pmod{p}$. All this communication takes place in the well known default mode.
- After normal demodulation and decoding of z, the receiver calculates the value: $z^a \pmod{p} = (g^b)^a \pmod{p} = s$

Both transmitter and receiver re-initialize themselves using the same secret key s.

Possible Interleavers

The S-type interleaver design method was described in [27] as follows: each randomly selected integer is compared to s previously selected integers. If the current selection is equal to any s previous selections within a distance of +/-s, then the current selection is rejected. This process is repeated until all N integers are selected. The duration of this search depends on s and usually converges very quickly for $s < \sqrt{N/2}$. The choice of s also depends on the performance desired for the codec. For example, the degradation in performance is less than 0.3 dB if the interleaver size is changed from 32,768 to 16,384. Therefore, if this penalty is acceptable, s can take any value between $\sqrt{8,192} \approx 90$ and $\sqrt{16,384} = 128$.

All possible permutations of a block of N bits is N! However, for a pseudo-random interleaver with a fixed separation threshold s, there are fewer possibilities. For the first mapping operation from position i to position $P(i)$ there are N options available, including the identity mapping. The mapping of position $i + 1$ can be done in any of the $(N - 1 - 2s)$ positions. This is because the mapping of two adjacent bits should not occur at a distance less than s, the separation threshold. The mapping of $i + 2$ can be done in any of the $(N - 2 - 4s)$ positions (worst case). Furthermore, any neighbors at up to s positions before and after the current position i, have to be mapped at a distance s from their neighbour. Therefore, the mapping of the $k - th$ bit, where $i - s < k < i + s$, can be done in any of the $(N - k - 2ks)$ positions (worst case). If s is made variable in a certain range, the total number of interleavers of size N can be increased accordingly.

11.5.2 Joint Source-channel Coding

Significant advances in the area of joint source and channel coding were presented in [29]. Some of the results and figures from the cited references are reproduced here courtesy of Dr Wei Xiang (`www.usq.edu.au/users/xiangwei`). Shannon's classical separation result states that the optimization of the end-to-end system design can be done by separately optimizing the source encoder-decoder pair and the channel encoder-decoder pair. However, this result holds only in the limit of infinite source code dimension and infinite channel code block length. For practical systems, a joint source and channel code design may reduce distortion, as well as complexity and delay.

State Transition Probabilities of VLC-Trellis

The source information is usually encoded in a variable length coded (VLC) sequence which can be represented by a trellis structure proposed by Balakirsky in [30]. Thus, a bit-level soft-in/soft-out (SISO) decoding module for VLCs can be derived by applying the BCJR (known also as the MAP) algorithm to such a VLC-trellis. Therefore, an a posteriori probability (APP) VLC

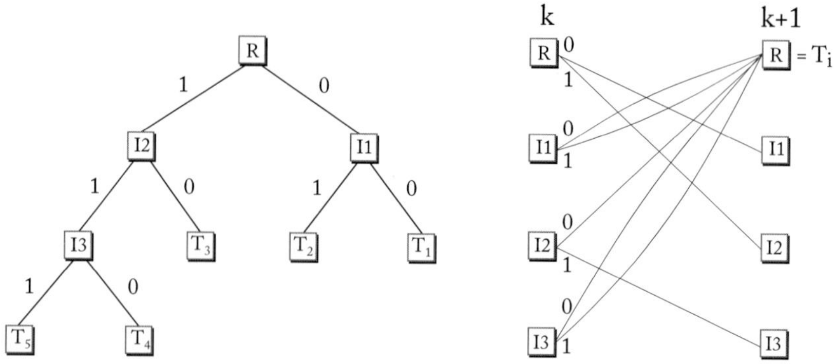

Fig. 11.12. Tree representation and VLC trellis with parallel transitions.

source decoder can be implemented to generate soft outputs and exchange extrinsic information between the source VLC APP decoder and channel APP decoder. For a variable length coded sequence, the symbol or codeword probabilities are usually assumed to be known. The state transition probabilities (STP) [29] can be considered as source a priori information based on the bit-level trellis representation of variable length codes.

The advantage of STPs is that they can be naturally incorporated into the maximum a posteriori (MAP) algorithm to improve system performance. For example, the Huffman code in [30] can be represented as a variable length code C = 00, 01, 10, 110, 111. The corresponding codeword probabilities are P $= p_1, p_2, p_3, p_4, p_5$. Fig. 11.12 shows the tree representation and the bit level trellis of the variable length code C. The nodes in the tree are subdivided into a root node (R), internal nodes (I) and terminal nodes (T).

The VLC trellis can be treated as an irregular trellis in comparison to the trellis of a recursive systematic convolutional (RSC) code. The trellis of an RSC code has nice systematic properties. Each state has two paths emanating and two paths merging with the same probabilities. In contrast to the systematic trellis of an RSC code, each state in the VLC trellis might have an arbitrary number of emanating or merging paths or even have parallel paths emanating and merging between states. Moreover, different paths might emanate or merge to a state with different probabilities from each other because the codeword probabilities of VLCs are in general different. The codeword probabilities of VLCs can be utilized to derive the state transition probabilities which are used as a priori information to facilitate iterative source channel decoding.

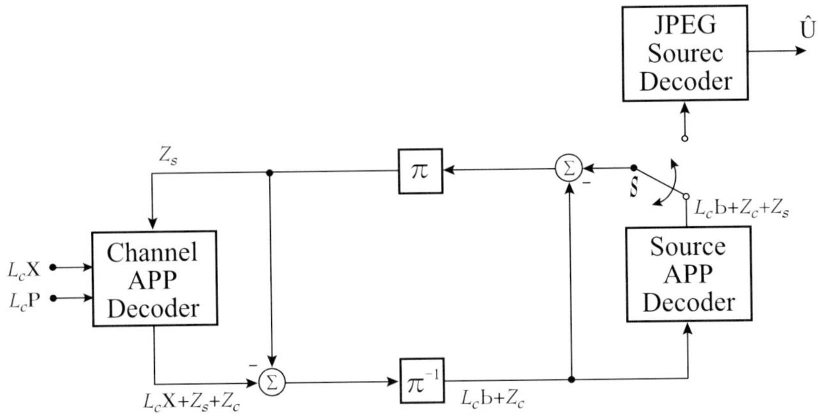

Fig. 11.13. Iterative source and channel decoder.

Iterative Decoding using RVLC

Codeword probabilities of VLCs can be used to derive the state transition probabilities as a priori information to facilitate iterative source-channel decoding. The error-correcting capability of a VLC is dominated by the free distance metric. As far as soft decoding of VLCs is concerned, reversible VLCs (RVLCs) with large free distance usually have better performance in terms of symbol error rate. Therefore, the free distance is a critical design criterion one has to take into consideration to construct an RVLC from an existing Huffman code. A block diagram of an iterative source and channel decoder is shown in Fig. 11.13.

It was interesting to observe that, when Huffman codes are used in this iterative decoding process, no significant gain is achieved. This can be explained by the fact that Huffman codes are too good, they do not leave enough redundancy to kick-start the iterative decoding process. If the Huffman codes are replaced with less efficient codes with a higher average codeword length, the iterative decoding process performs very well. The optimum case observed was for codes with 20% higher average codeword length than the Huffman code, as C_2 in Table 11.1.

The system that was simulated used a variable channel coding rate that would decrease in the same proportion with the increase in the average codeword length of the source, thus having no bandwidth expansion in the system. Simulation results shown in Fig. 11.14 confirm that iterative source-channel decoding of JPEG coded images provides a significant coding gain. Since most of the existing image/video standards use Huffman codes as the entropy code, the iterative source-channel decoding can be extended to those standards.

Table 11.1. Proposed variable length codes.

DC Symbol	Occurrence Probability $\mathcal{P}$	JPEG Lumin DC Huffman Code $\mathcal{C}_{\mathcal{H}}$	Symetric RVLC C_1	Asym. RVLC C_2 with $d_f = 2$	Asym. VLC C_3 with PfxDist = 2
0	0.371745	00	00	00	000
1	0.071615	010	111	111	011
2	0.102214	011	010	010	10100
3	0.147135	100	101	1101	11000
4	0.132812	101	0110	1011	1011100
5	0.124349	110	1001	0110	1101100
6	0.049479	1110	11011	10101	1011111
7	0.000651	11110	01110	11001	1101111
8	0.0	111110	10001	110001	
9	0.0	1111110	011110	1001001	
10	0.0	11111110	100001	1010010100	
11	0.0	111111110	110011	10011001100	
Average code length		2.6790	2.9857	3.1228	4.7279

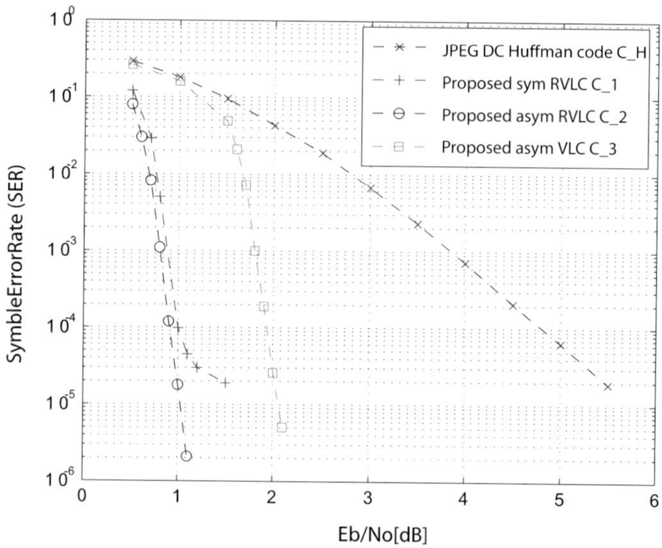

Fig. 11.14. BER curves for joint source channel decoding.

Table 11.2. New 8PSK bit mapping.

bit 1	bit 2	bit 3	bit 4	Symbol	I	Q
0	0	0	1	$S_1 = (0,0,0)$	$-A\sin(22.5^o)$	$-A\cos(22.5^o)$
0	0	1	1	$S_2 = (0,0,1)$	$-A\sin(22.5^o)$	$A\cos(22.5^o)$
0	1	0	0	$S_3 = (0,1,0)$	$-A\cos(22.5^o)$	$-A\sin(22.5^o)$
0	1	1	0	$S_4 = (0,1,1)$	$-A\cos(22.5^o)$	$A\sin(22.5^o)$
1	0	0	1	$S_5 = (1,0,0)$	$A\sin(22.5^o)$	$-A\cos(22.5^o)$
1	0	1	1	$S_6 = (1,0,1)$	$A\sin(22.5^o)$	$A\cos(22.5^o)$
1	1	0	0	$S_7 = (1,1,0)$	$A\cos(22.5^o)$	$-A\sin(22.5^o)$
1	1	1	0	$S_8 = (1,1,1)$	$A\cos(22.5^o)$	$A\sin(22.5^o)$

11.5.3 Higher Order Modulations for Satellite Systems

Until the invention of turbo codes, the use of higher order modulations in satellite systems was hampered by the need to operate in the linear region of a power amplifier. This operating mode was much less efficient than in the case of a QPSK or 8PSK modulation which could be used with the amplifier operating in saturation mode. However, with the advent of turbo codes, the satellite industry started to shift towards a more bandwidth efficient modulation which, when used with turbo codes, exhibited a significant improved performance from the standard convolutional codes. A new 8PSK mapping technique and 16QAM specific issues are discussed in detail in the following paragraphs.

8PSK

This section describes a particular bit to symbol mapping that can be applied to any conventional 8PSK modulator. This new technique allows for two of the three bits that are mapped on an 8PSK symbol to achieve an error probability close to the BPSK modulation while the soft demapper can perform a combining technique to estimate the third bit, therefore improving the overall system performance by 0.3 to 0.5 dB. This new mapping is based on independent I and Q modulation used in 16QAM and described in [2]. This mapping technique is applied to produce an 8PSK symbol as if it belongs to a 16QAM constellation where the fourth bit is the second bit inverted as described in Table 11.2 and shown in Fig. 11.15. The 8PSK symbol is defined by the first three bits. Bit 1 and bit 2 define the value for the I component whereas bit 3 and bit 4 (which is bit 2 inverted) define the value for the Q component.

This particular mapping allows independent I and Q mapping, that is each of them can be seen to be identical with a 4AM signal (4 levels amplitude modulation) [2]. In the case of the I signal, bit one is mapped to the most protected position of the 4AM symbol and bit two is mapped to the least protected position of the 4AM symbol. In the case of the Q signal, bit three is mapped in the most protected position of the 4AM symbol and bit four is mapped in the least protected position of the 4AM symbol. Every group of three bits to be transmitted is first transformed in a group of four bits, the fourth bit being the inverted version of the second bit. Then the four bits are mapped to a symbol as described above. The I and Q values associated with that symbol are sent through the channel.

The soft demapper calculates first the L_1 and L_2 values based on the received I component. Then the soft demapper calculates the L_3 and L_4 values based on the received Q component. The soft demapper makes its final decision as follows: bit 1 value is the L_1, bit 3 value is the L_3, and bit 2 value is $(L_2 - L_4)/2$ for equal gain combining.

For a 16QAM scheme, the symbol$u^k(u_1^k, u_2^k, u_3^k, u_4^k)$, is sent through the channel at time k and the point r_k in a two dimensional space is received. The 16QAM symbol is defined by the I and the Q components which are 4AM signals. Each signal is modulated independently by two bits. It is assumed that at time k, u_1^k and u_2^k modulate the I component, as in Fig. 11.16, and u_3^k and u_4^k modulate the Q component of a 16QAM scheme, as in Fig. 11.17. At the receiver, the I and Q signals are treated independently in order to take advantage of the simpler formulae for the 4 bit-LLR values.

For an AWGN channel, the following two LLRs need to be evaluated from the received I signal. These represent soft estimates for each bit of the 4AM symbol.

$$LLR(u_1^k) = log\frac{\sum_{i=1}^{2} exp[-\frac{1}{2\sigma^2}(I^k - a_{1,i}^k)^2]}{\sum_{i=1}^{2} exp[-\frac{1}{2\sigma^2}(I^k - a_{0,i}^k)^2]}$$

$$= log\frac{exp[-\frac{1}{2\sigma^2}(I^k - A_2)^2] + exp[-\frac{1}{2\sigma^2}(I^k - A_3)^2]}{exp[-\frac{1}{2\sigma^2}(I^k - A_0)^2] + exp[-\frac{1}{2\sigma^2}(I^k - A_1)^2]}$$

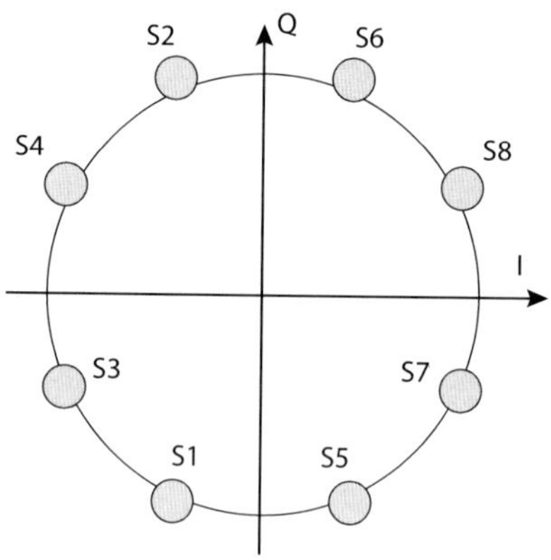

Fig. 11.15. New 8PSK symbol mapping.

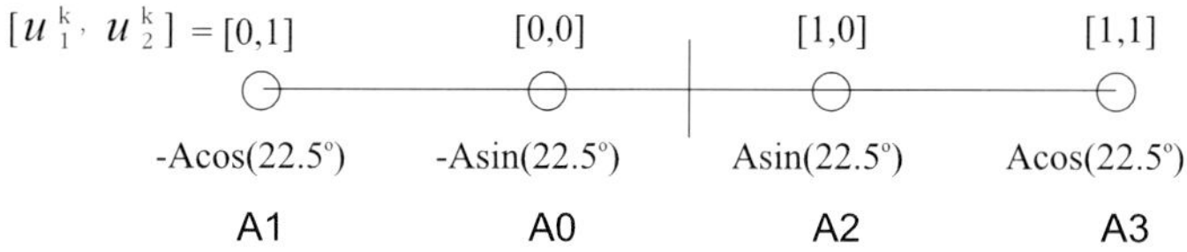

Fig. 11.16. Mapping in the I dimension.

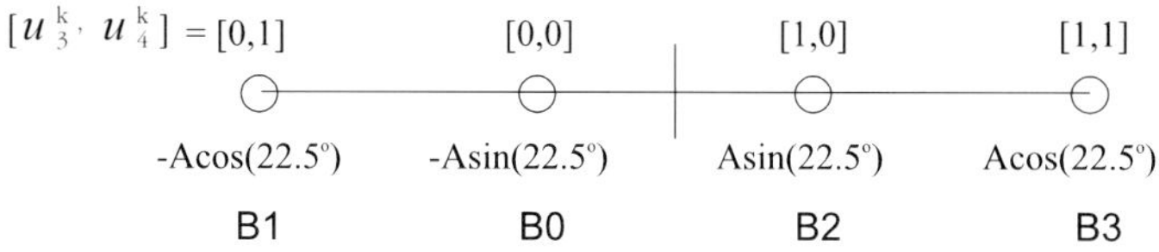

Fig. 11.17. Mapping in the Q dimension.

$$LLR(u_2^k) = log \frac{\sum_{i=1}^{2} exp[-\frac{1}{2\sigma^2}(I^k - a_{1,i}^k)^2]}{\sum_{i=1}^{2} exp[-\frac{1}{2\sigma^2}(I^k - a_{0,i}^k)^2]}$$

$$= log \frac{exp[-\frac{1}{2\sigma^2}(I^k - A_1)^2] + exp[-\frac{1}{2\sigma^2}(I^k - A_3)^2]}{exp[-\frac{1}{2\sigma^2}(I^k - A_0)^2] + exp[-\frac{1}{2\sigma^2}(I^k - A_2)^2]}$$

The above LLRs are renamed L_1 and L_2. The L_1 estimate is more reliable than the L_2 estimate and the performance of the turbo decoder depends on these reliabilities. Similar, for the Q dimension, the mapping is shown in Fig. 11.17.

For an AWGN channel, the following two LLRs need to be evaluated from the received Q signal:

$$LLR(u_3^k) = log \frac{\sum_{i=1}^{2} exp[-\frac{1}{2\sigma^2}(Q^k - a_{1,i}^k)^2]}{\sum_{i=1}^{2} exp[-\frac{1}{2\sigma^2}(Q^k - a_{0,i}^k)^2]}$$

$$= log \frac{exp[-\frac{1}{2\sigma^2}(I^k - B_2)^2] + exp[-\frac{1}{2\sigma^2}(Q^k - B_3)^2]}{exp[-\frac{1}{2\sigma^2}(Q^k - B_0)^2] + exp[-\frac{1}{2\sigma^2}(Q^k - B_1)^2]}$$

$$LLR(u_4^k) = log \frac{\sum_{i=1}^{2} exp[-\frac{1}{2\sigma^2}(Q^k - a_{1,i}^k)^2]}{\sum_{i=1}^{2} exp[-\frac{1}{2\sigma^2}(Q^k - a_{0,i}^k)^2]}$$

$$= log \frac{exp[-\frac{1}{2\sigma^2}(I^k - B_1)^2] + exp[-\frac{1}{2\sigma^2}(Q^k - B_3)^2]}{exp[-\frac{1}{2\sigma^2}(Q^k - B_0)^2] + exp[-\frac{1}{2\sigma^2}(Q^k - B_2)^2]}$$

The above $LLRs$ are renamed L_3 and L_4. These four values, L_1, L_2, L_3 and L_4 are used in making the final estimates for the three bits for every received 8PSK symbol.

nel, simulation results have indicated that more protection of the parity bits is required.

5. The parallel concatenation is known to perform closer to capacity than the serial concatenation, at similar complexity/delay and for relatively low BER with the crossover point around 1E-6 or 1E-7 [25]. The extra coding gain of the proposed parallel scheme gives around 0.3 to 0.5 dB improvement over serial concatenation schemes at the BER of interest. This scheme achieves a target BER of 1E-7 at as close as 0.8 dB from capacity for relative small block sizes.

Trellis coded modulation (TCM) can be applied to both PCCC and SCCC [26]. Turbo trellis coded modulation (TTCM) only performs marginally better than a Gray coded QAM structure with rate 1/2 turbo coding in an AWGN channel. The turbo TCM transmitter and receiver are more complicated to implement than the much simpler 16QAM scheme with the rate 1/2 turbo decoder proposed in [2].

TTCM structures have also been proposed in [26]. They offer improvements over conventional TCM techniques.The same authors investigated TTCM schemes for 8PSK, 16QAM and 64QAM modulation schemes with varying overall bandwidth efficiencies. The simulation results are only for BER higher than 1E-6 and show an "error floor" that can not be lowered below the 1E-6 threshold. The independent mapping of I and Q, each being treated as a 4AM signal, was presented for the first time in [2] achieves a very good performance using a less complex decoder than in the case of TTCM structures. Further more, the BER curve doesn't exhibit an error floor even for very low BERs below 1E-10.

A thorough investigation was performed in the non-linearities introduced by the high power amplifier (HPA) on the mobile terminal. Simulations and hardware measurements have shown that with minimum power amplifier back-off, optimal performance can be achieved with the 16QAM modulation. Of particular importance was the extra 0.5 dB coding gain that can be gained in higher order modulation schemes by mapping the information bits to the most protected positions in the 16QAM symbol.

11.5.4 Decoder-assisted Synchronization

The most power efficient coding schemes operate at very low SNRs that might exceed the ability of the receiver to achieve initial synchronization using traditional techniques. In [33], the error correction decoder is used to aid frame synchronization using the concept of list-synchronization, i.e., concatenated synchronization in which the first module generates a list of potential frame boundaries and the second module makes a hard decision. The same idea is applied to a turbo coded system in [34]. Instead of a traditional header, synchronization bits are embedded in a mid-amble of the information packet. A standard synchronization method is used to provide a list of potential delays.

The turbo decoder is trying various potential frame starting points while monitoring the anticipated state sequences. The probability of the decoder selecting the correct frame starting position improves with the number of iterations.

An even more bandwidth efficient technique that doesn't use any synchronisation words was presented in [35]. It relies on monitoring the amplitude of the MAP decoder output. Stream-oriented turbo codes [36] make use of non-block interleavers with low periodicity which present lower synchronization ambiguity than block interleavers. A maximum likelihood frame synchronisation technique that exploits both the presence of pilot symbols and the soft information provided by the MAP decoder was presented in [37]. The decoder can also assist other blocks in the demodulator, e.g., carrier phase recovery blocks, to achieve performance close to Cramer-Rao bound which would not be possible in "code-unaware" systems at very low signal-to-noise ratio [38–40].

11.5.5 Turbo Codes for Frequency-Hopped Spread Spectrum

Performance of a turbo coded system in non-coherent FH-SS with partial band interference was investigated in [41]. It shows a gain of more than 6 dB when compared with a RS-CC scheme described in [42]. However, there are some differences between the two systems that could not be eliminated, so the actual gain might be smaller. A dynamic power allocation among different dwells in a turbo coded system was presented in [43]. A new iterative algorithm for channel variance and carrier phase estimation (side information) was shown to provide superior performance to the case where no side information is available.

A turbo trellis coded modulation in conjunction with continuous phase modulation was used in a frequency-hop packet radio system as discussed in [44]. The interleaving was done across the hop frequencies to improve resistance to frequency-selective fading and partial-band interference. Diagonal interleaving was used to provide protection to co-site interference that may produce periodic bursts of noise.

11.5.6 Turbo Codes for Jammed Channels

It was shown in [45] that M-ary orthogonal signalling with coding can achieve error-free communications over the worst case partial band jamming. The minimum E_b/N_j is shown to be 4ln2 (4.43 dB) in the case of soft outputs and the optimum code rate that minimizes E_b/N_j is 1/2. Turbo code performance in partial band jamming was investigated in [46]. Simulation results have indicated that when side information is available, the turbo code is able to operate without degradation due to partial band jamming under various jammer occupancies and thermal noise levels. A similar conclusion was reached in [47] where Rayleigh fading was added to the channel model. If channel state estimates can be calculated accurately, turbo coded systems can mitigate the

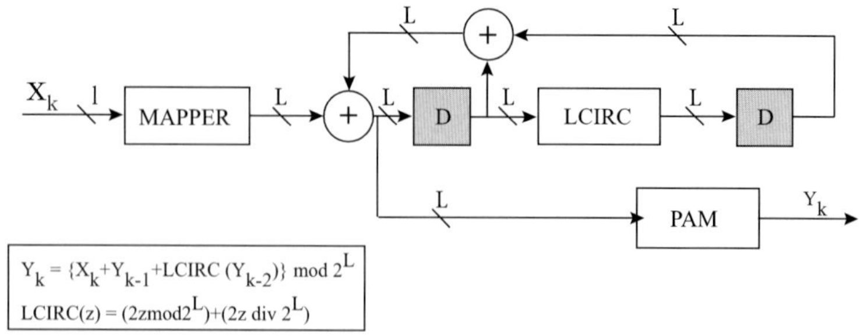

Fig. 11.19. Chaotic encoder.

effects of jamming and fading turbo codes were also considered a good candidate for medium data rates waveform in [48]. It was shown that even for short frame oriented systems, which are likely to be used in future satellite communications systems, turbo codes can provide better performance than convolutional codes by 1.5 to 2.5 dB for FH-SS.

11.5.7 Chaotic Turbo Codes

Chaos theory dates back to early 60's with the discovery of attractors within meteorological prediction systems. Analog systems were characterized in terms of their stability measured by Lyapunov Exponents. Recently, linear digital filters were designed to exhibit a "chaotic" behavior with a stable Lyapunov exponent using a nonlinear element [49]. The nonlinear element is a left circular shift register that is left to overflow. These digital filters can be used to produce "chaotic digital modulation" or can very easily be combined in a chaotic encoder scheme [50] as shown in Fig. 11.19.

The main advantage of these schemes is that the output of the encoder has a broadband noise-like spectrum. Another advantage is that the initial states of the decoder must be the same as at the encoder (otherwise the decoder will not work). The mapper is a serial to parallel converter. The delay elements and the adders have dimension L. The output from the shift register is converted to a PAM signal with 2L levels. For almost all choices of input sequences, two identical encoders initialized with two different but arbitrarily close states, will produce divergent outputs. This outputs are uncorrelated to one another.

A chaotic turbo encoder could be constructed as shown in Fig. 11.20.

11.5.8 Analog Decoders

Analog decoders have significantly less complexity than their digital relatives. Very simple transistor circuits were used in [51, 52] to implement the basic engines of iterative decoders. A decoder network was designed as an analog network in [53] using 940 BJTs and 650 PMOS transistors in a standard 0.8

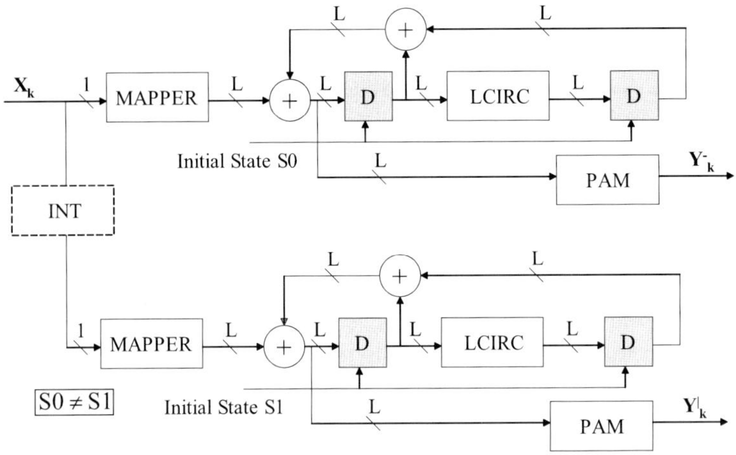

Fig. 11.20. Chaotic turbo encoder.

microns BiCMOS technology. Its power consumption is 98 mW. With 90 ns decoding time per codeword, the analog decoder can achieve a data rate of 100 Mbit/s.

An analog tail-biting MAP decoder was implemented in analog VLSI using 0.25 microns BiCMOS technology [55]. It achieves a speed of 160 Mbit/s and the power consumption is 20 mW. It is estimated that the digital equivalent decoder would have a 3.3 times increase in overall performance degradation, assuming 8 bits resolution, 8 times higher power dissipation and 5.2 times larger area. The analog decoding of high rate tail-biting codes using the dual code was presented in [56].

It was shown that the complexity of a decoder based on the dual trellis grows linearly (from a certain number of inputs) compared to an exponential growth in complexity for a decoder operating on the original trellis. An example is given for a rate 9/10 code that would require 25,000 transistors compared with 2,800 for the dual code implementation.

Analog decoding of product codes was presented in [57] with the potential to use larger code sizes. This can be achieved using currents to encode probabilities and differential voltages to encode log-likelihood ratios in the same circuits. Therefore differential sample-and-hold circuits can be used to store a stream of analog signals which can later be input in parallel to the decoder. Analog decoders have the potential to achieve very high speeds at very low power consumption. They can easily be integrated in the front end of the receiver together with similar circuits like adaptive equalizers. They are best suited for inter-chip communications, regardless of their sensitivity to noise, temperature and component variations.

It was observed in [58] that the physics of CMOS transistors in subthreshold mode is very similar to that of nervous cells. This points towards very

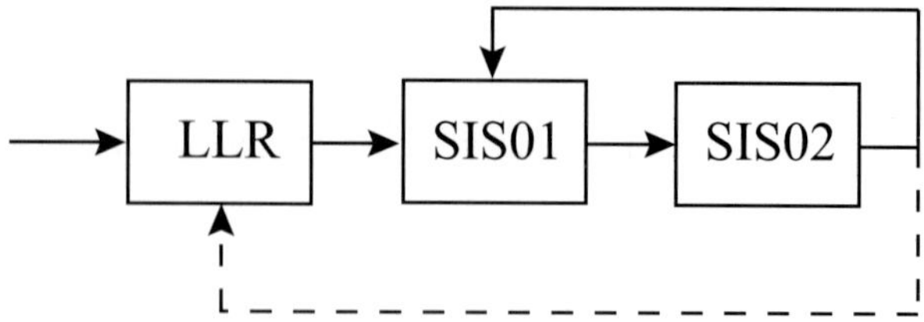

Fig. 11.21. Iterative demapping - decoding loop.

interesting research in reproducing in silicon the highly power efficient information processing that occurs in living organisms. Although the CMOS technology is slow, it has the advantage of occupying a small area. This allows for larger blocks to be used in the code structure. All implementations rely on the translinear principle: the current is proportional with the exponential voltage. Therefore the voltage sum becomes a sum of logarithms of current which in turn, represents current multiplication.

The work on decoder implementation is extended to eliminate the stand alone conventional A/D converter and to integrate it into the turbo decoder chip. Using differential sample and hold techniques, billions of bits can be held for decoder processing. The average power consumption for 1 Mbit/s throughput is below 1 mW, increasing to 4 mW for more than 100 Mbit/s throughput. Currently, only very small codes are being used: 24 bit LDPC, (8, 4) Hamming codes or $(16, 11)^2$ product codes.

A novel circuit topology for CMOS analog decoders which eliminates the need for bias voltages and can operate at lower supply voltages (e.g., 0.4 V) than typical Gilbert-style multipliers was described in [59]. The power consumption can be reduced even further using a special class of reversible LDPC codes which can use the decoder architecture for encoding thus implementing a time multiplexed analog codec [60].

11.5.9 De-mapping and Decoding

In the case of higher order modulation, the receiver needs to estimate the individual bit probabilities for each received symbol. These bit probabilities are then sent to the soft input soft output (SISO) outer decoder. Once the whole block was received, the outer and inner SISO decoders start their iterative process. The performance of the decoder depends on how accurate the bit probabilities are calculated. Simulation results for a linear channel and equiprobable constellations have indicated small performance improvements (e.g., 0.3 to 0.6 dB) if the symbol to bit log-likelihood ratio (LLR) estimator module is included in the iterative decoding loop as shown in Fig. 11.21.

Based on this approach, significant higher coding gains can be achieved in nonlinear channels or non-equiprobable constellations as described in the next section.

11.5.10 Performance in Nonlinear Channels

Sec. 11.5.3 introduced the specific issues that have to be considered when turbo codes are used in conjunction with higher order modulations, e.g., 16QAM. This section will focus on the specific non-linear channel that models most popular types of satellite transponders.

Equiprobable Constellations

It is well known that 16QAM and higher order modulations are sensitive to any non-linearity, e.g., the AM-AM and AM-PM transfer characteristics of a solid-state power amplifier (SSPA) in a satellite transponder. This section describes the excellent performance of the SCCC scheme even when the SSPA is operated at as little as 3 dB input back off (IBO) using 16QAM.

The AM-AM and AM-PM characteristics of an SSPA can be defined by two four-parameter algebraic formulas as described in [61]. The model used provided results which are reasonably close to the measurements of a GaAs FET power amplifier over the whole input range. The modelling method is based on a curve fitting technique; it is quite simple and allows very accurate characterization of any type of SSPA or travelling wave tube amplifier (TWTA). For the transmitted rectangular 16QAM constellation considered here, the average energy, E_{av} is 10. The outer constellation points, have energy 18, that is an increase of $10\log(18/10) = 2.55$ dB to the right of the average energy operating point. This means that for an IBO of 3 dB relative to the average energy, E_{av}, the outer constellation points are only 0.45 dB from saturation. The distortion introduced by the SSPA is shown in Fig. 11.22 for IBO of 3 dB. The distorted constellation is scaled back to have the same average energy as the rectangular constellation. This factor is the output back off (OBO) and will be considered later on in the final link budget. Perfect synchronization is assumed in this scenario. Fig. 11.23 shows the performance of the coding scheme only, in a linear channel, function of the number of iterations.

The SCCC rate 1/2 codec used a sliding window of 100 bits, an overlapping block of 4 bits and a channel interleaver of 256 bits. The SCCC interleaver was 8,192 bits. Increasing the interleaver size to 32,768 bits or increasing from 8 to 32 iterations, another 0.4 dB coding gain is expected. Fig. 11.24 shows the performance of the coding scheme function of the IBO after 8 iterations. Simulation results for a BER of 1E-10 show that for an IBO of 10 dB, the penalty with reference to the linear case is only 0.2 dB. Reducing the IBO to 6 dB, the SCCC needs 0.7 dB more E_b/N_o to achieve the same performance as in the linear case. The lower the IBO the higher the required E_b/N_o to achieve the same performance as in the linear case.

The final link budget has to consider the OBO mentioned before which is calculated for different IBO values in Table 11.4. It can be seen that there is an optimum operating point, IBO, for a given transponder characteristic. For this particular SSPA, the optimum operating point is at an IBO = 4 dB.

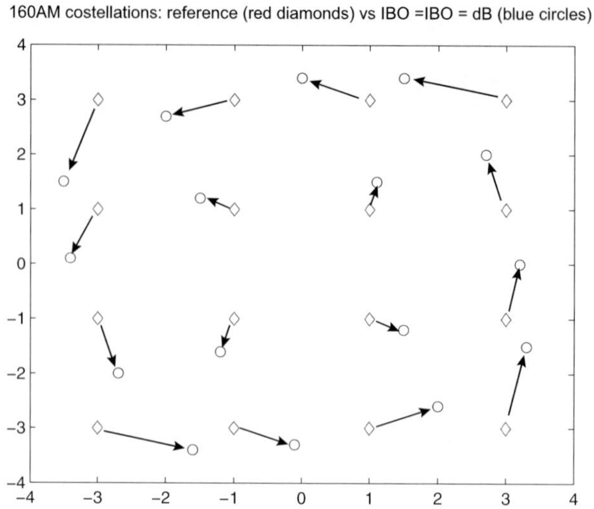

Fig. 11.22. Distortion of rectangular 16QAM constellation at an IBO 3 dB.

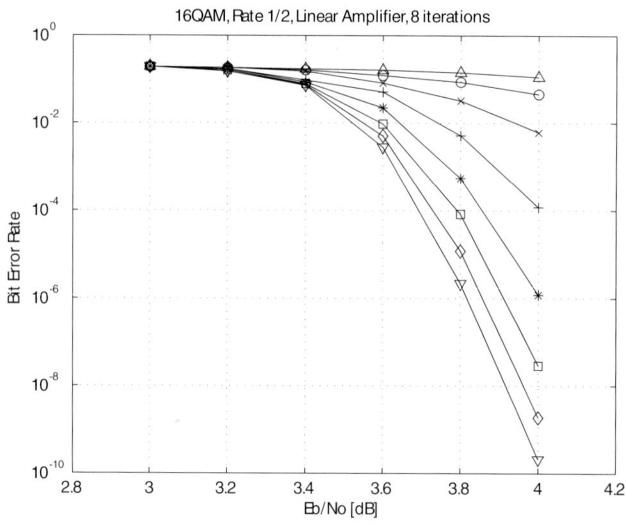

Fig. 11.23. BER in a linear channel after 8 iterations.

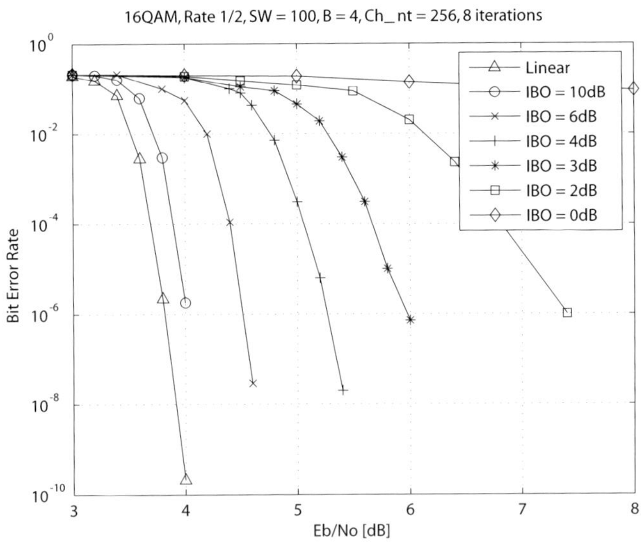

Fig. 11.24. Degradation of BER function of IBO.

In the final link budget that accounts for the OBO loss, the required E_b/N_o is 6.6 dB. Adding 0.2 dB to reach 1E-10 and 0.2 dB due to ACI caused by spectral re-growth, the final Eb/N0 is 7.0 dB, a 2.5 dB increase from the IF loop back result. The value of $E_b/N_o = 7.0$ dB is the common standard value used by rate 1/2 convolutional based QPSK modems to achieve a BER less than 1E-10 over a satellite channel. Hardware measurements using a development hardware platform called the Iterative Satellite Transceiver, (`www.itr.unisa.edu.au/rd/epg/overview.html`), have shown that, for the same data rate, the bandwidth can be halved without any increase in the old link budget used for industry standard rate 1/2 QPSK modems. In real applications, the OBO is changing function of the number of carriers on the transponder. The main advantage offered by the Iterative Satellite Transceiver is that it can identify the current OBO used by the satellite transponder and estimate the nonlinear distortion of the transmitted constellation.

Table 11.4. Optimal operating point.

Required E_b/N_o for BER $= 10^{-6}$							
IBO [dB]	0	2	3	4	5	6	10
OBO [dB]	0	0.77	1	1.36	1.75	2.2	4.4
E_b/N_o for same average energy[dB]	–	7.4	5.9	5.24	4.9	4.5	4.05
Final budget [dB]	–	8.17	6.9	6.6	6.65	6.7	8.45

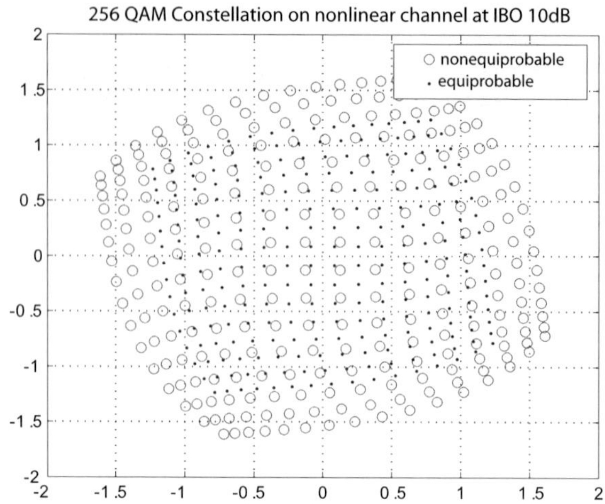

Fig. 11.25. 256QAM in nonlinear channels at IBO of 10 dB.

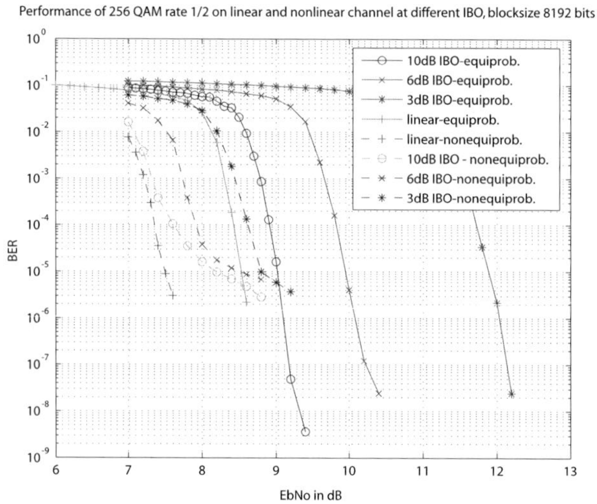

Fig. 11.26. 256QAM BER performance.

Nonequiprobable Constellations

Constellation shaping techniques for Gaussian channels are well known to increase the performance of the communication system. The constellation shaping refers to a Gaussian like distribution of the frequency with which each signal is used, leading to a reduced average energy.

To achieve the shaping gain, there were a number of proposed solutions. Trellis shaping was proposed in [62] where 1 dB gain can be achieved with a simple 4-state shaping code. Another approach proposed in [63] using variable rate codes or divide signal constellation into sub-region with different size as in [64]. The idea of using uniform space constellation with nonuniform distribution of signal point was explored in [65] or in [66], the authors proposed a solution where distance among signal points are varied while probability transmission of each signal point is the same. A recent publication of [67] follows the method presented in [63] and achieved a gain of 0.6 dB in linear channels.

This section applies nonequiprobable constellations to nonlinear channels. Depending on the IBO, a gain of almost 3 dB can be achieved. More details can be found in [68]. The LLRs output from the soft-demapper after being deinterleaved and demultiplexed are input into a normal binary turbo decoder using log-MAP algorithm. The comparison is made for the same spectral efficiency of 4 bits/s/Hz per 2-dimensional symbol. This requires the code rates for the equiprobable and nonequiprobable systems are different. The first scheme uses rate 1/3 turbo codes in conjunction with a 256QAM nonequiprobable constellation. The second scheme uses a rate 1/2 turbo code in conjunction with a 256QAM equiprobable constellation. The distortion of both constellations at an IBO = 10 dB is shown in Fig. 11.25. The performance of the two schemes is shown in Fig. 11.26. A few conclusions can be drawn:

- In a linear channel, the nonequiprobable scheme performs with almost 1 dB better than the equiprobable scheme.
- In the nonlinear channel, the gain of the nonequiprobable scheme increases to 1.5 dB at 10 dB IBO and to 3 dB at 3 dB IBO relative to the equiprobable scheme.
- An obvious flatenning of the BER curve is observed in the case of the nonequiprobable scheme. More research is required to reduce further the error floor in order to take advantage of the significant coding gains.

11.6 A New Turbo Hat?

Turbo-like codes are clearly the best coding technology developed at the end of the 20th century. In conjunction with the latest advances in antenna design, they will be soon used in wearable devices for personal satellite communications. An example is the microstrip antenna designed at the National Institute of Information and Communications Technology, Koanei-shi, Japan. This is a very flexible antenna which can be sewn into clothing and hats and allows communications in S-band with the Engineering Test Satellites (ETSs). It is estimated that it can receive data rates of 512 kbit/s with a low noise amplifier with NF of 1.5 dB and 0.5 dB feed cable loss which would result in a G/T of -19.6 dB/K. It will also be able to transmit data rates up to 7 kbit/s using a power amplifier of 0.25W and feed cable loss of 1 dB which will generate an EIRP of the terminal of -2.52 dBW. The same antenna can be used for GPS receivers thus opening the area of new applications related to health monitoring services.

The ever increasing need for communications anytime and everywhere using smaller and faster devices will rely more and more on turbo-like technologies capable to approach the channel capacity. In the near future satellite communications will no longer require large, bulky dishes but will be achieved with coin size antennas and will become a fashion statement!

This is indeed a wonderful new turbo world!

References

1. Berrou C, Glavieux A, Thitimajshima P (1993) "Near Shannon limit error-correcting coding and decoding: turbo-codes". *ICC 1993*, Geneva, Switzerland 1064–1070
2. Barbulescu SA, Farrell W, Gray P, Rice M (1997) "Bandwidth Efficient Turbo Coding for High Speed Mobile Satellite Communications". *International Symposium on Turbo Codes and Related Topics*, Brest, France 119–126
3. Feldman H, Ramana DV (1999) An "Introduction to Inmarsat's New Mobile Multimedia Service". *The Sixth International Mobile Satellite Conference*, Ottawa 226–229.
4. Trachtman E, Hart T (1999) "Research elements Leading to the Development of Inmarsat's New Mobile Multimedia Services". *The Sixth International Mobile Satellite Conference*, Ottawa 209–212
5. Broughton J, Nemes J (1999) "Multimedia Services for Aeronautical Mobile Satellite Applications". *The Sixth Int. Mobile Satellite Conference*, Ottawa 1–4
6. ETSI PR EN 301 790 (2000) *Digital video broadcasting (DVB): Interaction channel for satellite distribution systems* (V1.1.1, Apr. 2000)
7. ETSI EN 300 421 (1997) *Digital Video Broadcasting (DVB): Framing Structure, Channel Coding and Modulation for 11/12 GHz satellite services.* (V1.1.2, Aug. 1997)
8. ETSI EN 301 210 (1999) Digital *Video Broadcasting (DVB): Framing Structure, Channel Coding and Modulation for Digital Satellite News Gathering (DSNG) and other contribution applications by satellite.*(V1.1.1, Mar. 1999)
9. ETSI EN 300 744 (1997) *Digital video broadcasting (DVB-T); Framing structure, channel coding and modulation for digital terrestrial television.* (V1.1.2, August 1997)
10. Douillard C, Jezequel M, Berrou C, Brengarth (2000) "The Turbo Code Standard for DVB-RCS." *The 2nd International Symposium on Turbo Codes*, Brest, France, 535–538

11. Pyndiah R (1997) "Iterative decoding of product codes". *Proceedings International Symposium on Turbo Codes and Related Topics*, Brest, France, 71–79
12. Lodge J, Young R, Hoeher P, Hagenauer J(1993) "Separable MAP filters for the decoding of product and concatenated codes." *IEEE ICC'93*, Geneva, Switzerland 1740–1745
13. Pyndiah R, Glavieux A, Picart A, Jack S (1994) "Near optimum decoding of product codes." *IEEE GLOBECOM'94*, New Orleans, USA, 1/3:339–343
14. Hagenauer j, Offer E, Papke L (1996) "Iterative decoding of binary block and convolutional codes." *IEEE Transactions on Information Theory*, Vol. 42, 429–445
15. Barbulescu SA (1995) *Iterative Decoding of Turbo Codes and Other Concatenated Codes.* Ph.D. Dissertation, University of South Australia, 23–24
16. Pyndiah R, Glavieux A, Picart A (1995) "Performance of block turbo coded 16-QAM and 64-QAM modulations." *GLOBECOM'95*, Singapore, 1039–1043
17. Lentmaier M (1997) *Soft iterative decoding of generalised low-density parity-check codes based on MAP decoding of component Hamming codes.* Diploma thesis, University of Ulm, Sweden
18. Rice M (1993) *Soft-decision Reed-Solomon decoder.* Workshop on coding for radio channels, Adelaide, Australia
19. Benedetto S, Montorsi G, Divsalar D, Pollara F (1996) "Serial concatenation of interleaved codes: performance analysis, design and iterative decoding." *JPL TDA Progress Report* 42–126
20. Proakis JG (1989) Digital Communications. New York, McGraw-Hill
21. Krachkovsky VY, Lee YX (1998) "Decoding of parallel Reed-Solomon codes with applications tp product and concatenated codes." *1998 IEEE International Symposium onInformation Theory,* MIT, Cambridge, MA, USA 55
22. Ponnampalam V, Vucetic B (1998) "Maximum likelihood decoding of Reed-Solomon codes." *1998 IEEE International Symposium onInformation Theory,* MIT, Cambridge, MA, USA, 368
23. Rice M, Gray P, Barbulescu SA (1997) "Coding and modulation techniques for high speed data services by satellite." *ICICS'97*, Singapore
24. Benedetto S, Montorsi G (2000) "Versatile Bandwidth-efficient Parallel and Serial Turbo-Trellis-Coded Modulation." *The 2nd International Symposium on Turbo Codes*, Brest, France, 201–208
25. Barbulescu SA, Pietrobon SS (1999) "Turbo Codes: A Tutorial on a New Class of Powerful Error Correcting Coding Schemes." *Journal of Electrical and Electronics Engineering,* Australia, Vol.19, No.3, 129–152
26. Robertson P, Woerz T (1996) "A novel bandwidth efficient coding scheme employing turbo codes." *Proceedings ICC'96*, 962–967
27. Divsalar D, Pollara F (1996) "MultipleTurbo Codes for Deep-Space Communications." *TDA Progress Report*, Jet Propulsion Laboratory, California, USA, 42–121
28. Sodha J (2002) "Turbo code frame synchronisation." *Signal Processing Journal (Elsevier),* 82:803–809
29. Xiang W, Pietrobon SS, "Barbulescu SA (2003) Joint source-channel codec: a more powerful efficient technology for image transmission." *3rd International Symposium on Turbo Codes and Related Topics,* Brest, France, 59–62
30. Balakirsky VB (1997) "Joint source-channel coding with variable codes." *Proceedings of IEEE ISIT*, Ulm, Germany

31. Xiang W, Pietrobon SS, Barbulescu SA (2003) "Iterative decoding of JPEG coded images with channel coding." *Proceedings of IEEE International Conference on Telecommunications,* Papeete, Tahiti, 1356–1360
32. Guivarch L, Carlach L, Siohan P (2000) "Joint source-channel soft decoding of Huffman codes with turbo codes." *Proceedings of IEEE Data Compression Conference,* Snowbird, UT
33. Howlander MMK, Woerner BD (2000) "Decoder-assisted frame synchronisation of convolutionally encoded packet transmission." *MILCOM'00*
34. Howlander MMK, Woerner BD (2001) "Decoder-assisted frame synchronisation for turbo coded systems." *Electronics Letters,* 37:6, 362–363
35. Pietrobon SS (1996) "Efficient omplementation of continuous MAP decoders and a synchronisation techniquefor turbo decoders." *ISITA'96*
36. Hall EK, Wilson SG (2001) "Stream Oriented Turbo Codes." *IEEE Transactions on Information Theory,* 47:5
37. Wymeersch H, Moeneclaey M (2003) "Code-aided frame synchronizers for AWGN channels." *3rd International Symposium on Turbo Codes and Related Topics,* Brest, France, 243–246
38. Noels N, Steendam H, Moeneclaey M (2003) "Carrier Phase Recovery in turbo receivers: Cramer-Rao bound and synchronizer performance." *3rd International Symposium on Turbo Codes and Related Topics,* Brest, France, 323–326
39. Sutskover I, Shamai S, Ziv J (2003) "A novel approach to iterative joint decoding and phase estimation." *3rd International Symposium on Turbo Codes and Related Topics,* Brest, France, 83–86
40. Zhang L, Burr A (2003) "Application of turbo principle to carrier phase recovery in turbo encoded bit-interleaved coded modulation system." *3rd International Symposium on Turbo Codes and Related Topics,* Brest, France, 87–90
41. Kang JH, Stark WE (1998) "Turbo codes for noncoherent FH-SS with partial band interference." *IEEE Transactions on Communications,* Vol. 46, 1451–1458
42. Frank C, Pursley M (1996) "Concatenated coding for frequency-hop spread-spectrum with partial band interference." *IEEE Transactions on Communications,* Vol. 44, 377–387
43. El Gamal H, Geraniotis E (1998) "Turbo codes with channel estimation and dynamic power allocation for anti jam FH/SSMA." *MILCOM'98*
44. Grass JH, Curry PJ, Langford CJ (1999) "An application to turbo trellis coded modulation to tactical communications." *MILCOM'99*
45. Chu MJ, Stark WE (1998) "Asymptotic performance of a coded communication system with orthogonal signalling in partial band jamming." *MILCOM'98*
46. Jordan MA (1998) "Turbo code performance in partial band jamming." *MILCOM'98*
47. Kang JH, Stark WE (1998) "Performance of turbo coded FH-SS with partial band interference and Rayleigh fading." *MILCOM'98*
48. Levannier G, Bailly H (1999) "Comparison of convolutional codes against turbo codes for a frequency hopped SDPSK receiver in partial band jamming." *MILCOM'99*
49. Frey DR (1993) "Chaotic digital encoding: an approach to secure communications." *IEEE Transactions on Circuit Systems 40:10,* 660–666
50. Barbulescu SA, Guidi A, Pietrobon S (2000) "Chaotic Turbo Codes". *IEEE International Symposium on Information Theory,* Sorrento, Italy

51. Loeliger HA, Helfenstein M, Lustenberger F, Tarkoy F (1998) "Probability Propagation and Decoding in Analog VLSI." *IEEE International Symposium on Information Theory,* Cambridge, MA, 146
52. Hagenauer J, Winklhofer M (1998) "The Analog Decoder." *IEEE International Symposium on Information Theory,* Cambridge, MA, 145
53. Lustenberger F, Helfenstein M, Loeliger HA, Tarkoy F, Moschytz GS (1999) "An Analog VLSI Decoding Technique for Digital Codes." *IEEE International Symposium on Circuits and Systems,* II-424-II-427
54. Hagenauer J, Moerz M, Offer E (2000) "Analog Turbo-Networks in VLSI: The Next Step in Turbo Decoding and Equalization." *The 2nd International Symposium on Turbo Codes,* Brest, France, 209–218
55. Moerz M, Gabara T, Yan R, Hagenauer J (2000) "An Analog 0.25mm BiCMOS Tailbiting MAP Decoder." *2000 IEEE International Solid-State Circuits Conference,* 356–357
56. Moerz M, Schaefer A, Offer E, Hagenauer J (2001) "Analog decoders for high rate convolutional codes." *Information Theory Workshop,* Cairns, Australia, 128–130
57. Winstead C, Myers C, Schlegel C, Harrison R (2001) "Analog decoding of product codes." *Information Theory Workshop,* Cairns, Australia, 131–133
58. Mead C (1989) *Analog VLSI and Neural Systems.* Addison-Wesley
59. Winstead C, Nguyen N, Gaudet VC, Schlegel C (2003) "Low-voltage CMOS circuits for analog decoders." *3rd International Symposium on Turbo Codes and Related Topics,* Brest, France, 271–274
60. Haley D, Winstead C, Grant A, Schlegel C (2003) "An analog LDPC codec core." *3rd International Symposium on Turbo Codes and Related Topics,* Brest, France, 391–394
61. Saleh AAM (1981) "Frequency independent and frequency dependent nonlinear models of TWT amplifiers." *IEEE Transactions in Communications,* COM-29:11, 1715–1720
62. Forney G (1992) "Trellis Shaping." *IEEE Transaction on Information Theory,* IT-38:281–300
63. Kschischang F, Pasupathy S (1993) "Optimal Nonuniform Signalling for Gaussian Channels." *IEEE Transactions on Information Theory,* 39:3
64. Livingston J (1992) "Shaping Using Variable-Size Regions." *IEEE Transactions on Information Theory,* 38:4
65. Calderbank A, Orazou L (1990) "Nonequiprobable Signalling on Gaussian Channels." *IEEE Transactions on Information Theory,* 36:4
66. Fragauli C, Wesel RD, Sommer D, Fettweis GP (2001) "Turbo Codes with Nonuniform Constellation." *IEEE International Conference on Communication,* 1:70–73
67. Raphaeli D, Gurevitz A (2004) "Constellation Shaping for Pragmatic Turbo-Coded Modulation With High Spectral Efficiency." *IEEE Transaction on Communications,* 52:3
68. Ngo N, Barbulescu SA, Pietrobon SS (2005) "Performance of Nonuniform M-ary QAM constellation on nonlinear channels." *Australian Communications Theory Workshop*

Chapter 12

Turbo and LDPC Codes for Digital Video Broadcasting

Matthew C. Valenti, Shi Cheng, and Rohit Iyer Seshadri

West Virginia University, USA

The Digital Video Broadcasting (DVB) Project was founded in 1993 by the European Telecommunications Standards Institute (ETSI) with the goal of standardizing digital television services. Its initial standard for satellite delivery of digital television, dubbed DVB-S, used a concatenation of an outer (204,188) byte shortened Reed Solomon code and an inner constraint length 7, variable rate (r ranges from 1/2 to 7/8) convolutional code [1].

The same infrastructure used to deliver television via satellite can also be used to deliver Internet and data services to the subscriber. Internet over DVB-S is a natural competitor against cable modem and DSL technology, and its universal coverage allows even the most remote areas to be served. Because DVB-S only provides a downlink, an uplink is also needed to enable interactive applications such as web browsing. The uplink and downlink need not be symmetric, since many Internet services require a faster downlink.

One alternative for the uplink is to use a telephone modem, but this does not allow for always-on service, has modest data rates, and can be costly in remote areas. A more attractive alternative is for the subscriber equipment to transmit an uplink signal back to the satellite over the same antenna used for receiving the downlink signal. However, given the small antenna aperture and requirement for a low-cost, low-power amplifier, there is very little margin on the uplink. Therefore, strong FEC coding is desired. For this reason, the DVB Project has adopted turbo codes for the satellite return channel in its DVB-RCS (Return Channel via Satellite) standard [2].

At the same time that the DVB Project was developing turbo coding technology for the return channel, it was updating the downlink with modern coding technology. The latest standard, called DVB-S2, replaces the concatenated Reed-Solomon/convolutional coding approach of DVB-S with a concatenation of an outer BCH code and inner low density parity check (LDPC) code [3]. The result is a 30% increase in capacity over DVB-S. In this chapter, the coding strategies used by both DVB-RCS and DVB-S2 are discussed.

K. Sripmanwat (ed.), Turbo Code Applications: a journey from a paper to realization, 301–319.

12.1 DVB-RCS

The DVB-RCS turbo code was optimized for short frame sizes and high data rates. Twelve frame sizes are supported raging from 12 bytes to 216 bytes, including a 53 byte frame compatible with ATM and a 188 byte frame compatible with both MPEG-2 and the original DVB-S standard. The return link supports data rates from 144 kbps to 2 Mbps and is shared among terminals by using muti-frequency time-division multiple-access (MF-TDMA) and demand-assigned multiple-access (DAMA) techniques. Eight code rates are supported, ranging from $r = 1/3$ to $r = 6/7$.

Like the turbo codes used in other standards, a pair of constituent RSC encoders is used along with log-MAP or max-log-MAP decoding [4]. The decoder for each constituent code performs best if the encoder begins and ends in a known state, such as the all-zeros state. This can be accomplished by independently terminating the trellis of each encoder with a tail which forces the encoder back to the all-zeros state. However, for the small frame lengths supported by DVB-RCS, such a tail imposes a non-negligible reduction in code rate and is therefore undesirable. As an alternative to terminating the trellis of the code, DVB-RCS uses circular recursive systematic convolutional (CRSC) encoding [5], which is based on the concept of *tailbiting* [6]. CRSC codes do not use tails, but rather are encoded in such a way that the ending state matches the starting state.

Most turbo codes use binary encoders defined over GF(2). However, to facilitate faster decoding in hardware, the DVB-RCS code uses *duobinary* constituent encoders defined over GF(4) [7]. During each clock cycle, the encoder takes in two data bits and outputs two parity bits so that, when the systematic bits are included, the code rate is $r = 2/4$. In order to avoid parallel transitions in the code trellis, the memory of the encoder must exceed the number of input bits, and so DVB-RCS uses constituent encoders with memory three (a constraint length of four).

There are several benefits to using duobinary encoders. First, the trellis contains half as many states as a binary code of identical constraint length (but the same number of edges) and therefore needs half as much memory and the decoding hardware can be clocked at half the rate as a binary code. Second, the duobinary code can be decoded with the suboptimal but efficient max-log-MAP algorithm at a cost of only about 0.1-0.2 dB relative to the optimal log-MAP algorithm. This is in contrast with binary codes, which lose about 0.3-0.4 dB when decoded with the max-log-MAP algorithm [8]. Additionally, duobinary codes are less impacted by the uncertainty of the starting and ending states when using tailbiting and perform better than their binary counterparts when punctured to higher rates.

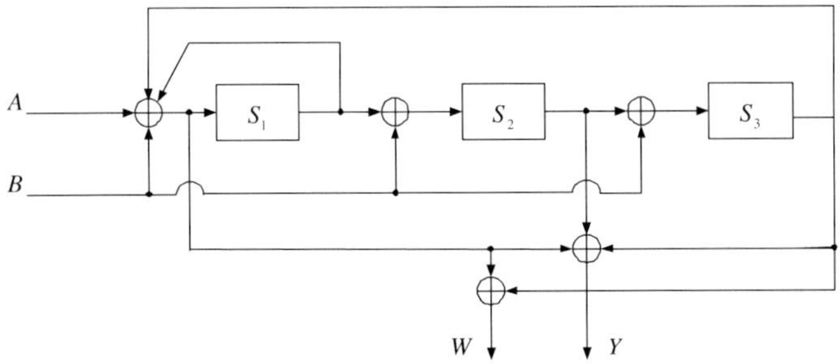

Fig. 12.1. Duobinary CRSC constituent encoder used by DVB-RCS.

12.1.1 Encoding

The CRSC constituent encoder used by DVB-RCS is shown in Fig. 12.1. The encoder is fed blocks of k message bits which are grouped into $N = k/2$ *couples*. The number of couples per block can be $N \in \{48, 64, 212, 220, 228, 424, 432, 440, \ 752, 848, 856, 864\}$. The number of bytes per block is $N/4$. In Fig. 12.1, A represents the the first bit of the couple, and B represents the second bit. The two parity bits are denoted W and Y. For ease of exposition, subscripts are left off the figure, but below a single subscript is used to denote the time index $k \in \{0, ..., N-1\}$ and an optional second index is used on the parity bits W and Y to indicate which of the two constituent encoders produced them.

Let the vector $\mathbf{S}_k = [S_{k,1} \ S_{k,2} \ S_{k,3}]^T$, $S_{k,m} \in \{0, 1\}$ denote the state of the encoder at time k. Note that although the inputs and outputs of the encoder are defined over GF(4), only binary values are stored within the shift register and thus the encoder has just eight states. The encoder state at time k is related to the state at time $k-1$ by

$$\mathbf{S}_{k+1} = \mathbf{G}\mathbf{S}_k + \mathbf{X}_k \tag{12.1}$$

where

$$\mathbf{X}_k = \begin{bmatrix} A_k + B_k \\ B_k \\ B_k \end{bmatrix} \tag{12.2}$$

and

$$\mathbf{G} = \begin{bmatrix} 1 \ 0 \ 1 \\ 1 \ 0 \ 0 \\ 0 \ 1 \ 0 \end{bmatrix} \tag{12.3}$$

Because of the tailbiting nature of the code, the block must be encoded twice by each constituent encoder. During the first pass at encoding, the encoder is initialized to the all-zeros state, $\mathbf{S}_0 = [0\ 0\ 0]^T$. After the block is encoded, the final state of the encoder $\mathbf{S}_N$ is used to derive the *circulation state*

$$\mathbf{S}_c = \left(\mathbf{I} + \mathbf{G}^N\right)^{-1} \mathbf{S}_N \tag{12.4}$$

Where the above operations are over GF(2). Note that the matrix $\mathbf{I} + \mathbf{G}^N$ is not invertible if N is a multiple of the period of the encoder's impulse response (which is seven for this encoder). However, this is not a problem because none of the permitted values of N are multiples of seven. In practice, the circulation state $\mathbf{S}_c$ can be found from $\mathbf{S}_N$ by using a lookup table (which is specified in the standard). Once the circulation state is found, the data is encoded again. This time, the encoder is set to start in state $\mathbf{S}_c$ and will be guaranteed to also end in state $\mathbf{S}_c$.

The first encoder operates on the data in its natural order, yielding parity couples $\{W_{k,1}, Y_{k,1}\}$. The second encoder operates on the data after it has been interleaved. Interleaving is performed on two levels. First, interleaving is performed within the couples, and second, interleaving is performed between couples. Let $\{A'_k, B'_k\}$ denote the sequence after the first level of interleaving and $\{A''_k, B''_k\}$ denote the sequence after the second level of interleaving. In the first level of interleaving, every other couple is reversed in order, i.e. $(A'_k, B'_k) = (B_k, A_k)$ if k is even, otherwise $(A'_k, B'_k) = (A_k, B_k)$. In the second level of interleaving, couples are permuted in a pseudorandom fashion. The exact details of the second level permutation can be found in the standard [2].

After the two levels of interleaving, the second encoder (which is identical to the first) encodes the sequence $\{A''_k, B''_k\}$ to produce the sequence of parity couples $\{W_{k,2}, Y_{k,2}\}$. As with the first encoder, two passes of encoding must be performed, and the second encoder will have its own independent circulation state. To create a rate $r = 1/3$ turbo code, a codeword is formed by first transmitting all the uninterleaved data couples $\{A_k, B_k\}$, then transmitting $\{Y_{k,1}, Y_{k,2}\}$ and finally transmitting $\{W_{k,1}, W_{k,2}\}$. The bits are transmitted using QPSK modulation, so there is a one-to-one correspondence between couples and QPSK symbols. Alternatively, the code word can be transmitted by exchanging the parity and systematic bits, i.e. $\{Y_{k,1}, Y_{k,2}\}$, followed by $\{W_{k,1}, W_{k,2}\}$ and finally $\{A_k, B_k\}$.

Code rates higher than $r = 1/3$ are supported through the puncturing of parity bits. To achieve $r = 2/5$, both encoders maintain all the Y_k but delete odd-indexed W_k. For rate 1/2 and above, the encoders delete all W_k. For rate $r = 1/2$, all the Y_k bits are maintained, while for rate $r = 2/3$ only the even-indexed Y_k are maintained, and for rate $r = 4/5$ only every fourth Y_k is maintained. Rates $r = 3/4$ and 6/7 maintain every third and sixth Y_k respectively, but are only exact rates if N is a multiple of three (otherwise the rates are slightly lower).

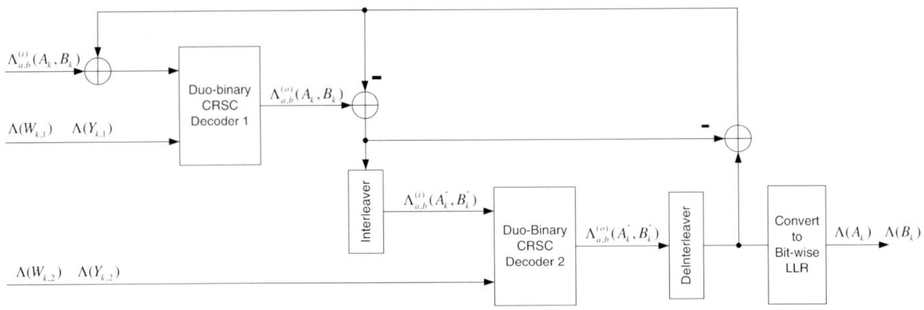

Fig. 12.2. A decoder for the DVB-RCS code.

12.1.2 Decoding

Decoding of the DVB-RCS code is complicated by the fact that the constituent codes are duobinary and circular. As with conventional turbo codes, decoding involves the iterative exchange of extrinsic information between the two component decoders. While decoding can be performed in the probability domain, the log-domain is preferred since the low complexity max-log-MAP algorithm can then be applied [4]. Unlike the decoder for a binary turbo code, which can represent each binary symbol as a single log-likelihood ratio, the decoder for a duobinary code requires *three* log-likelihood ratios. For example, the likelihood ratios for message couple (A_k, B_k) can be represented in the form

$$\Lambda_{a,b}(A_k, B_k) = \log \frac{P(A_k = a, B_k = b)}{P(A_k = 0, B_k = 0)} \tag{12.5}$$

where (a, b) can be $(0, 1)$, $(1, 0)$, or $(1, 1)$.

An iterative decoder that can be used to decode the DVB-RCS turbo code is shown in Fig. 12.2. The goal of each of the two constituent decoders is to update the set of log-likelihood ratios associated with each message couple. In the figure and in the following discussion, $\{\Lambda_{a,b}^{(i)}(A_k, B_k)\}$ denotes the set of LLRs corresponding to the message couple at the input of the decoder and $\{\Lambda_{a,b}^{(o)}(A_k, B_k)\}$ is the set of LLRs at the output of the decoder. Each decoder is provided with $\{\Lambda_{a,b}^{(i)}(A_k, B_k)\}$ along with the received values of the parity bits generated by the corresponding encoder (in LLR form). Using these inputs and knowledge of the code constraints, it is able to produce the updated LLRs $\{\Lambda_{a,b}^{(o)}(A_k, B_k)\}$ at its output.

As with binary turbo codes, extrinsic information is passed to the other constituent decoder instead of the raw LLRs. This prevents the positive feedback of previously resolved information. Extrinsic information is found by simply subtracting the appropriate input LLR from each output LLR, as indicated in Fig. 12.2.

The extrinsic information that is passed between the two decoders must be interleaved or deinterleaved so that it is in the proper sequence at the

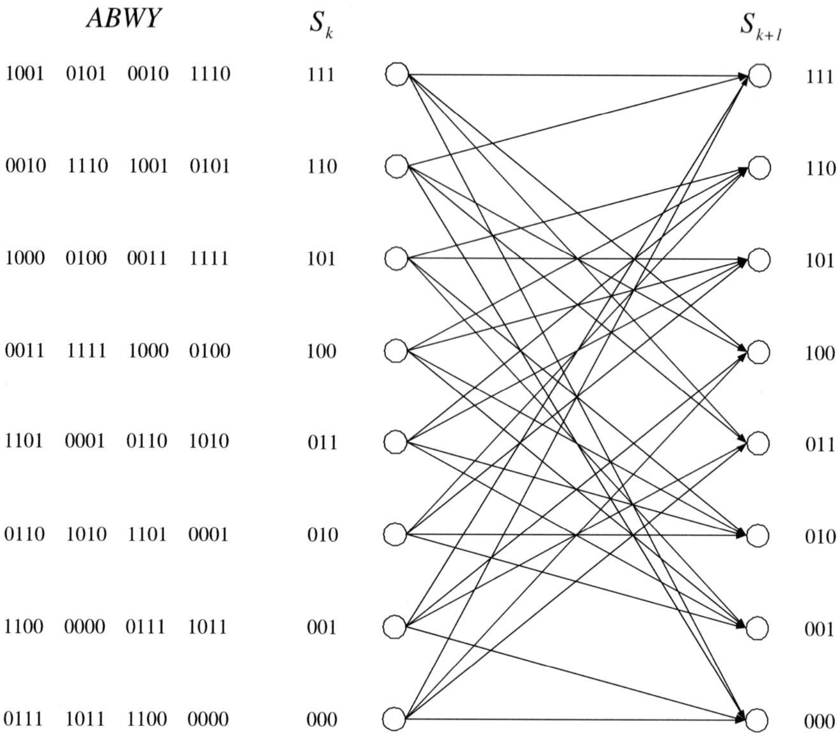

Fig. 12.3. Trellis associated with the duobinary CRSC constituent encoder used by DVB-RCS. The numbers on the left indicate the labels (A, B, W, Y) of the branches exiting each state. From left to right, the groups of numbers correspond to the exiting branches from top to bottom.

input of the other decoder. Interleaving and deinterleaving between the two constituent decoders must be done on a symbol-wise basis by assuring that the three likelihood ratios $\{\Lambda_{0,1}(A_k, B_k), \Lambda_{1,0}(A_k, B_k), \Lambda_{1,1}(A_k, B_k)\}$ belonging to the same couple are not separated.

The trellis for the duobinary constituent code is as shown in Fig. 12.3. The trellis contains eight states, with four branches entering and exiting each state. Note that this is in contrast with a conventional binary code which only has a pair of branches entering and exiting each state. The trellis contains two 4 by 4 butterflies, and because these two butterflies are independent, they can be processed in parallel. In the following, the ith state is denoted by $\mathbf{S}_i$ where $i \in \{0, ..., 7\}$ for DVB-RCS. Note that the subscript i takes on a slightly different connotation depending on whether encoding or decoding is being discussed. When discussing encoding, the subscript was used to indicate a time step, but when discussing decoding the subscript indicates a particular state.

The extension of the log-MAP and max-log-MAP algorithms [4] to the duobinary case is fairly straightforward. Each branch must be labeled with the log-likelihood ratios corresponding to the systematic and parity couples associated with that branch. Because QPSK modulation is orthogonal, the LLR of message couple (A, B) can be initialized prior to being fed into the first decoder as $\Lambda^{(i)}_{a,b}(A_k, B_k) = a\Lambda(A_k) + b\Lambda(B_k)$, where $\Lambda(C) = \log[P(C = 1)/P(C = 0)]$. Because extrinsic information about the parity bits is not exchanged, the parity bits can always be decomposed in a similar manner. However, for the systematic bits, the three likelihood ratios defined in Eq. 12.5 must be calculated during each iteration and exchanged between the decoders.

Let $\gamma_k(\mathbf{S}_i \rightarrow \mathbf{S}_j)$ denote the branch metric corresponding to state transition $\mathbf{S}_i \rightarrow \mathbf{S}_j$ at time k. The branch metric depends on the message and parity couples that label the branch along with the channel observation and extrinsic information at the decoder input. In particular, if transition $\mathbf{S}_i \rightarrow \mathbf{S}_j$ is labelled by $(A_k, B_k, W_k, Y_k) = (a, b, w, y)$ then

$$\gamma_k(\mathbf{S}_i \rightarrow \mathbf{S}_j) = \Lambda^{(i)}_{a,b}(A_k, B_k) + w\Lambda(W_k) + y\Lambda(Y_k) \tag{12.6}$$

As with binary codes, the constituent decoder must perform a forward and a backward recursion. Let $\alpha_k(\mathbf{S}_i)$ denote the normalized forward metric at trellis stage k and state $\mathbf{S}_i$, while $\alpha'_{k+1}(\mathbf{S}_j)$ is the forward metric at trellis stage $k + 1$ and state $\mathbf{S}_j$ prior to normalization. The forward recursion is

$$\alpha'_{k+1}(\mathbf{S}_j) = \max_{\mathbf{S}_i \rightarrow \mathbf{S}_j} * \left\{\alpha_k(\mathbf{S}_i) + \gamma_k(\mathbf{S}_i \rightarrow \mathbf{S}_j)\right\} \tag{12.7}$$

where the max$*$ operation[1] is performed over the four branches $\mathbf{S}_i \rightarrow \mathbf{S}_j$ leading into state $\mathbf{S}_j$ at time $k + 1$. While the log-MAP algorithm uses the exact definition of max$*$, the max-log-MAP algorithm uses the approximation $\max*(x, y) \approx \max(x, y)$.

After computing $\alpha'_{k+1}(\mathbf{S}_j)$ for all $\mathbf{S}_j$ at time $k+1$, the forward metrics are normalized with respect to the metric stored in state zero

$$\alpha_{k+1}(\mathbf{S}_j) = \alpha'_{k+1}(\mathbf{S}_j) - \alpha'_{k+1}(\mathbf{S}_0) \tag{12.8}$$

Similarly, let $\beta_{k+1}(\mathbf{S}_j)$ denote the normalized backward metric at trellis state $k+1$ and state $\mathbf{S}_j$ and $\beta'_k(\mathbf{S}_i)$ denote the backward metric at trellis state k and state $\mathbf{S}_i$ prior to normalization. The backward recursion is

$$\beta'_k(\mathbf{S}_i) = \max_{\mathbf{S}_i \rightarrow \mathbf{S}_j} * \left\{\beta_{k+1}(\mathbf{S}_j) + \gamma_k(\mathbf{S}_i \rightarrow \mathbf{S}_j)\right\} \tag{12.9}$$

where max$*$ is over the four branches $\mathbf{S}_i \rightarrow \mathbf{S}_j$ exiting state $\mathbf{S}_i$ at time k. As with α, the β's are normalized with respect to the metric stored in state zero

[1] The max$*$ operation is defined in [9] as $\max*(x, y) = \max(x, y) + \log(1 + e^{-|x-y|})$. Multiple arguments imply a recursion of pairwise operations, i.e. $\max*(x, y, z) = \max*(x, \max*(y, z))$.

$$\beta_k(\mathbf{S}_i) = \beta'_k(\mathbf{S}_i) - \beta'_k(\mathbf{S}_0) \tag{12.10}$$

Because the encoders are circular, special care must be taken to initialize the forward and backward recursions. Since the starting and stopping states are identical, the code trellis can be visualized as a cylinder (see, for example, Fig. 1 in [10]). The forward recursion can be interpreted as going around the cylinder in the clockwise direction and the backward recursion as going around the cylinder in the counter-clockwise direction.

Several algorithms are presented in [10] for decoding circular/tailbiting convolutional codes. The most practical algorithm, called algorithm A3 in [10], begins by initializing the decoder so that all initial states are equally likely. The forward recursion is initialized so that $\alpha_0(\mathbf{S}_j) = 0, \forall \mathbf{S}_j$. The forward recursion then cycles through the entire trellis in the clockwise direction. If the encoder was terminated in a known state, then the forward recursion could halt once it reaches the end of the trellis. However, since the starting and ending states are not known, the forward recursion continues around the cylinder a second time. During the second cycle, the new value of the α_k's are compared against the same value computed during the first cycle, and the new α_k's are used to replace the ones from the last cycle. Once all the α_k's are close to the value from the last cycle, the forward recursion halts. The number of extra trellis sections beyond the first cycle around the cylinder is called the *wrap depth*. For long frames, the wrap depth is typically smaller than the frame length (so an entire second cycle does not need to be run). However, for short frames, a third or fourth cycle around the trellis cylinder could be required, i.e. the wrap depth could exceed N.

The backward recursion is executed in similar manner, with $\beta_N(\mathbf{S}_i) = 0, \forall \mathbf{S}_i$ and the decoder cycling around the cylinder in the counter-clockwise direction. After making one lap around the cylinder, the algorithm continues until the β_k's closely match the values computed during the previous lap.

After the forward and backward recursions have been completed, a full set of $\{\alpha_k\}$ and $\{\beta_k\}$ metrics will be stored in memory. The next step is for the decoder to use these metrics to compute the LLRs given by Eq. 12.5. This is accomplished by first computing the likelihood of each branch

$$Z_k(\mathbf{S}_i \to \mathbf{S}_j) = \alpha_k(\mathbf{S}_i) + \gamma_k(\mathbf{S}_i \to \mathbf{S}_j) + \beta_{k+1}(\mathbf{S}_j) \tag{12.11}$$

Next, the likelihood that message pair $(A_k, B_k) = (a, b)$ is calculated using

$$t_k(a, b) = \max_{\mathbf{S}_i \to \mathbf{S}_j : (a,b)}{*} \{Z_k\} \tag{12.12}$$

where the max* operator is over the eight branches labelled by message couple (a, b). Finally, the LLR at the output of the decoder is found as

$$\Lambda^{(o)}_{a,b}(A_k, B_k) = t_k(a, b) - t_k(0, 0) \tag{12.13}$$

where $(a, b) \in \{(0, 1), (1, 0), (1, 1)\}$.

After the turbo decoder has completed a fixed number of iterations or met some other convergence criterion, a final decision on the bits must be made. This is accomplished by computing the LLR of each bit in the couple (A_k, B_k) according to

$$\begin{aligned}\Lambda(A_k) &= \max* \left\{ \Lambda_{1,0}^{(o)}(A_k, B_k), \Lambda_{1,1}^{(o)}(A_k, B_k) \right\} \\ &\quad - \max* \left\{ \Lambda_{0,0}^{(o)}(A_k, B_k), \Lambda_{0,1}^{(o)}(A_k, B_k) \right\} \\ \Lambda(B_k) &= \max* \left\{ \Lambda_{0,1}^{(o)}(A_k, B_k), \Lambda_{1,1}^{(o)}(A_k, B_k) \right\} \\ &\quad - \max* \left\{ \Lambda_{0,0}^{(o)}(A_k, B_k), \Lambda_{1,0}^{(o)}(A_k, B_k) \right\}, \end{aligned} \tag{12.14}$$

where $\Lambda_{0,0}^{(o)}(A_k, B_k) = 0$. The hard bit decisions can be found by comparing each of these likelihood ratios to a threshold.

12.1.3 Simulation Results

In this section, simulation results are presented that illustrate the performance of the DVB-RCS turbo code. Fig. 12.4 shows the frame error rate (FER) of several decoding algorithms when using blocks of $N = 212$ message couples (53 bytes) and code rate $r = 1/3$. One problem with using circular constituent codes is that the circulation state is unknown at the decoder. The two lowermost curves in Fig. 12.4 show the impact of the unknown circulation state when using eight iterations of log-MAP decoding. The lowermost curve was created by using a genie-aided decoder that knows the exact circulation state of the encoder. While this decoder is not feasible in practice, it serves as a bound for more practical decoders. The second curve shows the performance when the circulation state is not known and algorithm A3 from [10] is used to compensate for the unknown circulation state. Note that the loss due to imperfect knowledge of the circulation state is only about 0.02 dB at a FER of 10^{-4}.

The uppermost curve in Fig. 12.4 shows the performance when using eight iterations of the max-log-MAP algorithm along with algorithm A3 from [10] to handle the unknown circulation state. At a frame error rate (FER) of 10^{-4}, the loss due to using max-log-MAP is only about 0.16 dB. This is in contrast with the $0.3 - 0.4$ dB losses that are incurred when decoding binary turbo codes with the max-log-MAP algorithm, and for this reason many DVB-RCS decoder implementations use max-log-MAP [11].

Fig. 12.5 shows the influence of the block size. Frame error rate results are shown for blocks of $N = \{48, 64, 212, 432, 752\}$ message couples, or correspondingly $\{12, 16, 53, 108, 188\}$ bytes. In each case, the code rate is $r = 1/3$, the circulation state is unknown at the decoder, and eight iterations of max-log-MAP decoding are performed. The SNR required to achieve a FER of 10^{-4} is $\mathcal{E}_b/N_o = \{3.02, 2.77, 1.86, 1.65, 1.44\}$ dB for $N = \{48, 64, 212, 432, 752\}$, respectively.

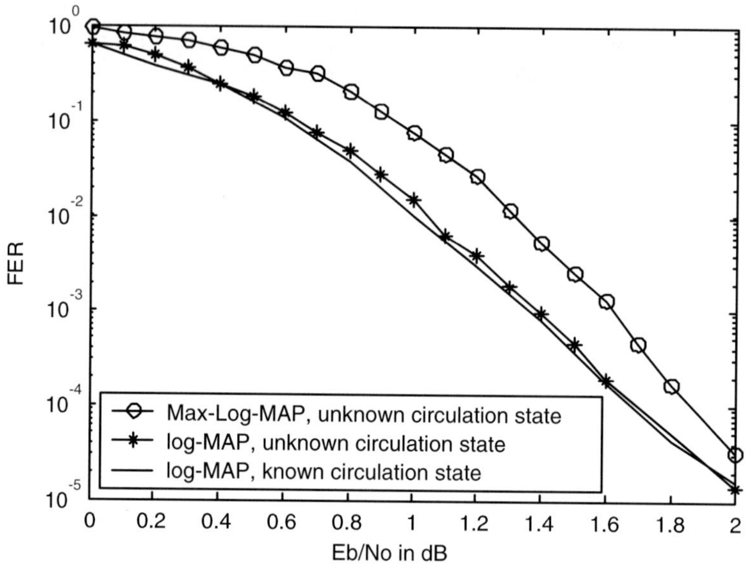

Fig. 12.4. Influence of decoding algorithm on the performance of the DVB-RCS turbo code. The frame length is $K = 212$ (53 bytes) and code rate is $r = 1/3$. Eight iterations of decoding are performed. The curve with the best performance shows the performance of the log-MAP algorithm if the circulation state used by the encoder were to be known by the decoder. The other two curves show the performance of log-MAP and max-log-MAP decoding when the decoder does not know the circulation state and uses algorithm A3 from [10].

Fig. 12.6 shows the influence of the code rate. Frame error rate results are shown for all seven code rates when the block size is $N = 212$ message couples. As with Fig. 12.5, eight iterations of max-log-MAP decoding are performed and the circulation state is unknown at the decoder. The SNR required to achieve a FER of 10^{-4} is $\mathcal{E}_b/N_o = \{1.86, 2.03, 2.37, 3.29, 3.96, 4.57, 5.21\}$ dB for $r = \{1/3, 2/5, 1/2, 2/3, 3/4, 4/5, 6/7\}$, respectively.

12.2 DVB-S2

The DVB-S2 standard uses a serial concatenation of two binary linear codes: an outer BCH code and an inner low density parity check (LDPC) code [3]. With binary linear block codes, a k bit message $\mathbf{d}$ is encoded into a n bit codeword $\mathbf{c}$ according to

$$\mathbf{c} = \mathbf{dG}, \tag{12.15}$$

where $\mathbf{G}$ is the k by n *generator matrix* and the matrix multiplication is modulo-2. The *parity-check matrix* $\mathbf{H}$ is a $m = n - k$ by n matrix that spans the null space of $\mathbf{G}$ and therefore satisfies $\mathbf{GH}^T = \mathbf{0}$.

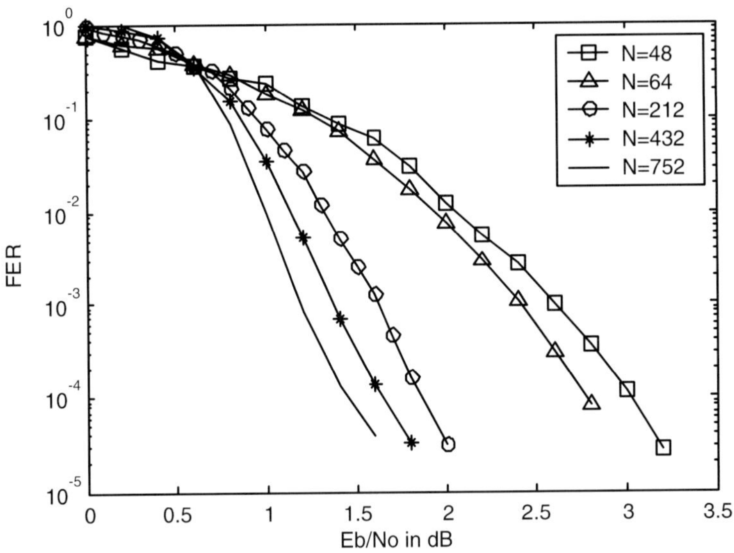

Fig. 12.5. Influence of block size on the performance of the DVB-RCS turbo code. The code rate is $r = 1/3$, block size is N message couples, and eight iterations of max-log-MAP decoding are performed.

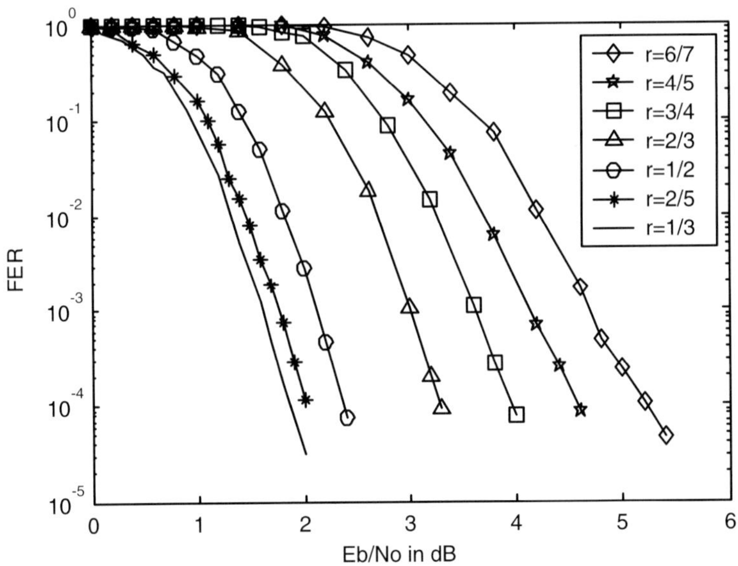

Fig. 12.6. Influence of code rate on the performance of the DVB-RCS turbo code. The code rate is r, block size is $N = 212$ message couples, and eight iterations of max-log-MAP decoding are performed.

Table 12.1. The input k' and output n' word sizes of the outer BCH code used by normal DVB-S2 frames. Also listed is the error correcting capability t of the BCH code.

rate	k'	n'	t
9/10	58192	58320	8
8/9	57472	57600	8
5/6	53840	54000	10
4/5	51648	51840	12
3/4	48408	48600	12
2/3	43040	43200	12
3/5	38688	38880	12
1/2	32208	32400	12
2/5	25728	25920	12
1/3	21408	21600	12
1/4	16008	16200	12

LDPC codes are characterized by having very sparse parity check matrices. An LDPC code is said to be *regular* if all rows in **H** have the same weight (number of ones) and all columns in **H** have the same weight; otherwise the code is *irregular*. LDPC codes were originally proposed by Gallager in his 1960 dissertation [12] along with an iterative process for decoding. Although Gallager proved that the codes were good in theory, they were largely ignored until the advent of turbo codes because the decoder was thought to be too complex. However, after turbo codes showed the practicality of iterative decoding, interest in LDPC codes was soon renewed. In the mid-1990's, MacKay rediscovered LDPC codes [13, 14] and showed that they are capable of approaching the Shannon limit. Soon afterwards, Richardson and Urbanke [15] and Luby et al [16] showed that long irregular LDPC codes can be superior to turbo codes of the same length and can approach the Shannon capacity by a fraction of a decibel [17].

12.2.1 Encoding

The DVB-S2 channel encoder begins by first encoding a length k' binary message into a n' bit systematic BCH codeword. The $k = n'$ BCH codeword is then encoded into a n bit systematic LDPC codeword. The codeword length n can be either 64,800 or 16,200 bits long, producing *normal* and *short* frames, respectively. Note that unlike DVB-RCS, which fixes the value of the encoder *input*, DVB-S2 fixes the length of the encoder *output*. Because of this, the length k' of the input to the BCH encoder and the length k of the input to the LDPC encoder (which equals the length n' of the output of the BCH encoder) are variable and depend on the rate r of the LDPC code. Normal frames can be encoded at eleven different code rates, as shown in Table 12.1. Short frames can be encoded at all the same code rates except for rate $r = 9/10$, which is not supported, as shown in Table 12.2. For short frames, the "rates"

Table 12.2. The input k' and output n' word sizes of the outer BCH code used by short DVB-S2 frames. Each of these BCH codes has error correcting capability $t = 12$. Also listed is the actual rate $r = k/16200$ of the inner LDPC code.

rate	k'	n'	actual r
8/9	14232	14400	8/9
5/6	13152	13320	37/45
4/5	12432	12600	7/9
3/4	11712	11880	11/15
2/3	10632	10800	2/3
3/5	9552	9720	3/5
1/2	7032	7200	4/9
2/5	6312	6480	2/5
1/3	5232	5400	1/3
1/4	3072	3240	1/5

$1/4, 1/2, 3/4, 4/5$, and $5/6$ are misnomers because the actual rates, as defined by $r = k/n = k/16200$ are $1/5, 4/9, 11/15, 7/9$ and $37/45$, respectively.

For short frames, the BCH code can correct $t = 12$ errors at the output of the LDPC decoder, while for normal frames it can correct between $t = 8$ and $t = 12$ errors, depending on rate of the LDPC code. The BCH code requires 168 parity bits for short frames and between 128 and 192 parity bits for long frames. Thus the overall rate r' of the concatenated BCH/LDPC code is slightly lower than that of the LDPC code alone. For instance, short frames encode messages of length $k' = k - 168 = 16200r - 168$ and thus have overall rate $r' = r - 168/16200$.

Unlike with turbo codes, the encoding of LDPC codes can be very complex if the code is not designed with encoding in mind [18]. This is especially true for systematic LDPC codes, because a sparse $\mathbf{H}$ matrix could require a dense $\mathbf{G}$ matrix. In order to facilitate systematic encoding and produce irregular LDPC that are especially suitable for high code rates, DVB-S2 uses a class of LDPC codes called *extended irregular repeat accumulate* (eIRA) codes [19]. With eIRA codes, the parity check matrix is constrained to be in the form

$$\mathbf{H} = [\mathbf{H}_1 \ \mathbf{H}_2] \tag{12.16}$$

where $\mathbf{H}_1$ is a sparse m by k matrix and $\mathbf{H}_2$ has the form

$$\mathbf{H}_2 = \begin{bmatrix} 1 & & & & & \\ 1 & 1 & & & & \\ & 1 & 1 & & & \\ & & & \cdots & & \\ & & & & 1 & 1 & \\ & & & & & 1 & 1 \end{bmatrix} \tag{12.17}$$

Given the constraint on the $\mathbf{H}$ matrix, the generator matrix can be expressed in systematic form as

$$\mathbf{G} = [\mathbf{I} \ \mathbf{P}] \tag{12.18}$$

where $\mathbf{P} = \mathbf{H}_1^T \mathbf{H}_2^{-T}$ and the form of $\mathbf{H}_2^{-T}$ is

$$\mathbf{H}_2^{-T} = \begin{bmatrix} 1\,1 \cdots & 1\,1 \\ 1 \cdots & 1\,1 \\ \ddots & \vdots\;\vdots \\ & 1\,1 \\ & 1 \end{bmatrix} \tag{12.19}$$

The matrix $\mathbf{H}_2^{-T}$ is actually the generator matrix for a differential encoder (also called an accumulator). Thus the encoding of the DVB-S2 LDPC code can be accomplished in two stages. First the output of the BCH encoder $\mathbf{d}$ is multiplied by the sparse matrix $\mathbf{H}_1^T$, yielding the intermediate result $\mathbf{p}' = \mathbf{d}\mathbf{H}_1^T$. The standard specifies the matrix $\mathbf{H}_1^T$ in the form of a table that lists the locations of the ones in the sparse matrix $\mathbf{H}_1^T$ for each code rate and length. Next, the intermediate result is differentially encoded yielding the set of parity bits $\mathbf{p} = \mathbf{p}'\mathbf{H}_2^{-T}$. Finally the parity bits and message are combined into the systematic codeword as $\mathbf{c} = [\mathbf{d}\ \mathbf{p}]$.

The modulation can be either QPSK, 8PSK, 16APSK, or 32APSK. The 16APSK constellation consists of two concentric rings with 4 uniformly spaced symbols on the inner ring and 12 uniformly spaced symbols on the outer ring. The 32APSK adds a third ring outside the 16APSK constellation, with 16 equally spaced symbols along the third ring. For the higher order modulations (everything except QPSK), a bit interleaver is placed between the channel encoder and the modulator, and thus the system uses bit interleaved coded modulation (BICM) [20].

12.2.2 Decoding

An LDPC code can be decoded iteratively using a message passing algorithm [14] over a graphical representation of the code's parity check matrix called a Tanner graph [21]. A Tanner graph is a bipartite graph consisting of n *variable* nodes (v-nodes) and m *check* nodes (c-nodes). Variable node y_j is connected to check node f_i if and only if the $(i,j)^{th}$ entry of $\mathbf{H}$ is equal to one. As an example, consider the systematic $(7,4)$ Hamming code with parity check matrix:

$$\mathbf{H} = \begin{bmatrix} 1\,1\,1\,0\,1\,0\,0 \\ 1\,1\,0\,1\,0\,1\,0 \\ 1\,0\,1\,1\,0\,0\,1 \end{bmatrix} \tag{12.20}$$

The corresponding Tanner graph for this code is shown in Fig. 12.7.

In the message passing algorithm, messages (in the form of extrinsic information) flow up from the variable nodes to the check nodes and down from the check nodes to the variable nodes. A full description of the decoding algorithm is presented in [22], so here only the main results are presented. The

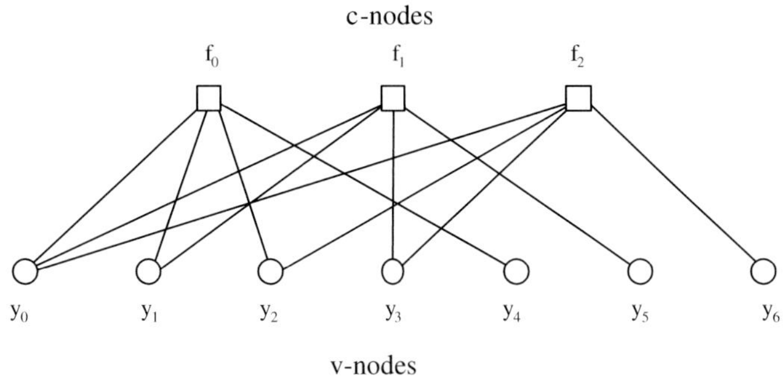

Fig. 12.7. Tanner graph for the $(7, 4)$ systematic Hamming code. Each row of $\mathbf{H}$ is represented by a check node and each column of $\mathbf{H}$ is represented by a function node.

goal of the decoder is to compute the LLR of the ith code bit, $\Lambda^{(o)}(c_i)$, from which the systematic bits can be extracted and a hard decision made.

In the following, let $q_{i,j}$ represent the message passed from v-node i to c-node j and $r_{j,i}$ represent the message passed from c-node j to v-node i. Let $C_i = \{j : h_{j,i} = 1\}$ denote the row locations of the 1's in the ith column of $\mathbf{H}$ and likewise $R_j = \{j : h_{j,i} = 1\}$ the column locations of the 1's in the jth row. The set $R_{j\backslash i}$ is equal to R_j with i excluded.

Initially, the messages passed from the v-nodes to the c-nodes are set to the channel likelihood values, i.e. $q_{i,j} = \Lambda^{(i)}(c_i)$. Then each c-node computes the messages that it sends to every v-node it is attached to according to:

$$r_{j,i} = \left(\prod_{i' \in R_{j\backslash i}} \text{sign}(q_{i',j}) \right) \phi \left(\sum_{i' \in R_{j\backslash i}} \phi\left(|q_{i',j}|\right) \right) \tag{12.21}$$

where the nonlinear $\phi(x)$ function is defined as

$$\begin{aligned} \phi(x) &= -\log \tanh\left(\frac{x}{2}\right) \\ &= \log\left(\frac{e^x + 1}{e^x - 1}\right). \end{aligned} \tag{12.22}$$

Next, each v-node updates the LLR of its corresponding code bit according to

$$\Lambda^{(o)}(c_i) = \Lambda^{(i)}(c_i) + \sum_{j \in C_i} r_{j,i} \tag{12.23}$$

and then generates the output message sent to every c-node that it is connected to

$$q_{i,j} = \Lambda^{(o)}(c_i) - r_{j,i} \tag{12.24}$$

Table 12.3. The $\mathcal{E}_b/N_o$ required to achieve $FER = 10^{-3}$ for the LDPC codes used in DVB-S2. Values marked with an asterisk (*) are extrapolated from Fig. 12.9.

rate	short	normal
9/10	N/A	3.78 dB
8/9	3.78 dB	3.68 dB
5/6	3.06 dB	3.03* dB
4/5	2.72 dB	2.68 dB
3/4	2.33 dB	2.18 dB
2/3	1.95 dB	1.86* dB
3/5	1.59 dB	1.36 dB
1/2	0.93 dB	0.85* dB
2/5	0.55 dB	0.54 dB
1/3	0.37 dB	0.22* dB
1/4	0.25 dB	0.13 dB

A hard decision can be made by simply comparing $\Lambda^{(o)}(c_i)$ to a threshold. Since LDPC codes are linear block codes, they have a built in error detecting mechanism. When the estimated codeword $\hat{\mathbf{c}}$ is a valid codeword, then $\hat{\mathbf{c}}\mathbf{H}^T = \mathbf{0}$ and thus the decoder can halt. If $\hat{\mathbf{c}}\mathbf{H}^T \neq \mathbf{0}$, then the decoder can iterate by having each c-node compute new messages to send to the v-nodes according to Eq. 12.21. Once a maximum number of iterations is reached, the decoder will quit.

12.2.3 Simulation Results

In this section, simulation results are presented that illustrate the performance of the DVB-S2 LDPC code. Fig. 12.8 shows the frame error rate (FER) performance of the short frame size and Fig. 12.9 shows the FER performance of normal frame size. In each case, up to 100 iterations of the log-domain sum-product algorithm [14] described in the last subsection are executed. Table 12.3 shows the $\mathcal{E}_b/N_o$ required to achieve a FER of 10^{-3} for each rate and frame size. Because of the large size of the normal frame size code and the steepness of the corresponding FER curve, results could not be simulated all the way down to a FER 10^{-3} for every code rate. Thus, extrapolated results are given for rates $r = 1/3, 1/2, 2/3$ and $5/6$. Note that the results presented here are only for the LDPC code. The outer BCH code used by DVB-S2 helps to clean up additional errors at the output of the LDPC decoder and will improve the overall performance (mainly be reducing an error floor that begins to emerge below the shown error rates).

12.3 Putting It All Together

The turbo principle stands to revolutionize the delivery of digital content via satellite. The future of the DVB project hinges upon turbo-like coding

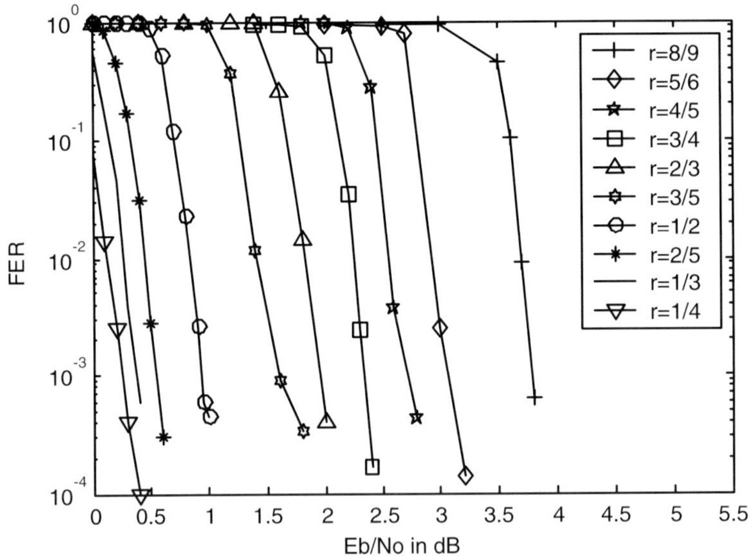

Fig. 12.8. Frame error rate performance of the $n = 16,200$ bit (short frame) LDPC code used in DVB-S2. The decoder uses 100 iterations of the log-domain sum-product algorithm.

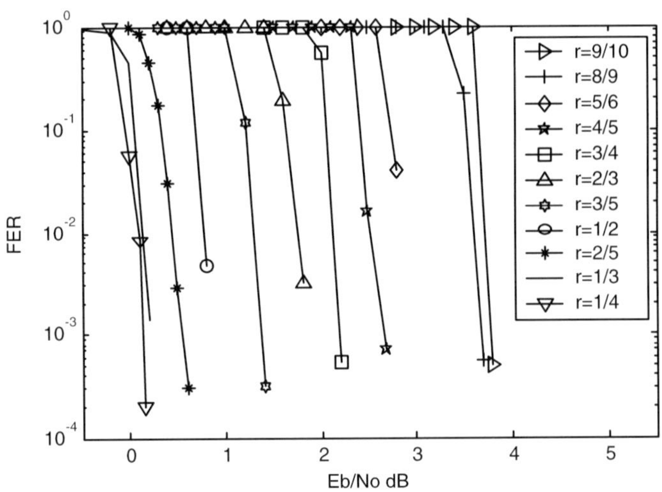

Fig. 12.9. Frame error rate performance of the $n = 64,800$ bit (normal frame) LDPC code used in DVB-S2. The decoder uses 100 iterations of the log-domain sum-product algorithm.

techniques. The DVB-RCS standard, which uses a circular duobinary turbo code, provides a return channel for Internet services, thereby instantly making satellite a serious competitor to cable modems and DSL. The DVB-S2 standard, which uses LDPC codes, represents a significant improvement in the satellite downlink. However, for these technological improvements to be a complete success several hurdles remain. Turbo and LDPC codes are still more complex than their convolutional and Reed Solomon brethren, and therefore significant advances in implementation must still come to fruition. In addition, iteratively decodable codes are more sensitive to channel estimation and synchronization errors and therefore these issues must be dealt with carefully.

12.4 About the Simulations

The software to generate the plots in this chapter has been made available to the public at the http://www.iterativesolutions.com website. The software runs within matlab, but the key encoding and decoding functions are written in c for rapid execution and called as c-mex functions from matlab.

References

1. European Telecommunications Standards Institute. "Digital broadcasting system for television, sound, and data services." *ETS 200 421*, 1994.
2. European Telecommunications Standards Institute. "Digital video broadcasting (DVB); interaction channel for satellite distribution systems;." *ETSI EN 301 790 V1.2.2 (2000-12)*, 2000.
3. European Telecommunications Standards Institute. 'Digital video broadcasting (DVB) second generation framing structure, channel coding and modulation systems for broadcasting, interactive services, news gathering and other broadband satellite applications." *DRAFT EN 302 307 DVBS2-74r15*, 2003.
4. P. Robertson, P. Hoeher, and E. Villebrun. "Optimal and sub-optimal maximum a posteriori algorithms suitable for turbo decoding." *European Trans. on Telecommun.*, 8(2):119–125, Mar./Apr. 1997.
5. C. Berrou, C. Douillard, and M. Jezequel. "Multiple parallel concatenation of circular recursive convolutional (CRSC) codes." *Annals of Telecommunication*, 54(3-4):166–172, Mar.-Apr. 1999.
6. H. H. Ma and J. K. Wolf. "On tail biting convolutional codes." *IEEE Trans. Commun.*, 34:104–111, May 1986.
7. C. Berrou and M. Jezequel. "Non binary convolutional codes for turbo coding." *IEE Electronics Letters*, 35(1):39–40, Jan. 1999.
8. M. C. Valenti and J. Sun. "The UMTS turbo code and an efficient decoder implementation suitable for software defined radios." *Int. J. Wireless Info. Networks*, 8:203–216, Oct. 2001.
9. A. J. Viterbi. "An intuitive justification and a simplified implemetation of the MAP decoder for convolutional codes." *IEEE J. Select. Areas Commun.*, 16(2):260–264, Feb. 1998.

10. J. B. Anderson and S. M. Hladik. "Tailbiting MAP decoders." *IEEE J. Select. Areas Commun.*, 16:297–302, Feb. 1998.
11. M. R. Soleymani, Y. Gao, and Y. Vilaipornsawai. *Turbo Coding for Satellite and Wireless Communications.* Kluwer Academic Publishers, Dordrecht, the Netherlands, 2002.
12. R. G. Gallager. *Low-Density Parity-Check Codes.* PhD thesis, Massachusetts Institute of Technology, Cambridge, MA, 1960.
13. D. J. C. MacKay and R. M. Neal. "Near Shannon limit performance of low density parity check codes." *IEEE Electronics Letters*, 32:1645–1646, Aug. 1996.
14. D. J. C. MacKay. "Good error correcting codes based on very sparse matrices." *IEEE Trans. Inform. Theory*, 45:399–431, Mar. 1999.
15. T. Richardson, A. Shokrollahi, and R. Urbanke. "Design of capacity approaching irregular low density parity check codes." *IEEE Trans. Inform. Theory*, 47:618–637, Feb. 2001.
16. M. Luby, M. Mitzenmacher, A. Shokrollahi, and D. Spielman. "Improved low-density parity-check codes using irregular graphs." *IEEE Trans. Inform. Theory*, 47:585–598, Feb. 2001.
17. S. Y. Chung, G. D. Forney, T. J. Richardson, and R. Urbanke. "On the design of low-density parity-check codes within 0.0045 dB of the shannon limit." *IEEE Commun. Letters*, 5:58–60, Feb. 2001.
18. T. Richardson and R. Urbanke. "Efficient encoding of low-density parity-check codes." *IEEE Trans. Inform. Theory*, 47:638–656, Feb. 2001.
19. M. Yang, W. E. Ryan, and Y. Li. "Design of efficiently encodable moderate-length high-rate irregular LDPC codes." *IEEE Trans. Commun.*, 52:564–571, Apr. 2004.
20. G. Caire, G. Taricco, and E. Biglieri. "Bit-interleaved coded modulation." *IEEE Trans. Inform. Theory*, 44:927–946, May 1998.
21. R. M. Tanner. "A recursive approach to low complexity codes." *IEEE Trans. Inform. Theory*, 27:533–547, Sept. 1981.
22. W. E. Ryan. An introduction to LDPC codes. In B. Vasic, editor, *Handbook for Coding and Signal Processing for Recording Systems.* CRC, New York, 2004.

Chapter 13

Turbo Code Applications on Telemetry and Deep Space Communications

Gian Paolo Calzolari[1], Franco Chiaraluce[2], Roberto Garello[3] and Enrico Vassallo[1]

[1] European Space Agency, Germany
[2] Università Politecnica delle Marche, Italy
[3] Politecnico di Torino, Italy

Telemetry from space is regulated by the CCSDS (Consultative Committee for Space Data Systems) recommendations, used worldwide by international space Agencies (CNES, DLR, ESA, JAXA, NASA, RSA, etc.). These recommendations establish common basis for all aspects of space telemetry communication links, including their physical layers, data formats and protocols. Applications, ranging from space research, deep space and near-Earth missions, to Earth-observation, meteorological, tracking and data-relay satellites, cover a wide range of different requirements.

The CCSDS Telemetry Channel Coding Recommendation [1] establishes a common framework and provides a standardized basis for the coding schemes used by CCSDS Agencies for space telemetry data communications. This standard traditionally provides the benchmark for new and emerging coding technologies. Turbo codes were chosen as a new option for this standard in 1999, only 6 years since their official presentation to the international community: this was the first international standard including turbo codes. The reason was the significant improvement in terms of power efficiency assured by turbo codes over the old codes of the standard. At that time, the standard included these coding options:

- **RS**: a (255,223) Reed-Solomon code with 8-bit symbols and Error Correction Capability of 16 symbols.
- **CC**: a 64-state, rate-1/2 binary convolutional code.
- **SC**: the serial concatenation of the Reed-Solomon code **RS** (outer code) and the convolutional code **CC** (inner code) through an interleaver of length (255 I) bytes, with I = 1,2,3,4, or 5.

Turbo codes demonstrated an additional coding gain larger than 2 dB with respect to the concatenated scheme **SC**. This significant gain was considered essential for new missions with constrained link budgets as those for deep space missions.

K. Sripmanwat (ed.), Turbo Code Applications: a journey from a paper to realization, 321–344.

In this chapter various aspects of CCSDS turbo codes performance are presented and discussed, including simulation error rate curves, comparison with theoretical limits, minimum distance and error floor behavior, and symbol synchronization properties. Space missions using turbo codes are reviewed, and some hints on future development of CCSDS standard discussed.

13.1 Theory

Let us denote by $\mathbf{d}$ the k-bit input frame of the CCSDS turbo encoder. With specific reference to packet telemetry [2], the input frame consists of:

- The frame primary header: 48 header bits introduced for various uses (spacecraft identification, virtual channel identification, frame count, etc.).
- The frame data field: $(k - 64)$ data bits.
- The frame error control field: 16 parity check bits added by a $(k, k - 16)$ Cyclic Redundancy Check (CRC) encoder with the generator polynomial $G(D) = 1+D^5+D^{12}+D^{16}$. (Since the CRC encoder is preset to an all "1" state prior to encoding, the parity check bits are inverted with respect to a usual CRC encoding. As an example, $(k-16)$ all "0" input bits generate 16 all "1" parity check bits.)

(Other than packet telemetry systems, the CCSDS turbo codes are also applicable, for example, to Advanced Orbiting Systems (AOS) [3]. In this case, the structure of the input frame is very similar.)

When turbo codes are used, the input frame length k must assume one of these values:

$$k = 1784, 3568, 7136 \text{ or } 8920 \quad \text{bits}$$

(These lengths were chosen for commonality with other codes of the standards and therefore to re-use existing telemetry frame generators.)

The structure of the CCSDS turbo encoder is depicted in Fig. 13.1, and consists of the parallel concatenation of:

- Two equal binary, linear, systematic, recursive convolutional encoders E_1 and E_2 with rate-1/4 and 16 states. Their feedback polynomial is $(1 + D^3 + D^4)$, while their forward polynomials are $(1 + D + D^3 + D^4)$, $(1 + D^2 + D^4)$,$(1 + D + D^2 + D^3 + D^4)$.
- A block interleaver with length k, characterized by a permutation law $\pi : \mathbf{Z}_k \rightarrow \mathbf{Z}_k$. The input bits of E_1 are not permuted, whereas those of E_2 are permuted by π. Therefore, given $\mathbf{d}$, the i-th input bit is d_i for E_1 and $d_{\pi(i)}$ for E_2.

The interleaver permutation laws $\pi : \mathbf{Z}_k \rightarrow \mathbf{Z}_k$ were proposed by Berrou, and are generated by the following algorithm [1]:

- First express k as $k = k_1 k_2$ where $k_1 = 8$ and $k_2 = k/8$.

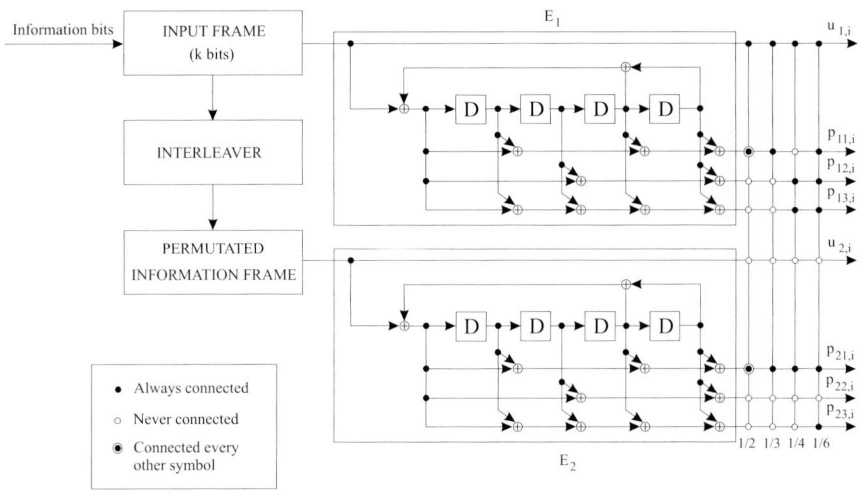

Fig. 13.1. Encoder for the CCSDS turbo code family.

- Then, for each s from 0 to $(k-1)$, compute $\pi(s)$ by using the following equations:

$$m = s \mod 2$$

$$i = \lfloor \frac{s}{2k_2} \rfloor$$

$$j = \lfloor \frac{s}{2} \rfloor - ik_2$$

$$t = (19i + 1) \mod 4$$

$$q = t \mod 8 + 1$$

$$c = (p_q j + 21m) \mod k_2$$

$$\pi(s) = 2t + 8c + 1 - m$$

where p_q denotes one of the following eight prime integers:

$$p_1 = 31;\ p_2 = 37;\ p_3 = 43;\ p_4 = 47;\ p_5 = 53;\ p_6 = 59;\ p_7 = 61;\ p_8 = 67.$$

It is worthwhile to mention that an algorithmically generated permutation law of turbo code interleavers is important for space applications to avoid on-board large memory storage. In space, cosmic rays may alter the memory content. In case of single event upset of this kind, it would be easier to re-transmit to the spacecraft the generating algorithm than the whole permutation law.

Four nominal code rates $R_c = 1/r$, for $r = 2$, 3, 4, and 6, can be obtained through the puncturing options described in Fig. 13.1. As an example, for the rate-1/3 code the output sequence is composed of k triplets of the form

Table 13.1. Parameters for the CCSDS turbo codes: codeword length n as a function of input frame length k and nominal code-rates.

	n [bits]			
k [bits]	R_c=1/2	R_c=1/3	R_c=1/4	R_c=1/6
1784	3576	5364	7152	10728
3568	7144	10716	14288	21432
7136	14280	21420	28560	42840
8920	17848	26772	35696	53544

$(u_{1,i}, p_{11,i}, p_{21,i})$ while, for the rate-1/2 code, the output sequence is composed of $(k/2)$ 4-tuples of the form $(u_{1,i}, p_{11,i}, u_{1,i+1}, p_{21,i+1})$.

The constituent convolutional encoders are terminated in four steps which generate $(4r)$ additional output bits. As a consequence of termination, the turbo code becomes a block code $C(n, k)$ with $n = (k + 4)r$. The resulting (n, k) pairs are summarized in Table 13.1.

For codeword synchronization at the receiver side, an attached sync marker (ASM) is appended to each codeword before transmission. The ASM is not a part of the input frame and it is not encoded. It is simply attached before the codeword at the output of the turbo encoder. The ASM length depends on the code-rate: it consists of $(32r)$ predefined bits [1].

CCSDS standards include a number of modulation options [4], like 2-PSK, 4-PSK, GMSK, and others. All the results presented in this chapter are referred to a Gray-labelled 4-PSK modulation.

13.2 CCSDS Turbo Codes Performance

In this section, several aspects of CCSDS turbo code performance are presented and discussed. In Sec. 13.2.1, simulated error rate curves are presented and compared against theoretical limits. The code minimum distance and the corresponding error floors are reviewed in Sec. 13.2.2. Symbol synchronization properties are discussed in Sec 13.3.

Throughout the remainder of the chapter, reference will be made to the Additive White Gaussian Noise (AWGN) channel. This choice is justified by the observation that the deep space channel, which is the main subject of the present analysis, can be considered to be AWGN at the frequencies currently in use. Possible interferences with other deep space probes are avoided thanks to predictions based on orbits and other mission parameters. These previsions are developed in large advance with respect to the launch date (space agencies strictly co-operate on this task). Possible interferences from other systems, like point-to-point and point-to-multipoint terrestrial systems or non deep-space missions, are also avoided by using very severe protection rules, continuously enforced in the Earth stations that receive deep-space data. In such a scenario, it is therefore justified that gaussian noise is practically the only significant disturbance contribution.

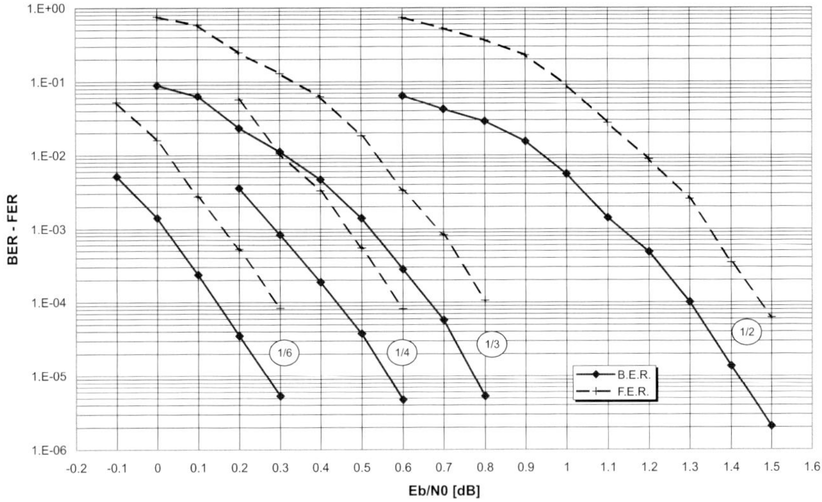

Fig. 13.2. BER and FER simulated performance of CCSDS turbo codes for k=1784 bits.

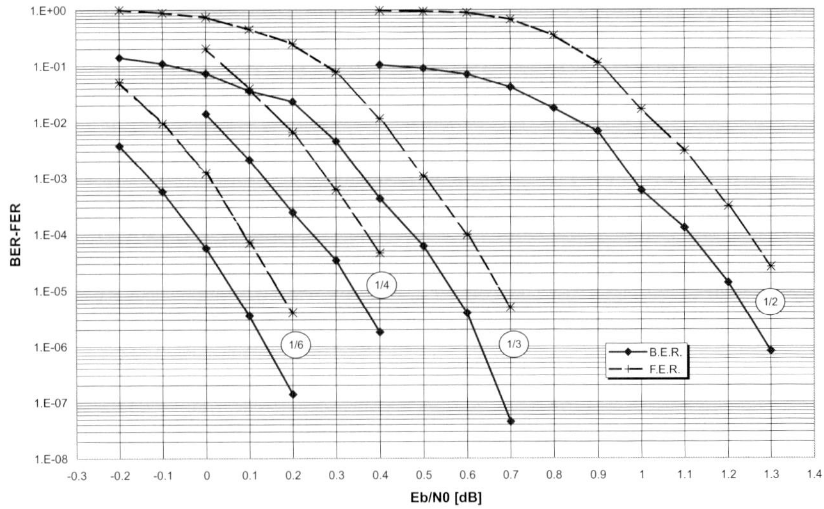

Fig. 13.3. BER and FER simulated performance of CCSDS turbo codes for k=3568 bits.

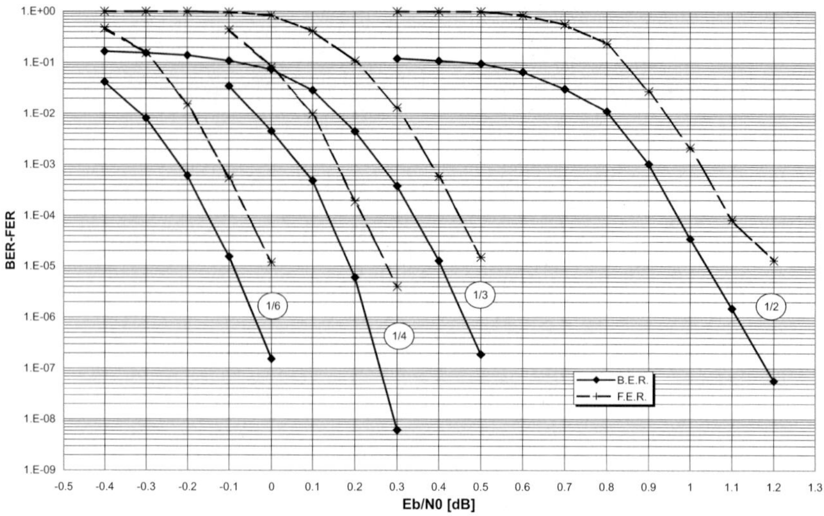

Fig. 13.4. BER and FER simulated performance of CCSDS turbo codes for k=7136 bits.

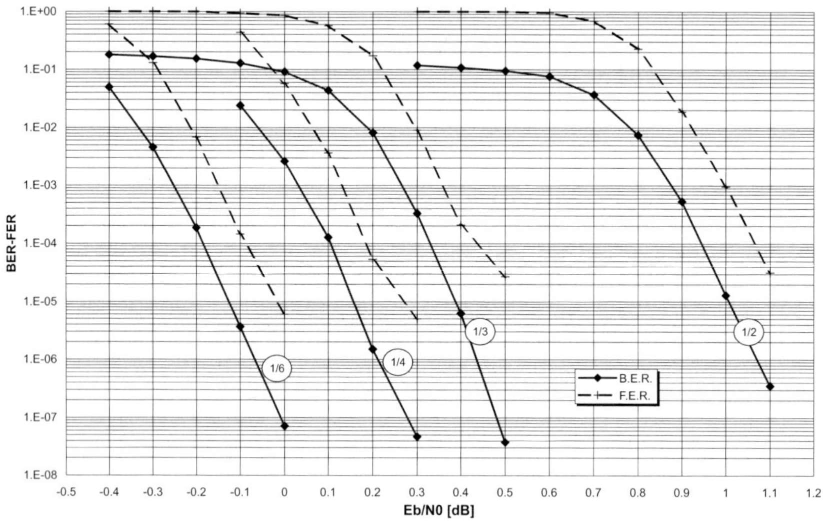

Fig. 13.5. BER and FER simulated performance of CCSDS turbo codes for k=8920 bits.

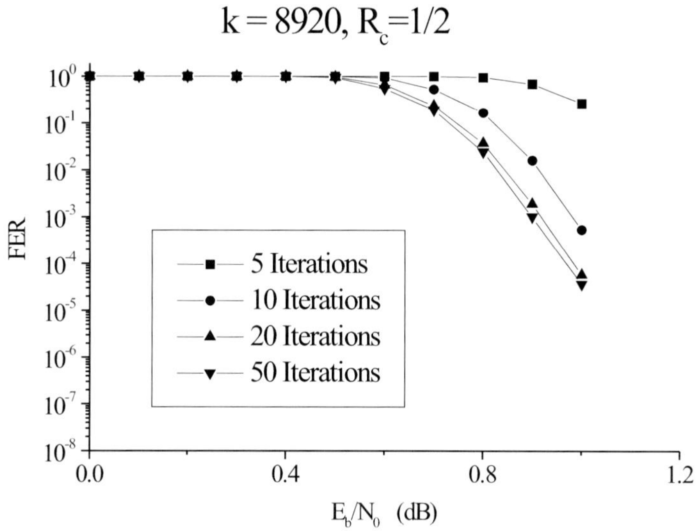

Fig. 13.6. FER performance of the CCSDS turbo code with k=8920 bits and $R_c = 1/2$ for different values of the number of iterations.

13.2.1 Error Rates Curves and Comparisons

The Bit Error Rate (BER) and Frame Error Rate (FER) simulated performance of the CCSDS turbo codes over the AWGN channel are reported in Fig. 13.2 (k=1784), Fig. 13.3 (k=3568), Fig. 13.4 (k=7136), and Fig. 13.5 (k=8920), by considering the four nominal code rates $R_c = 1/2, 1/3, 1/4$ and $1/6$ (written within circles in the figures). They have been obtained by 10 iterations of the SISO (Soft Input Soft Output) decoding algorithm.

The effect of different choices for the iteration number on the FER performance is shown in Fig. 13.6, by assuming k=8920 bits and $R_c = 1/2$ (similar behaviors hold for the other frame lengths and rates). By assuming 10 iterations, the FER performance in the water-fall region is less than 0.1 dB worse than assuming 20 iterations. On the other hand, performance with 50 iterations is only slightly better than with 20 iterations, whilst the processing time is much higher. The assumption of 10 iterations seems then a reasonable compromise, in the perspective to obtain good performance with limited complexity (*i.e.*, acceptable processing time).

To investigate the "goodness" of CCSDS turbo code performances, it is useful to compare them against the channel coding theoretical limits. For a fixed code-rate k/n and a specific constellation, the ideal spectral efficiency η (measured in bps/Hz) is computed by referring to ideal Nyquist base-band filters. For 4-PSK constellations the following expression results:

$$\eta = 2\frac{k}{n}.$$

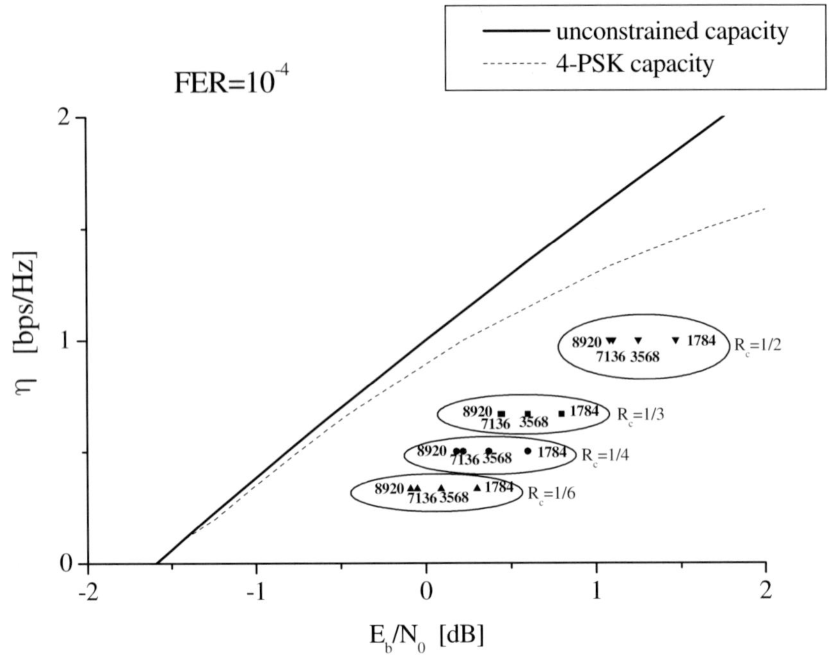

Fig. 13.7. CCSDS turbo codes FER simulated performance at FER=10^{-4} vs. the unconstrained capacity and the 4-PSK capacity.

Given η, it is possible to invoke the Shannon limit and compute the minimum E_b/N_0 ratio which allows vanishing error probability for that spectral efficiency. In Fig. 13.7, the E_b/N_0 ratios required by the CCSDS codes to achieve FER=10^{-4} are reported and compared against both the unconstrained capacity and the 4-PSK capacity curves.

These results show that the CCSDS turbo codes have good performance, not far from 4-PSK capacity at FER=10^{-4}. However, the capacity curves are achievable by codeword lengths growing to infinite. For taking into account of the unavoidable penalty introduced by codeword length finiteness, it is possible to compute the Sphere Packing Bound (SPB). Given η, the SPB provides a lower bound to the FER performance achievable by a code with finite codeword length [5], as a function of the E_b/N_0 ratio. For infinite codeword length, the SPB converges on the capacity limit for the considered constellation (4-PSK for CCSDS codes).

For a fixed Frame Error Rate value (for example FER=10^{-4}), the distance in dB between the code FER curve and the SPB curves is a fair indication about the "goodness" of the considered code, taking into account also the finiteness of codeword lengths. The comparisons between the FER simulated performance of the CCSDS turbo codes and the corresponding sphere packing

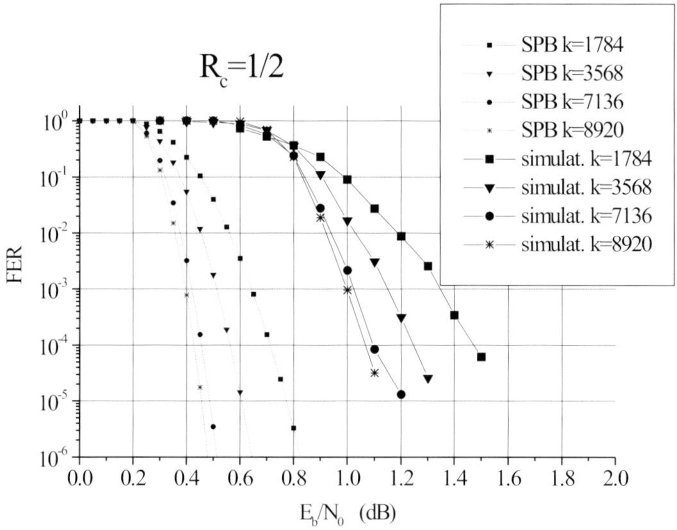

Fig. 13.8. CCSDS turbo codes FER simulated performance vs. the sphere packing lower bound: rate-1/2 codes.

lower bounds are reported in Fig. 13.8 (rate-1/2 codes), Fig. 13.9 (rate-1/3 codes), Fig. 13.10 (rate-1/4 codes) and Fig. 13.11 (rate-1/6 codes).

As a general comment, it is possible to observe that the CCSDS turbo code performances are very good. At FER $= 10^{-4}$, most of the codes stay within 0.8 dB from the SPB. Clearly, this gap is destined to increase for larger E_b/N_0 values (lower error rates).

13.2.2 Minimum Distances and Error Floors

Turbo codes are characterized by coding gains extremely large for low/medium signal-to-noise ratios (high/medium error rates), very close to the theoretical Shannon limits. However, it is well known that their performance may be not as good at very high signal-to-noise ratios (very low error rates), where the "error floor" phenomenon occurs. As will be shown in the following of this section (see Eq. 13.1 and 13.2 in particular), in the region of high signal-to-noise ratios, the performance of any binary code is dominated by its minimum distance $d_{\min}$ (the minimum Hamming distance between codewords, which coincides with the minimum Hamming weight of a nonzero codeword for linear codes) and its multiplicity values. In some cases, turbo codes may have very low minimum distances, despite very large interleaver lengths. This causes their performance curves to flatten according to the slope imposed by $d_{\min}$, after the typical "water-fall" decreasing at low signal-to-noise ratios. To estimate the turbo code performance in the region of very low error rates (for

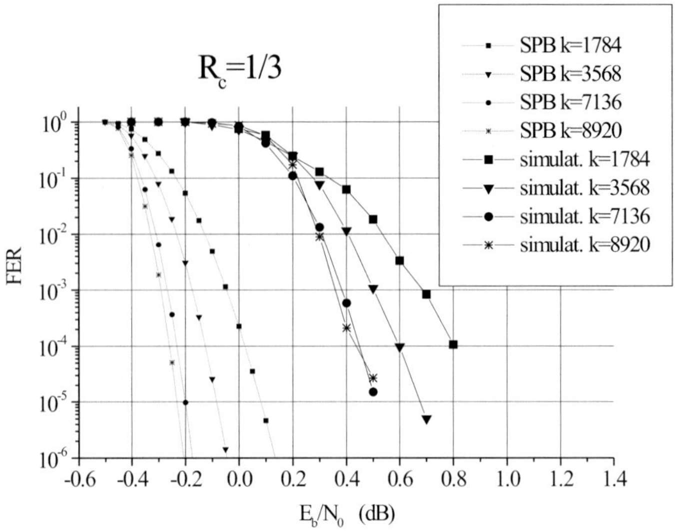

Fig. 13.9. CCSDS turbo codes FER simulated performance vs. the sphere packing lower bound: rate-1/3 codes.

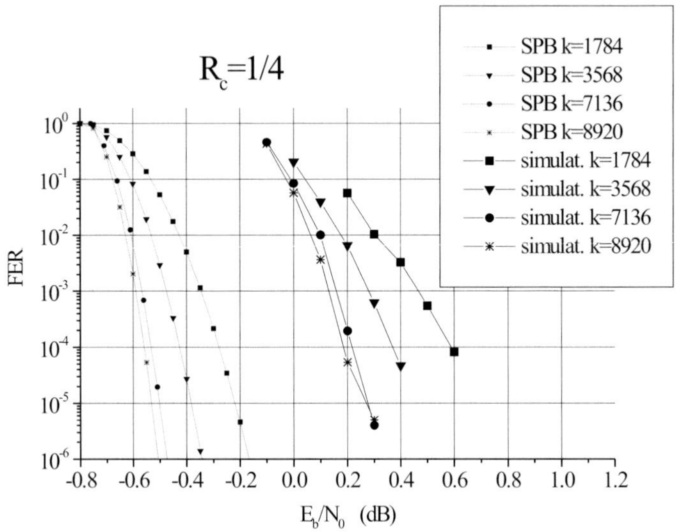

Fig. 13.10. CCSDS turbo codes FER simulated performance vs. the sphere packing lower bound: rate-1/4 codes.

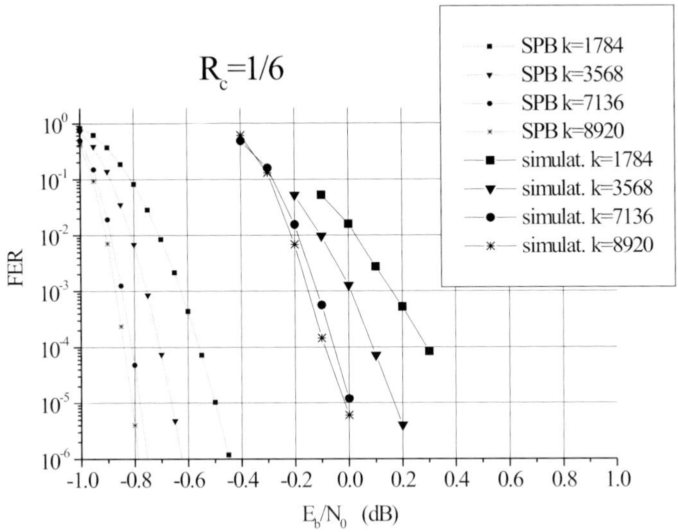

Fig. 13.11. CCSDS turbo codes FER simulated performance vs. the sphere packing lower bound: rate-1/6 codes.

example BER $\leq 10^{-9}$), extremely long simulations are needed, usually impossible without a dedicated hardware. As an alternative, an analytical approach can be applied, requiring the computation of the minimum distance.

Consider the transmission of a binary turbo code over the AWGN channel by a Gray labelled 4-PSK. At very high signal-to-noise ratios (SNR), that is very low error rates, the code performance practically coincides with the union bound, truncated to the contribution of the minimum distance [6]. The FER and BER code performance can then be approximated by:

$$\text{FER} \simeq \frac{1}{2} A_{\min} \ \text{erfc}\left(\sqrt{d_{\min} \frac{k}{n} \frac{E_b}{N_0}}\right) \tag{13.1}$$

$$\text{BER} \simeq \frac{1}{2} \frac{w_{\min}}{k} \ \text{erfc}\left(\sqrt{d_{\min} \frac{k}{n} \frac{E_b}{N_0}}\right) \tag{13.2}$$

where $A_{\min}$ is the code *dominant multiplicity* (number of codewords with weight $d_{\min}$), and $w_{\min}$ is the code *dominant information multiplicity* (sum of the Hamming weights of the $A_{\min}$ information frames generating the codewords with weight $d_{\min}$). When comparing the simulated curves with Eq. 13.1 and 13.2 a small fixed penalty (usually less than 0.25 dB for turbo codes) must be also taken into account, due to the sub-optimality of iterative decoding.

Eq. 13.1 and 13.2 allow to obtain a good approximation of the code performance in the region of very low error rates, without resorting to exhausting

Table 13.2. The minimum distances (and the multiplicities) of the CCSDS turbo codes, computed by the constrained subcode algorithm (k=1784 and k=3568) and the all-zero algorithm (k=7136 and k=8920).

k	rate	$d_{\min}$	$A_{\min}$	$w_{\min}$	rate	$d_{\min}$	$A_{\min}$	$w_{\min}$	rate	$d_{\min}$	$A_{\min}$	$w_{\min}$	rate	$d_{\min}$	$A_{\min}$	$w_{\min}$
1784	1/2	17	2	6	1/3	32	1	2	1/4	42	1	2	1/6	70	1	2
3568	1/2	20	1	1	1/3	40	3	9	1/4	56	2	6	1/6	93	2	6
7136	1/2	25	1	3	1/3	48	3469	20814	1/4	66	1	3	1/6	117	2	6
8920	1/2	25	1	3	1/3	48	1414	8484	1/4	72	1	3	1/6	108	1	1

simulations. However, in order to apply such an analytical approach for evaluating turbo code performance in this region, an efficient algorithm for turbo code minimum distance evaluation is required. The CCSDS turbo code distances and their multiplicity values were computed in [6] by applying the constrained subcode algorithm to all the codes with k=1784 and k=3568. This algorithm provides the exact values of the computed parameters. However, its application to larger frame lengths is limited by complexity problems. In this case, it is possible to apply the iterative all-zero algorithm [7], which was used to compute the CCSDS turbo code distances and their multiplicity values for k=7136 and k=8920. It is important to remark that, in general, the iterative all-zero algorithm may produce false results. In the case of turbo codes, however, its reliability is very high; as a consequence, for the codes here of interest, the probability that the true distances are lower, or the true multiplicities are larger, than those obtained via the all-zero algorithm is very small [7].

To summarize, the minimum distances and the multiplicities of the CCSDS turbo codes, computed by the constrained subcode (k=1784 and k=3568) and the all-zero algorithm (k=7136 and k=8920) are reported in Table 13.2.

The corresponding FER analytical error floor curves, computed by applying Eq. 13.1 are depicted in Fig. 13.12 (k=1784), Fig. 13.13 (k=3568), Fig. 13.14 (k=7136) and Fig. 13.15 (k=8920).

These analytical error floor curves provide useful information on the code behavior for very low error rates. When joined with the simulated performance curves at high/medium error rates presented in the previous subsection, they provide a complete analysis of the turbo code performance. As an example, the Frame Error Rate simulated performance and error floor curves for the rate-1/2 turbo code with data frame length k=1784 bits are reported in Fig. 13.16. It is clear that the error floor curve allows to foresee the behavior of the turbo code for low error rates, without resorting to exhausting simulations.

13.3 Symbol Synchronization Properties

In order to acquire and maintain symbol synchronization at the receiver side, the coded bits produced by the turbo encoder (simply called "symbols" in

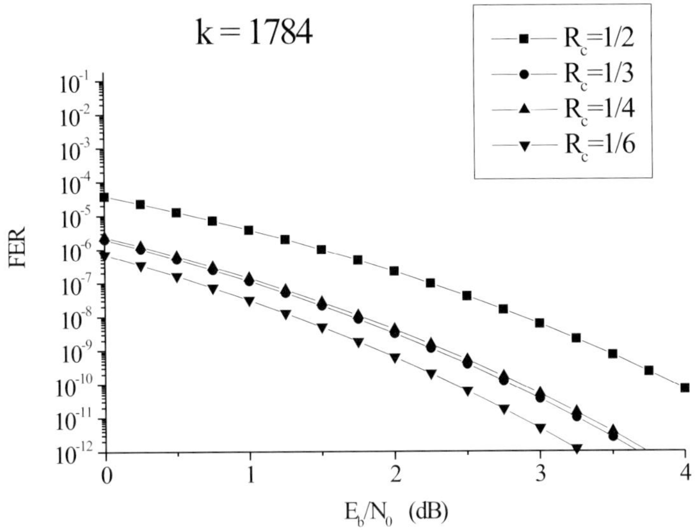

Fig. 13.12. CCSDS turbo codes FER error floor curves: k=1784 bits.

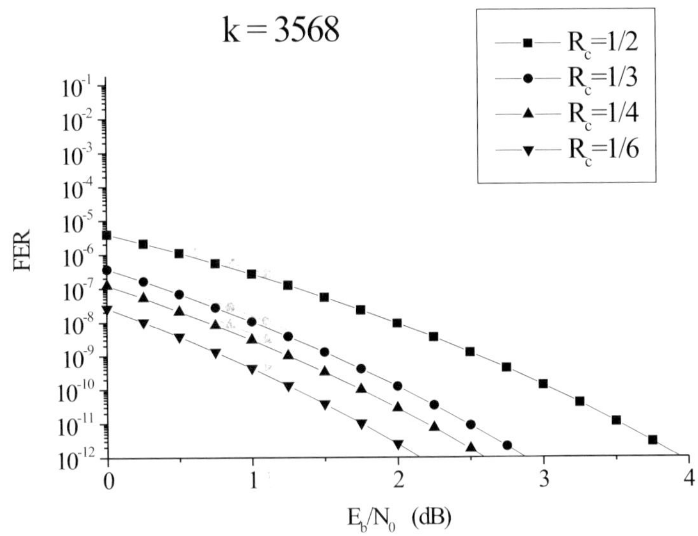

Fig. 13.13. CCSDS turbo codes FER error floor curves: k=3568 bits.

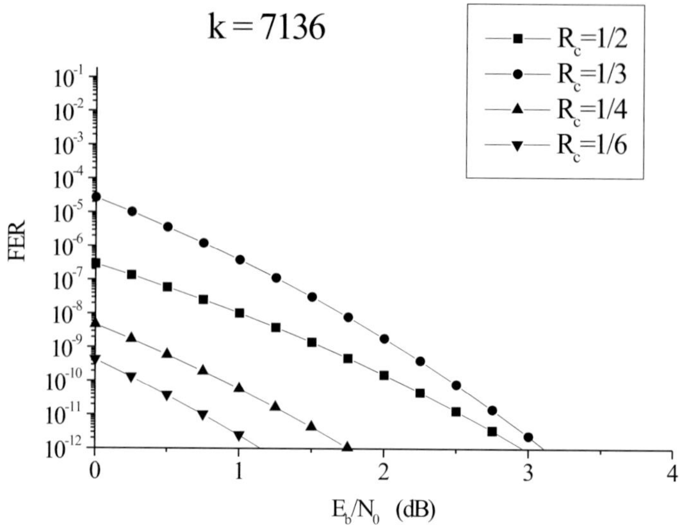

Fig. 13.14. CCSDS turbo codes FER error floor curves: k=7136 bits.

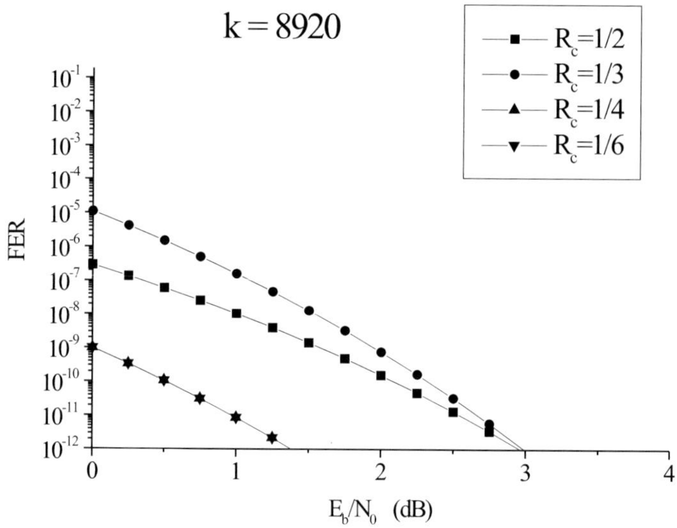

Fig. 13.15. CCSDS turbo codes FER error floor curves: k=8920 bits.

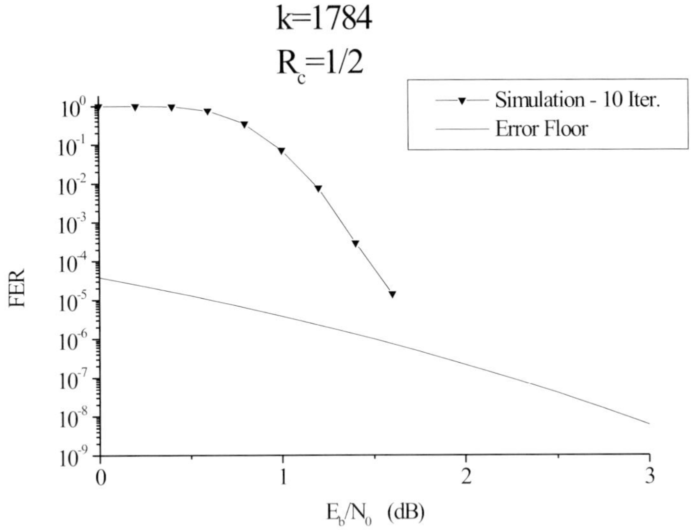

Fig. 13.16. FER performance of the CCSDS turbo code with rate-1/2 and k=1784 bits: simulation and error floor curves.

the following) and transmitted over the channel must have a sufficient symbol transition density. In fact, it is well known that practical demodulators encounter trouble during acquisition or tracking whenever the number of transitions is too small or when long sequences of equal symbols are received.

As a consequence, a certain number of quantities are important for synchronization ability:

- The transition probability **P(TR)** (probability that an encoded binary symbol is different from the previous one) and the no-transition probability **P(NTR)**.
- The symbol probabilities **P(0)** and **P(1)** (probability that a symbol is 0 or 1).
- The probability $\mathbf{P(RL} = i\mathbf{)}$ of having a run of length i (i equal consecutive symbols).
- The maximum run-length **MAX_RL** (maximum number of equal consecutive symbols).
- The probability $\mathbf{P(T1000} = i\mathbf{)}$ of having i transitions in a sequence of 1000 consecutive symbols.
- The minimum number **MIN_T1000** of transitions in any sequence of 1000 consecutive symbols.

These quantities must be compared against realistic requirements of practical circuits for symbol synchronization recovery. As a benchmark, CCSDS recommends [4]:

Table 13.3. Relevant parameters for symbol synchronization: CCSDS turbo code with k=8920 bits.

R_c	P(TR)	P(NTR)	P(0)	P(1)	MAX_RL	CCSDS	MIN_T1000	CCSDS
1/2	0.499943034	0.500056966	0.499999799	0.500000201	42	64	407	275
1/3	0.499923689	0.500076311	0.500000708	0.499999292	51	64	408	275
1/4	0.499929704	0.500070296	0.500000391	0.499999609	68	64	380	275
1/6	0.499888131	0.500111869	0.500000285	0.499999715	90	64	394	275

- A maximum run-length **MAX_RL** of 64 symbols.
- A minimum number of transitions **MIN_T1000** of 275.

The other quantities can instead be compared against the characteristics of an ideal random sequence of equiprobable and independent binary symbols, which is characterized by:

- **P(TR)** = **P(NTR)** = 0.5
- **P(0)** = **P(1)** = 0.5
- **P(RL** = i**)** = $1/2^i$ (1/2 of the runs are of length 1, 1/4 of length 2, and so on).

Similar to the error rate performance evaluation, the only way to properly analyze the synchronization properties of a turbo code employing a chosen deterministic interleaver is to simulate it with sufficient detail. Here, a number of 10,000,000 input frames have been simulated to obtain the results (corresponding, for a mission data rate $R_b = 1$ Mbit/s and a frame length $k = 8920$ bits, to about one day of continuous working). To perform this analysis some reasonable hypotheses have been undertaken. With reference to the telemetry systems input frame structure described in Sec. 13.1, the data bits have been supposed equally likely and independent one each other, and the 48 bits of the frame primary header have been assimilated to data bits. (An extended discussion on the analysis assumptions and their consequences can be found in [8].)

For the sake of brevity, in this section only the case k=8920 bits is presented (in [8] also smaller frame lengths were considered, by demonstrating they lead to quite similar results). In Table 13.3 the parameters **P(TR)**, **P(NTR)**, **P(0)**, **P(1)**, **MAX_RL**, **MIN_T1000** are listed for all the four nominal code-rates, and compared against the CCSDS requirements (simply noted by CCSDS in the table).

The probabilities **P(RL** = i**)** are plotted explicitly as a function of the run-length i in Fig. 13.17–13.20. As a reference, the behavior of an ideal binary random sequence is also depicted. The probabilities **P(T1000** = i**)** are reported in Fig. 13.21 for code-rate 1/2 (the curves for the other rates being very similar [8]).

Based on the presented results, it is possible to state that the binary sequences output by the CCSDS turbo encoder have excellent characteristics for symbol synchronization.

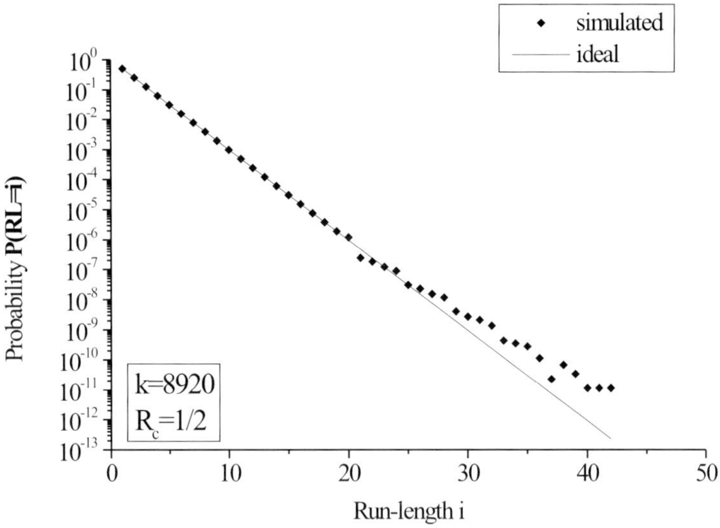

Fig. 13.17. **P(RL** $= i$**)** for CCSDS turbo code with k=8920 and code-rate 1/2, and ideal behavior.

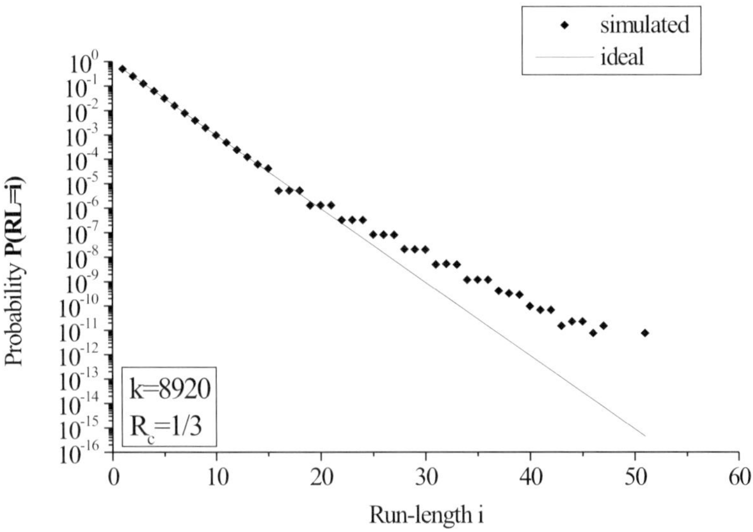

Fig. 13.18. **P(RL** $= i$**)** for CCSDS turbo code with k=8920 and code-rate 1/3, and ideal behavior.

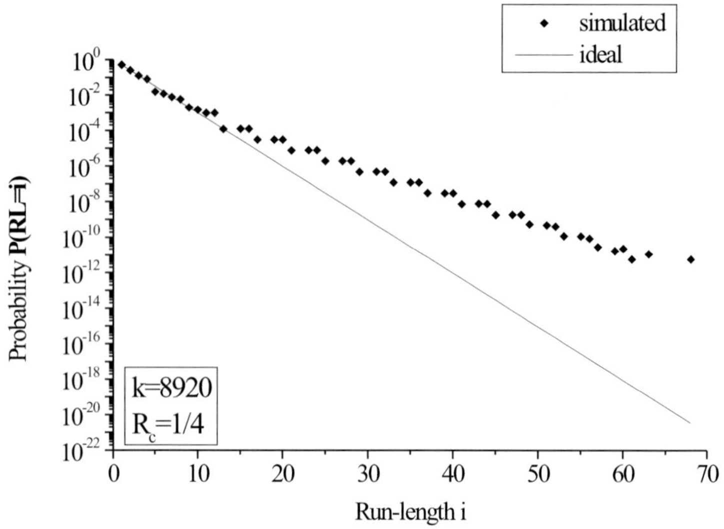

Fig. 13.19. **P(RL** = i) for CCSDS turbo code with k=8920 and code-rate 1/4, and ideal behavior.

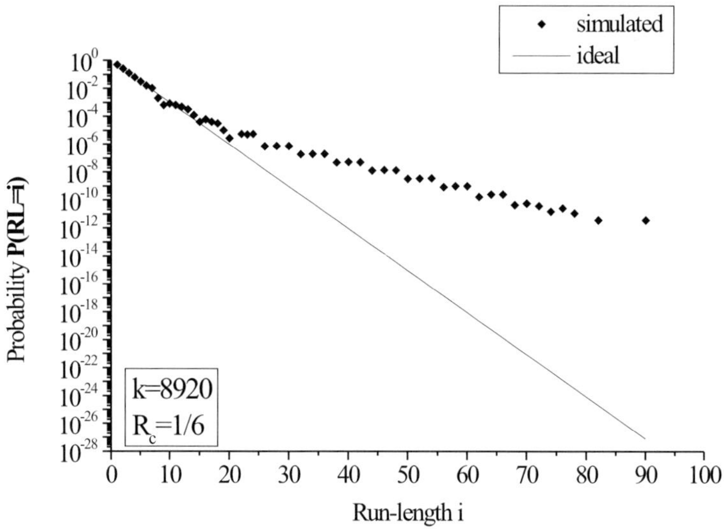

Fig. 13.20. **P(RL** = i) for CCSDS turbo code with k=8920 and code-rate 1/6, and ideal behavior.

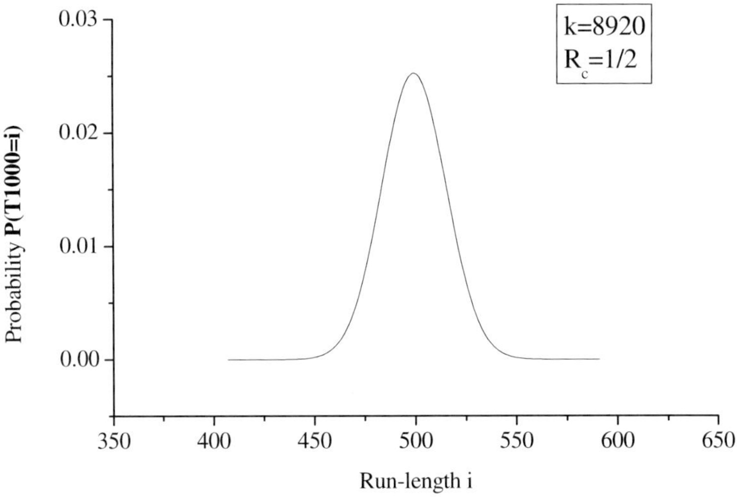

Fig. 13.21. P(T1000 $= i$) for CCSDS turbo code with k=8920 and code-rate 1/2.

The requirement on **MIN_T1000** is always satisfied: in any case the minimum values found were by far larger than the minimum (275) imposed by the practical design considerations settled by CCSDS recommendation [4]. On the contrary, it is possible to observe that the requirement on **MAX_RL** was satisfied for rates 1/2 and 1/3, but not for rates 1/4 and 1/6. In order to correctly interpret the results, it is necessary to stress they represent probabilistic outputs, obtained by simulating a very large number of information frames composed of ideal random information bits. As a matter of fact, they represent the probability of a given event in one-day continuous working of a transmission link working at a data rate $R_b = 1$ Mbps. So, they show that, in ideal working conditions, critical events on **MAX_RL** for code-rates 1/4 and 1/6 may occur, even if with very small probabilities.

As for the comparison of turbo codes against the ideal random sequence characteristics, there is a satisfactory match, not only in terms of transition/symbol probabilities, but also in terms of run-length probability, especially for rate-1/2 and rate-1/3 codes, extended down to probabilities less than 10^{-6}. For the sake of comparison, by evaluating the run-length distribution of usual block and convolutional codes, it is easy to verify that their match is significantly less pronounced, usually limited to probabilities not smaller than 10^{-4} [8]. For the other two code-rates, some run-lengths never appear, and this may explain less pronounced match with the ideal case. From an unusual point of view, these results give another support to the random coding interpretation of turbo codes, where their excellent performances are explained by their similarity to random binary codes.

As a general comment based on the presented results, it is possible to conclude that the CCSDS turbo encoder scheme has very good performance in terms of transition density with respect to both practical requirements and random behavior, with only a few exceptions. Since the CCSDS standard [1] calls for an optional pseudo-randomizer to be used in cases where data randomness at the telemetry generator output is not guaranteed, these exceptions are not a problem for practical use.

13.4 Applications: CCSDS Turbo Codes and Space Missions

As mentioned in introducing the chapter, the CCSDS Telemetry Channel Coding recommendation [1] is used worldwide by space Agencies for space communications links. Its primary application is the transmission between a remotely located data source to users located in space or on Earth. Typically, data are generated by scientific and engineering sensors on board a spacecraft, and transmitted to Earth. The categories of missions interested to use CCSDS turbo codes include deep space missions and near-Earth missions in highly elliptical orbits or at Lagrangian points (1.5 Mkm from Earth) for which communications margins are low.

Currently, ESA's SMART-1 mission is using turbo codes as an experiment. SMART-1 was launched to the Moon in September 2003. Propelled by solar electric propulsion, its main scope is the testing of miniaturization technology and new techniques while exploring the Moon from its orbit. Actually, in a recent experiment, the KaTE (Ka-band Transmission Experiment) instrument of SMART-1 has transmitted turbo encoded frames (rate 1/4) that have been received at the ESA Tracking of Villafranca del Castillo (Spain) and decoded by the ESA prototype turbo codes decoder developed by Space Engineering Rome (Italy) and Politecnico of Torino (Italy) on behalf of the European Space Agency.

Turbo codes are successfully being used also on NASA's MESSENGER, the first Mercury orbiter, launched on August 2004. The spacecraft is using the length 8920-bit, rate-1/6 turbo code, at a variety of rates up to 104 Kbits/second.

Other turbo-coded transmissions are planned for the near future. For ESA's Rosetta mission it will be possible to programme the on board computers to use turbo codes. The goal of Rosetta is to study the origin of comets, the relationship between cometary and interstellar material and its implications with regard to the origin of the Solar System. After postponement of the initial launch, its new target is Comet 67 P/Churyumov-Gerasimenko, with a 10 year journey to the comet. Future candidates for using turbo codes are also ESA's Gaia and BepiColombo missions. Gaia plans be launched in 2010: its goal is to chart a three-dimensional map of the Milky Way, in the process revealing the composition, formation and evolution of the Galaxy. BepiColombo, an ESA mission in cooperation with Japan, will explore Mercury, and plans to be launched in 2012.

Moreover, a number of NASA missions, recently launched, near to be launched or in the development phase have planned to use turbo codes. Among these missions, we can mention: Deep Impact (launched January 2005, to impact comet Tempel 1 and study the induced crater), Stereo (launch November 2005, to study the Sun-Earth system), New Horizons (launch January 2006, the first probe to reach Pluto, its moon Charon, and go deeper to visit one or more Kuiper Belt Objects), and Dawn (launch June 2006, to orbit Vesta and Ceres, two of the largest asteroids in the solar system).

13.5 Future Developments

Traditionally designed for deep-space missions, CCSDS standard coding options must cover new demanding applications like Earth observation missions, characterized by high data rates and very stringent requirements in terms of bandwidth and power. In particular, coding schemes realizing high spectral efficiency and large coding gains are strongly needed.

It is worthwhile to mention that, in the last years, a few other coding options with high code-rate have been added to [1], with respect to turbo codes and the other previous options listed at the beginning of the chapter:

- **RS8**: a (255,239) Reed-Solomon code with 8-bit symbols and Error Correction Capability of 8 symbols.
- **CCP**: a family of 64-state convolutional codes with code-rate 2/3, 3/4, 5/6, and 7/8, obtained by puncturing the mother rate-1/2 code **CC**.
- All possible serial concatenation of an outer standard Reed-Solomon code (**RS/RS8**) and an inner standard convolutional code (**CC/CCP**) through an interleaver of length (255 I) bytes, with I = 1,2,3,4, or 5. (Recently, the interleaving depth $I = 8$ has also been added to the options.)

Moreover, a number of coding schemes approaching the Shannon limit are currently studied by the CCSDS working groups for possible inclusion in the standard, like LDPC (Low Density Parity Check) Codes, Serial Turbo Codes and others. A few years ago, ESA investigated the performance of high rate punctured CCSDS turbo codes for possible inclusion, but this solution has currently been abandoned. It is, however, interesting to briefly review their characteristics, given the context of this book.

Punctured CCSDS turbo codes are obtained by simply puncturing the output of the CCSDS turbo encoder of Fig. 13.1. A set of nominal code-rates (3/4, 7/8, and 15/16) was largely studied in [9], [10].

The performance of rate-3/4 punctured CCSDS turbo codes is very good, especially for large data frame lengths. As an example, the FER performance of a rate-3/4 turbo code is depicted in Fig. 13.22, in terms of both simulation curve and error floor behavior. This code is obtained by puncturing the

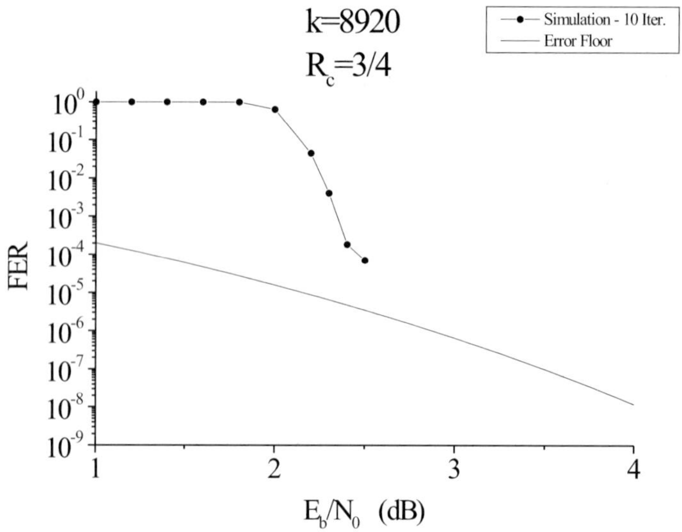

Fig. 13.22. Rate-3/4 turbo code with k=8920 bits obtained by puncturing the CCSDS turbo code: FER simulated performance and error floor.

CCSDS turbo code with k=8920 bits and code-rate 1/2. Its minimum distance (and multiplicity), computed by the constrained algorithm of [6] are equal to $(d_{\min}/A_{\min}/w_{\min}) = (10/29/169)$. Rate-3/4 punctured turbo codes are quite competitive with other solutions like LDPC and serial turbo codes.

The performances of punctured CCSDS turbo codes with higher code-rate are not so good, especially at low error rates, because puncturing causes a significant reduction of their minimum distance. To partially attenuate this drawback, nonsystematic puncturing can be employed, which usually allows to obtain larger minimum distances. However, this causes a worsening of the code performance for higher error rates. As an example, the FER performance of a rate-7/8 turbo code, obtained by puncturing the CCSDS turbo code with k=8920 bits and code-rate 1/2 is reported in Fig. 13.23, in terms of both simulation curve and error floor behavior. For this code, the minimum distance and its multiplicity are $(d_{\min}/A_{\min}/w_{\min}) = (5/79/316)$. In the same figure, the performance of an equivalent rate-7/8 code is reported, obtained by applying a nonsystematic puncturing pattern to the same mother code. This code achieves $(d_{\min}/A_{\min}/w_{\min}) = (6/3/18)$. Its performance for low signal-to-noise ratios are worse. However, its analytical error floor is lower: the nonsystematic puncturing code improves the systematic puncturing code at low error rates.

In general, the performances of punctured turbo codes with code-rate near to the unity (for example, $R_c > 3/4$) are not as good as for lower code-rates.

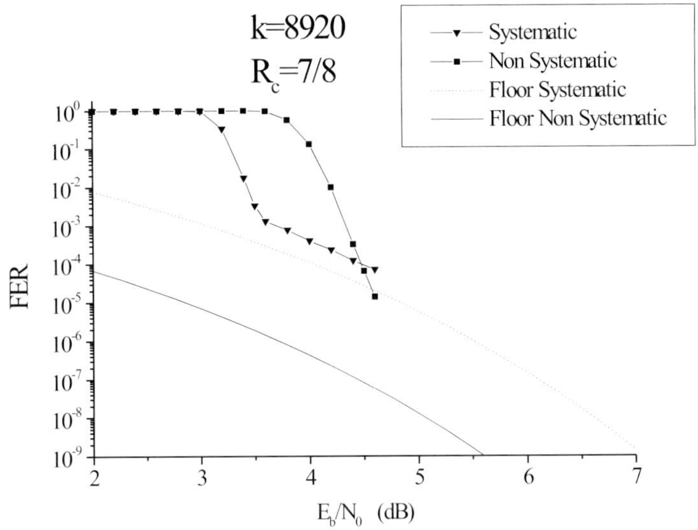

Fig. 13.23. Rate-7/8 turbo codes with k=8920 bits obtained by puncturing the CCSDS turbo code: FER simulated performance and error floor for systematic and nonsystematic puncturing.

The penalty introduced by puncturing is significant and other coding schemes, like LDPC or Hamming product codes, can outperform them [10].

13.6 Conclusive Remarks

In this chapter the turbo codes family of CCSDS standard, the first international standard including turbo codes in early 1999, has been analyzed. They are 16-state parallel turbo codes, with code-rates ranging from 1/6 to 1/2, and input frames from 1784 to 8920 bits. Their performances have been presented, in terms both of simulated error rates and analytical error floors computed from their minimum distances. It has been shown that they are indeed excellent codes, performing near the theoretical limits. CCSDS turbo codes also possess valuable properties for codeword synchronization at the receiver side. Analysis of punctured CCSDS turbo codes show that their performance, still rather good up to code-rate 3/4, becomes worse at higher rates due to puncturing action. CCSDS agencies will use turbo codes for space telemetry, where their very large coding gain will be extremely useful for missions with low communications margins.

13.7 Acknowledgment

CCSDS turbo codes were included in the standard as the result of a joint scientific effort of the international space agencies. Relevant contributions came from NASA-JPL, in particular through the work of Fabrizio Pollara and Dariush Divsalar, and from ESA, in co-operation with Claude Berrou (ENST), Sergio Benedetto and Guido Montorsi (Politecnico di Torino). Finally, a special contribution came from Warner H. Miller of NASA, who was Chairman of the Channel Coding working group and provided his excellent contribution to numerous standards for error control coding. This chapter is dedicated to his memory.

References

1. *Consultative Committee for Space Data Systems (2003) TM Synchronization and Channel Coding.* Blue Book 131.0-B-1
2. *Consultative Committee for Space Data Systems (2000) Packet Telemetry.* Blue Book 102.0-B-5
3. *Consultative Committee for Space Data Systems (2001) Advanced Orbiting Systems, Networks and Data Links: Architectural Specifications.* Blue Book 701.0-B-3
4. *Consultative Committee for Space Data Systems (2003) Radio Frequency and Modulation Systems - Part 1: Earth Stations and Spacecraft.* Blue Book 401.0-B-14
5. Beyer G, Engdahl K, Zigangirov K Sh (2001) "Asymptotic analysis and comparison of two coded modulation schemes using PSK signaling - Part I." *IEEE Transactions on Information Theory* 47:2782–2792
6. Garello R, Benedetto S, Pierleoni P (2001) Computing the free distance of turbo Codes and serially concatenated codes with interleavers. IEEE J. on Selected Areas on Communications 19:800–812
7. Garello R, Vila A (2004) "Turbo code minimum distance computation by the all-zero iterative decoding algorithm." In: *Proceedings of ICC'04* :800–812, Paris, France
8. Calzolari G P, Chiaraluce F, Garello R, Vassallo E (2001) "Symbol synchronization properties of CCSDS turbo codes." *International Journal of Satellite Communications* 20:379-390
9. University of Ancona (2000) *Study of Bandwidth-Efficient Coding Schemes for Near-Earth Applications.* ESA/ESOC Contract No. 14128/00/D/SW - Final Report
10. University of Ancona (2001) *Highly Efficient Channel Codes for High Data Rate Missions.* ESA/ESOC Contract No. 15048/01/D/HK(SC) - Final Report

Part V

Implementations

Chapter 14

VLSI for Turbo Codes

Guido Masera

Politecnico di Torino, Italy

The impressive developments of high performance digital CMOS technologies in the last years have made possible the implementation of very complex algorithms in low cost chips. Digital signal processors, programmable devices such as FPGAs and VLSI technologies have come to the point where the computing power and the memory required to execute several real time applications can be incorporated even in cheap portable devices. Among the several application fields that have been strongly developed by this technology progress, channel decoding is one of the most significant and interesting: as predicted by Shannon, the achievement of performance close to the theoretical limit implies the adoption of high complexity algorithms, so making the search for good decoding architecture a challenging and fascinating problem for many digital architects and VLSI designers. There are a number of aspects of turbo codes that make them so interesting also from an implementation point of view: first of all the algorithms that are used to implement the MAP decoding (such as the so called BCJR algorithm) are of great complexity; this complexity, coupled with the iterative nature of the whole decoding process, makes very difficult the accomplishment of throughput, latency, cost and energy dissipation constraints imposed by several modern multimedia and interactive applications. Moreover turbo decoders include large RAM memories that need to be organized and managed properly. Finally, the decoding algorithm can be parallelized in several ways, acting at different levels and the best solution for each application can only be selected by carefully exploring the space of design alternatives.

The chapter reviews all main achievements obtained in the last 10 years in the hardware implementation of turbo codes and particularly the critical aspects that designers have to deal with. The main addressed problems are the following.

- *Digital architectures for the implementation of basic BCJR equations, related to the forward and backward recursions*
  Starting from the straightforward mapping of the SISO equations to a

K. Sripmanwat (ed.), Turbo Code Applications: a journey from a paper to realization, 347–382.

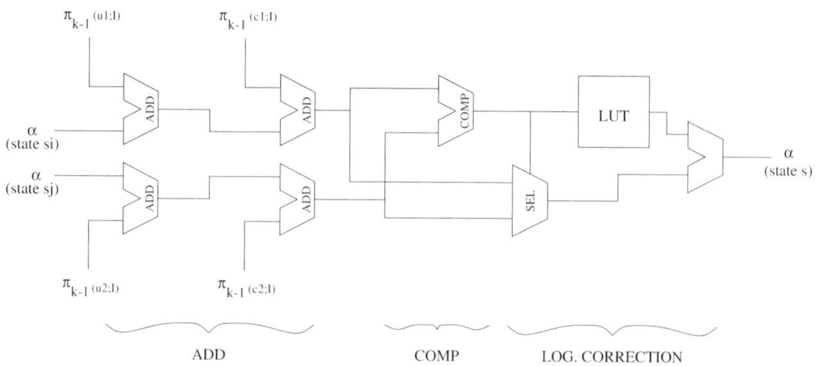

Fig. 14.3. Block scheme for a *max star* architecture

The two most widely known techniques to approximate Eq. 14.5 are the *max* and *max star* solutions:

(a) In the *max* approximation, given in Eq. 14.7, add operations are used to evaluate the logarithm of two or more product terms (indicated as λ_1 and λ_2 in Eq. 14.5), while the max operation, implemented as a compare and select sequence of steps, approximates the logarithm of a sum of exponential terms. This solution, known as *max-log-MAP*,

$$ln(exp(\lambda_1) + exp(\lambda_2)) \approx max(\lambda_1, \lambda_2) \tag{14.7}$$

is of course a great simplification from the implementation point of view, but is known to determine a performance loss of about 0.5 dB on the BER curve of a turbo coded system [9].

(b) The *max star* approximation achieves coding gains within 0.1 dB from the ideal MAP decoder by simply adding to the previously defined max operation a correction term based on the difference of the two λs:

$$ln(exp(\lambda_1)+exp(\lambda_2)) \approx \overset{*}{\max}(\lambda_1, \lambda_2) = max(\lambda_1, \lambda_2)+ln(1+exp(-|\lambda_1-\lambda_2|)) \tag{14.8}$$

A few bits are typically used to represent the correction term, which can then be stored in quite a small look-up-table.

The *max star* approximation is slightly more complicated than the simple *max* operator, requiring the addition of a correction term usually generated through a look-up table. The typical structure of the arithmetic unit implementing the computation of Eq. 14.1 or 14.2 by means of *max star* approximation is given in Fig. 14.3: the block diagram resembles the ACS (Add-Compare-Select) architecture traditionally adopted in Viterbi decoders, but is completed by the final adder that sum the correction term provided by the look-up table. The main drawback of this solution is mainly related to the delay contribution that is due to the final adder placed along the metric

Table 14.1. Comparison between *max* and *max star* approximations in a modified ACS unit

approximation	bit nr.	Virtex gates	Virtex frequency	Virtex2P gates	Virtex2P frequency
max ACS	6	660	91 MHz	600	114 MHz
	4			500	128 MHz
max star ACS	6	900	63 MHz	900	98 MHz
	4			700	104 MHz

computation path; this adder and the look-up table of course also extend the occupied area. It has been proved that the performance loss associated with the *max star* approximation is in the order of 0.1 dB relative to the exact MAP formulation, what is usually considered acceptable. Synthesis results are given in Table 14.1 for a modified ACS unit implemented with the two mentioned approximations; the reported figures refer to two different FPGA technologies and show that the *max star* version roughly costs 30% of additional area and delay with respect to the pure *max* structure. For this example, input LLRs are represented on 6 and 4 bits (second table column), while state metrics are numbers respectively on 9 and 7 bits (see Sec. 14.2.2 for a discussion on the fixed point representation of received and internal metrics).

The original formulation of the BCJR algorithm, based on a sum-product algorithm, is implemented in the special case of analog decoders, as an alternative to the logarithm probability domain; Gilbert multipliers are typically allocated to evaluate the probability products, achieving very high performance to complexity ratios [10] [11] [12], with reduced power dissipations. However the advantages of the analog turbo decoder implementation are at the moment offset by typical limitations of analog circuits, such as the sensitivity to temperature and process variation and poor scalability with technology.

14.2.2 Fixed-point Representation

One of the key points in the implementation of channel decoders is the fixed point representation for all quantities involved in the decoding algorithm. This design choice is usually based on acceptable trade-offs between error correcting performance on one hand, and cost and power dissipation on the other: both dynamic and precision adopted for representing data tend to affect heavily the decoder behaviour; the allocated number of bits also has a direct impact on cost and power dissipation, especially when dealing with hardware implementation for FPGA and VLSI technologies. The choices to be taken are related to

- finite precision representation of LLRs
- finite precision arithmetic for SISO internal processing

The finite precision representation of input LLRs has a strong impact from the hardware point of view as it directly affects the capacity of input buffers;

as a consequence the number of allocated bits has to be set to the minimum possible for a given bit error rate target. The log-likelihood ratio produced for a given coded bit c by an ideal, infinite precision demodulator is a real number that can be expressed as

$$\lambda = log\frac{p(y|c=1)}{p(y|c=0)} = 4\frac{y\sqrt{E_S}}{N_0}$$

where y is the output of the matched filter, E_S is the signal energy and $N_0/2$ is the two-sided noise power spectral density. According to the fixed point representation, in general d bits will be used to represent the integer part of λ (d is usually defined as the "dynamic") and p additional bits will be reserved for the decimal part ("precision"):

$$\lambda \quad \rightarrow \quad b_{d-1}2^{d-1} + ... + b_1 2^1 + b_0 2^0 + b_{-1}2^{-1} + ... + b_{-p+1}2^{-p+1} + b_{-p}2^{-p}$$

Simulations results have been used to understand the influence of both precision and dynamic on the performance of turbo codes and so to properly choose the LLRs representation. In general, when a low number of bits d is allocated for the dynamic, metrics saturate at low values and this tends to badly affect the "error floor" of the performance curve; on the contrary, an insufficient number of bits p of precision affects the decoder behaviour at low SNRs (convergence abscissa of the iterative decoding). As an example, in [13] and [14] extensive simulations have been performed on some significant cases (such as for example the UMTS turbo decoder) for different choices of p and d: from the performed simulations, two bits of precision and three of dynamic give performance very close to the ideal case, while reducing p to one single bit results in a loss around 0.1 dB at low to medium SNRs and setting $d = 2$ costs 0.3 dB

Concerning the fixed-point processing performed internally to the SISO, it must be considered that path metrics α and β increase their values with k, and some action is required in order to avoid an arithmetic overflow (or underflow). Three possible solutions have to be considered:

(a) subtracting the smallest metric from all the path metrics (α or β);
(b) subtracting the metric related to the same state at each trellis step;
(c) managing the overflow effects, rather than avoiding it.

The first solution implies that a set of comparators and subtracters are allocated to find the smallest metric and to perform the normalization; of course this choice results in both area and delay penalty.

In the second solution, instead of searching for the smallest metric at each step, a fixed state metric is subtracted from all path metrics, so avoiding the allocation of comparators; saturation must be used to prevent overflows.

The normalization of the α and β metrics at each updating step can be avoided by adopting a proper numerical representation of the metric differences.

- metrics are updated after relative comparisons performed by the ACS sections and the result does not change if the same constant is added to each α or β metric;
- all possible differences between pairs of path metrics are upper-bounded: path metrics may grow from one trellis step to the other, but their relative differences have an upper limit, Δ_{MAX}, which depends on the distance profile of the code and on the number of bits used to represent LLRs. An expression giving the largest difference assumed by the SISO path metrics has been derived in [14]:

$$\Delta_{MAX} = \min_{w}(wM_u + d_{min}(w)M_c)$$

  where $d_{min}(w)$ is the minumum weight of the code sequences generated by input sequences with weight w, $\pm M_u$ and $\pm M_c$ are the bounds for respectively extrinsic information and LLRs received by the soft demodulator: while $d_{min}(w)$ depends on the considered code, $\pm M_u$ and $\pm M_c$ are simply related to the number of allocated bits in the intrinsic and extrinsic information.

For the convolutional code of the UMTS-3GPP standard, $d_{min}(w) = 8$ for $w = 2$ and $d_{min}(w) = 6$ for $w = 3$ [14]; assume that $N_b = 6$ bits of quantization are used for both the received LLRs and extrinsic information, represented in two's complement format: the maximum absolute values for u and c symbols are

$$M_u = M_c = 2^{N_b - 1} = 32$$

the worst case is when all LLRs along the trellis path are equal to M_c; in this case, the maximum path metric difference is given by

$$\Delta_{MAX} = wM_u + d_{min}(w)M_c = 3 \times 32 + 6 \times 32 = 288$$

If path metrics are represented as positive integers using n bits, where n is given by

$$n = \lceil log_2 \Delta_{MAX} \rceil + 1 \tag{14.9}$$

(n is equal to 10 in this example) the metric difference has a range from $-2^{n-1} = -512$ to $2^{n-1} - 1 = 511$, so that it can be represented in general as a 2's complement value with n bits and the sign bit has weight $n - 1$.

As also done for the computation of metrics in Viterbi decoders [15] [16], the limited spread of path metrics can be exploited to achieve two important implementation results:

- to define the proper number of bits to be used to represent metrics
- to avoid the normalization step when updating the forward and backward metrics, so saving the implementation cost of subtractors.

Both choices are based on the fact that, if metrics are represented according to Eq. 14.9, the metric spread will always be smaller than the half of the

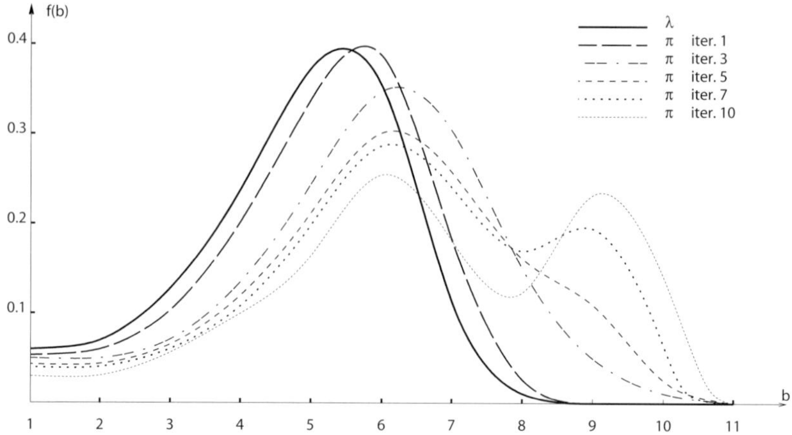

Fig. 14.4. Density of LLR values in a 8 state SISO with $N_{int} = 320$ and $E_b/N_0 = 0$ dB

maximum quantity that can be represented by the n bits; under this condition the result of a subtraction between two metrics is unaffected by the occurred overflows or underflows and as a consequence the normalization is not required.

For a given fixed point representation of LLRs, the number of bits required to represent path metrics can be reduced by slightly reducing the dynamic range of LLRs, with a minor impact on the decoder performance: for the UMTS-3GPP example, if LLRs are saturated to ± 21 instead of -32 and $+31$, we have:

$$Mc = 21 \qquad \Delta_{MAX} = 255$$

so the required number of bit is

$$n = \lceil log_2 \Delta_{MAX} \rceil + 1 = 9$$

A further choice to be taken in the hardware implementation of iterative decoders is related to the management of fixed-point quantities between one iteration and the other. During the iterative decoding process, the extrinsic informations generated at the output of the SISOs tend to modify their statistics in a way that depends on the signal to noise ratio: typically at high SNR the LLRs values show a fast growing, while at low SNR the growing is null or slow.

Fig. 14.4 reports the LLR density for the case of a 8 state SISO with $N_{int} = 320$ and $E_b/N_0 = 0$ dB: extrinsic information (π_i) and input LLR (λ_k) are described through the probability density function

$$f(b) = P[2^{b-2} \leq |\gamma| < 2^{b-1}]$$

A second example, again for a 8 state SISO with $N_{int} = 320$, is given for the case $E_b/N_0 = 2$ dB in Fig. 14.5.

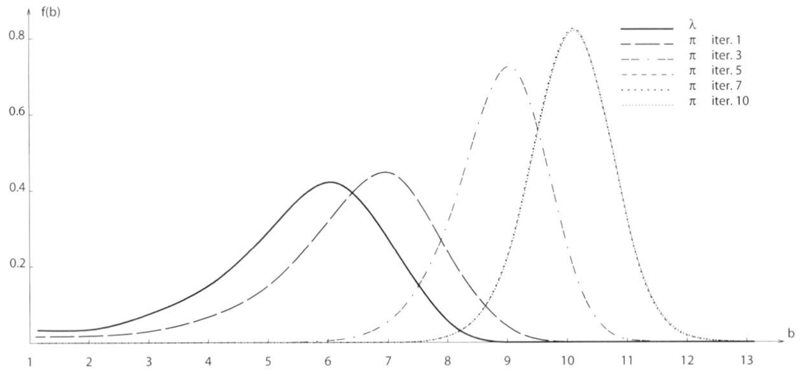

Fig. 14.5. Density of LLR values in a 8 state SISO with $N_{int} = 320$ and $E_b/N_0 = 2$ dB

To better accommodate this statistical distribution of extrinsic information along the different stages of the decoding process, the number of bits used to represent the integer part of LLRs (d) propagated from one SISO to the other can be increased, so improving the "error floor" on the performance curve. A simulation study reporting the effect of different choices for the UMTS turbo code is available in [14]: the work shows that the output LLR dynamic should be $d + 2$ when setting to d the dynamic of the input LLRs. If a single SISO hardware unit is shared among different stages, the given rule results in the constraint that all input and output metrics make use of the same representation; when multiple hardware units are allocated to execute in pipelining subsequent iterations, the number of bits can be changed from one unit to the other, gradually increasing the dynamic and reducing the precision. More in general, the fixed point representation may be made adaptive with the changing SNR measured on the channel: at high SNR, d can be reduced, by shifting the decimal point to the left, while at low SNR, p had better be increased (right shift). To support this function, a SISO unit needs to be completed with adaptive logic capable of modifying the point position according to provided SNR estimations (Fig. 14.6).

14.3 SISO Architecture

In the *sliding window* approximation, the block of data to be decoded is divided into a number of segments, or windows (length N_W), and the decoding algorithm is applied to each of them in sequence. It has been proved that the performance loss due to this approximation is negligible, provided that, in addition to the considered window, Eq. 14.1 and 14.2 are applied along a initialization portion of the trellis (length N_S), which is processed just with the purpose of obtaining reliable starting state metrics for the actual decoding of the current window (see Fig. 14.7). In order to properly concatenate

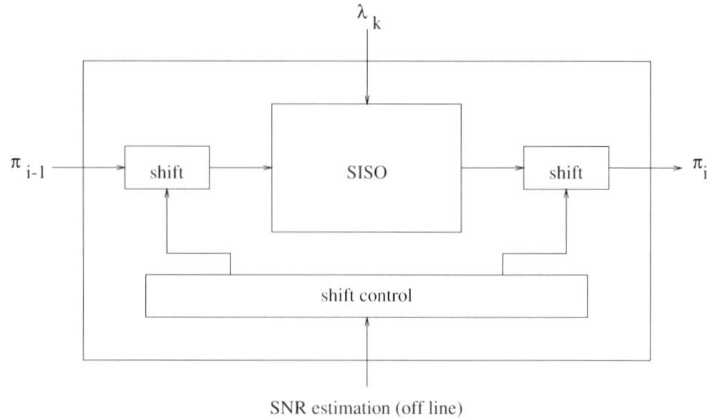

Fig. 14.6. Adaptive logic to control dynamic and precision

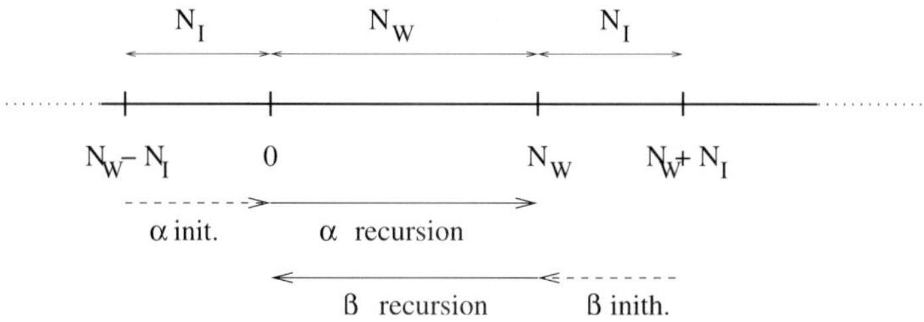

Fig. 14.7. Forward and backward recursions with initialization portions

the portions of updated metrics, windows have to be overlapped, resulting in some overhead of required hardware resources.

The lowest cost solution is to limit the window overlap to the backward recursion, obtaining the starting α metrics for a given window from the last metrics of the previous one. This solution, known as *single flow* structure and already introduced in [17] as an alternative to the storage of the entire state metric history, consists in performing a double β backward recursion: the first one starts with uniform LLR's and goes back along the trellis for N_I steps (say from step $N_W + N_I$ back to step N_W), without the need of storing intermediate β values, except for the last ones, that are generated at the trellis position N_W; these values become reliable initialization metrics for the second β backward recursion, going from step N_W back to step 0. During this second recursion, all generated β metrics are reliable enough to be used for decoding in Eq. 14.3 and 14.4. Three processing units are then required: one is needed to update α metrics, two must be allocated to support two simultaneous backward recursions, respectively along the currently decoded window and along the overlapped portion of the following one.

In practical implementations adopting the structure with three processing units, the overlapping portion is usually made equal to the full window length, since this choice results in the best performance to cost ratio and also tends to make simple the control unit. This can be easily shown by means of the following estimations.

Assume that f is the processing rate of the three units that calculate α and β metrics in the *single flow* structure, N_W is the window length and N_I the length of the overlapped portion ($N_I \leq N_W$); the SISO throughput will be given by

$$t = \frac{N_W}{N_W + N_I} f$$

The hardware complexity can be roughly estimated as

$$c = a \times N_W + b$$

where the first term, linear with N_W, is the cost due to memories that store window metrics and the second term includes the other contributions not related to the window length. For a fixed value of N_I, imposed by the expected decoder performance, the throughput to cost ratio will have a maximum for $N_W = N_I$. This conclusion, obtained through an oversimplified cost and performance model, is substantially confirmed also by more accurate investigations, adopting precise cost, delay and even energy consumption models [18]; it is worth noticing here that different optimization criteria for N_W and N_S have been derived in [19] when more general SISO structures are considered and the problem of joint minimization of decoding delay and metric storage is addressed.

A powerful while simple and elegant tool for the analysis of complex signal processing algorithms is represented by the data flow graph (DFG) [20], that has also been applied to various formulations of the *sliding window* SISO [21] [22] [23] [19]. A simplified DFG representation for a possible organization of α and β recursions in the SISO decoding algorithm is given in Fig. 14.8, where time measured along the trellis runs horizontally across data blocks, while processing time runs vertically from top to bottom; both the initialization portions and the updating windows are indicated for α and β processing. More DFG of the kind shown in Fig. 14.8 can be combined to obtain different scheduling schemes of the involved operations, and the DFG representation can be used to evaluate both decoder performance (throughput and latency) and hardware cost (in terms of number of processing units to be allocated and amount of required storage for branch and path metrics).

A popular structure that makes use of the three processing units mentioned above can be represented with the DFG of Fig. 14.9 (*single flow* structure), where the concatenation of three blocks of data is shown; in this structure, no initialization is required for α metrics, as they are inherited from the end of one block to the beginning of the following one, while two β processing units are always operating at the same time to perform β update and initialization along consecutive windows. The black filled square boxes placed along β recursion indicate the operation of output LLR generation, that needs forward and backward path metrics as well as branch metrics related to the current trellis step (α_{k-1}, β_k, $\pi_k(u)$ or $\pi_k(c)$).

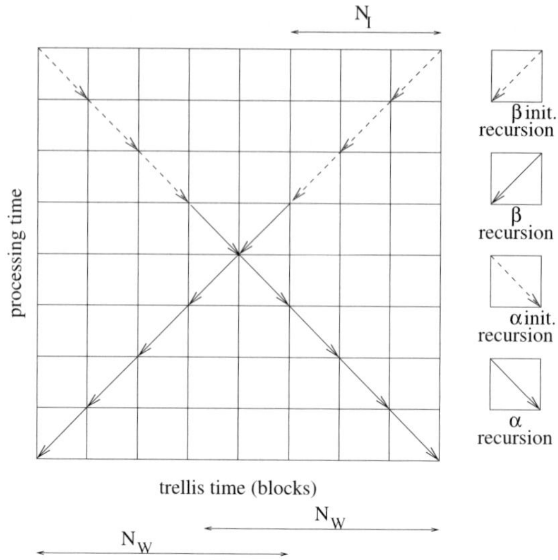

Fig. 14.8. DFG representation for α and β recursions and initializations

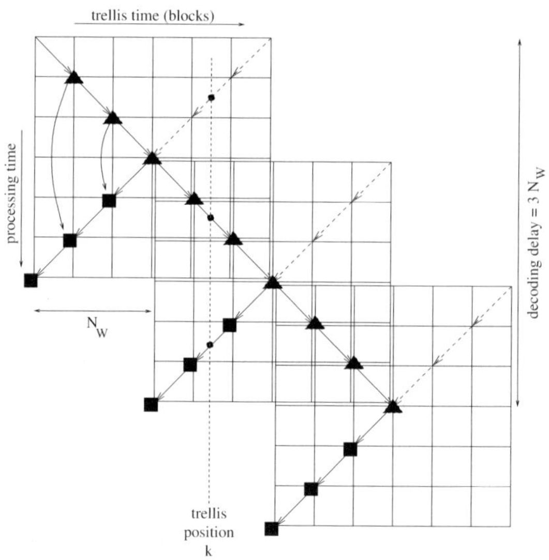

Fig. 14.9. DFG for the *single flow* structure

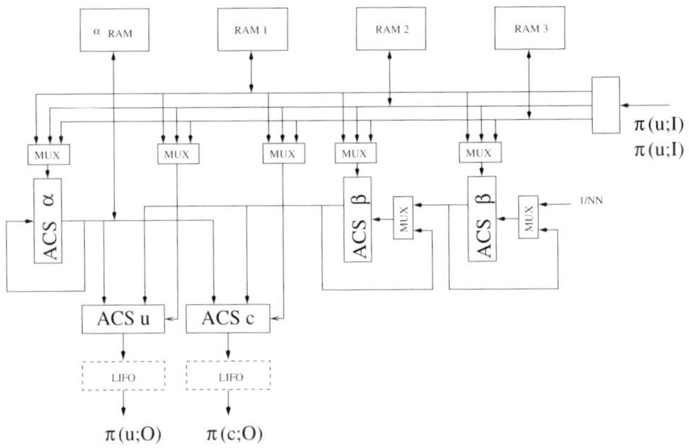

Fig. 14.10. *Single flow* SISO architecture

Memories are therefore required to store branch and path metrics and the amount of necessary memory can be estimated from the DFG analysis. If a vertical slice of the DFG in Fig. 14.9 is considered, it can be easily seen that three operations performed in different times are aligned on the same trellis position, so requiring the same input branch metric: these are the β init., α and β update operations indicated at the trellis position k by the filled black circles. Three buffer memories of size N_W are therefore required to store branch metrics along the three involved windows.

Additionally, the updated α metrics generated while processing a given window (these are indicated in Fig. 14.9 with black triangles) have to be stored up to the output LLR calculation in the following window (see arrowed curves going from triangles to black squares): a fourth memory of size N_W must be used to store these metrics.

From the DFG, the decoding delay can also be evaluated looking at the amount of time that the full decoding of a window takes along the vertical axis: for this example, the delay is given by $3 \times N_W$, as three complete window processing are required.

A general view of the architecture implementing the described processing structure is given in Fig. 14.10, where two further processing units are indicated as ACS u and ACS c: these blocks generate the output LLRs, $\pi(u)$ and $\pi(c)$; these outputs are evaluated simultaneously with the β backward recursion and as a consequence they are generated in the reversed order; the correct order can be restored either allocating LIFO memories at the SISO output or changing properly the interleaving law of the decoder.

The indicated multiplexers distribute the stored branch metrics to the different processing units that operate in a synchronous way, so that a single counter can be used as address generator for all memories.

The DFG representation together with the described method to evaluate decoding delay and required storage can be applied to different computational structures. For example, a *double flow* processing structure has been proposed in [18] as a low latency solution, capable of reducing the decoding delay to two time windows instead of the three required by the *single flow* case: in this solution, the same processing is applied to both β and α recursions, for a DFG similar to that given in Fig. 14.8.

Again the DFG method can be applied to the study of parallel SISO implementations, where more windows are processed in parallel rather than serially [23] [24] [25]: for parallel structures, more DFG's such as the two previously considered, have to be aligned laterally, so supporting the simultaneous decoding of multiple windows.

Although the practical implementation of high performance decoders also involves other important issues, such as the reduction of the dissipated energy or the search for parallel interleaving schemes, a clear and complete study on the many possible alternatives in the SISO internal organization has been recently published by M. Mansour and N. Shanbhag [19], who formally derived expressions for the optimization of decoding delay and metric memories in *single flow* and parallel SISO's.

A completely different approach, which has been recently proposed [26], is based on the idea of evaluating backward state metrics in the forward direction: in principle this is possible if one is able to invert the basic β updating equation, so expressing β_k as a function of β_{k-1}'s. The possible advantage offered by this approach comes from the elimination of the path metric memory in the SISO architecture: in fact the decoding algorithm, after a backward recursion with no storage of intermediate results, two forward processes can be started that simultaneously generate α and β metrics and immediately calculate the output LLR's with very reduced memory requirements.

Therefore the forward calculation of reverse state metrics is potentially able to reduce both SISO complexity and energy dissipation, and some implementations have already been proposed [27] [28] reporting improvements between 15% and 35%; however this approach seems to be critical for two main reasons: first of all, the inversion of the original β recursive equations poses problems of singularity and computational complexity; moreover the numerical precision in the reversed computation has an heavy impact on the BER performance of the code, so that techniques such as the periodic partial storing of traced back metrics are currently being studied as methods for controlling the BER degradation.

The use of radix-4 computation structures has also been explored [29]: similarly to what is routinely done in the implementation of Viterbi decoders [16], the trellis can be compressed in time, collapsing two trellis steps in a single one; this potentially doubles the throughput for a given clock frequency with respect to the radix-2 version, but on the other hand also doubles the edges for each state, so requiring the implementation of radix-4 log-MAP or max-log-MAP units with tree organization [30] [29]. In order to actually exploit the potential speed advantage, a careful optimization is required for the processing units of the SISO.

14.4 Parallel Architectures

Additionally to the widely known adoption of turbo codes for some important communication standards, such as for example DVB-RCS for digital video broadcasting, 3GPP UMTS and CCSDS for space telemetry, they have also been applied to some other different contexts, where very high data throughputs are required, such as 10 Gb/s optical communications and 1 Gb/s high performance storage [31]. In those contexts the hardware implementation of efficient turbo decoders represents a remarkable challenge and a parallel formulation of the decoding process is mandatory to achieve the wanted data rates.

There are a number of possibilities to introduce forms of parallelism in a turbo decoder. The most straightforward idea is to simply allocate more processing units capable of decoding in parallel a number of received data blocks. This approach offers the advantage of reusing an already available decoder and to keep low the latency of the parallel system, which will be the same as the latency of the single processing unit, while the throughput roughly increases linearly with the number of parallel units. Of course the cost paid for these advantages is the area increase, resulting from both allocated decoder instances and additional communication and storage resources: especially with large block sizes, the required buffering between pipelining stages and the replicated interleavers would lead to unacceptable die areas; moreover the complexity increase is accompanied by a proportional power dissipation increase.

Rather then having multiple complete decoders working at the same time, specific sub-parts of the decoding algorithm can be arranged for parallel processing; typically the levels of parallelism that are considered are related to the path metric update Eq. 14.1–14.4, to the data block and to the iteration process.

Parallelism is routinely introduced in the implementation of the ACS modules that implement Eq. 14.1–14.4: forward and backward updating recursions can be applied independently for each state by allocating multiple hardware units that work in parallel on the different states of the convolutional code.

The sliding window technique described in Sec. 14.3 enables the independent processing of consecutive sub-blocks and can therefore be seen as a method to perform parallel decoding at the block level. Although the independent decoding of sub-blocks implies a degradation of the performance in terms of error correcting capability, due to the fact that state-metrics at the borders of each sub-block are not known, in practice the effect on performance can be made negligible provided that the border metrics are estimated on overlapped initialization windows of extension typically equal to a few times the constraint length of the convolutional code [32]. For example, the data block can be divided in M sub-blocks and, instead of having a single hardware module for the whole forward metric updating, M modules can be used to start simultaneously the processing on each of the M sub-blocks; the initial metrics for each sub-block are computed on the tail of the previous sub-block and this can be achieved by simply overlapping the sub-blocks (*overlapping sliding window* method [33]).

An alternative method that avoids the throughput and latency penalty coming from this overlapping is to initialize the border metrics at the values that metrics assumed in the previous iteration of the decoding process.

The independent encoding of each sub-block has also been proposed [34] as a method to support block level parallelism without sub-block overlapping: this method, known as *multiple slice turbo encoding*, modifies the encoder, where separated sub-blocks are independently encoded with proper tail terminations; on the decoder side the additional overhead is limited to the estimation of starting and ending states for each sub-block.

Finally parallelism can be added at the iteration level, allocating multiple decoders (I) that work in a pipelined fashion: at a given time each of the I decoders performs a single iteration on a separated block, so supporting a decoding throughput I times higher than for a single decoder architecture. Similarly to the parallelization of whole decoders working on subsequent blocks, also the unfolding of the iterative decoding algorithm in a pipelined architecture requires additional storage for the multiple blocks that are simultaneously processed.

In order to speed-up the decoding process without excessive penalty in terms of memories to store blocks of data, the single decoder must be made faster, rather than simply putting more decoders in parallel or in pipelining. This approach implies that each SISO is replicated in M parallel units, again interconnected either in parallel or in pipelining. However this approach generates another obstacle to the parallelization of a turbo decoder that is currently subject of investigation by several researchers. In a turbo decoder, extrinsic information has to be exchanged among SISO units via interleavers or de-interleavers, at each decoding iteration; interleaver and deinterleaver are typically implemented as unique memories with proper addressing units. In the case of a parallel architecture, multiple SISOs generate a number of soft values that have to be carried in parallel to another set of SISO units in a scrambled order, that is through an interleaver: in the parallel structure, the interleaver cannot be mapped to a unique memory, because its access time would set a severe throughput limitation and the interleaver would become the system bottleneck. Instead more memories must be allocated, with the constraint that all simultaneous accesses are always referred to different memories.

It can be easily seen that meeting this constraint is not trivial and a random interleaver frequently generates access requests referred to the same memory: this event is usually known as *collision*. There are a number of techniques to avoid collisions. The most straightforward solution to this problem is to simply accept the collision and to stop the decoder when a collision occurs: in other words, when two SISOs need to read a data from the same memory, two subsequent access operations are scheduled and the decoding process is slowed down. This solution is rarely acceptable, due to its cost in terms of latency and throughput.

As the interleaver plays a key role in determining the error correcting capabilities of a turbo code, the design of a dividable interleaver that avoids any collision when mapped to a given number of parallel memories is not straightforward, but requires to jointly consider performance and collision avoidance constraints. Many works are available in the literature on the subject of collision management in turbo codes and the proposed techniques can be classified in two categories. The first broad category includes several methods of designing turbo codes capable of supporting up to a given amount of parallelism: in this case, the collision constraint has been inserted in the process of code design, so that the obtained codes are intrinsically collision-free and they can be supported by suitable parallel architectures. The second category on the other hand collects architectural approaches that are collision-free independently on the code, or at least capable of reducing the effects of collisions for all possible codes. Sec. 14.4.1 and 14.4.2 present the most interesting solutions proposed for the two categories.

14.4.1 Design of Collision-Free Turbo Codes

In principle the design of a collision-free turbo code implies a proper definition of the permutation law, where the two defined access sequences - the "in order" and the "scrambled" ones - are both supported by a parallel architecture. The problem of partitioning of the interleaver on more parallel memories can be represented by means of a graph in which each vertex represents a bit position in the block (and therefore an address in an undivided interleaver memory) and the edges connect vertexes that are accessed simultaneously: adjacent vertexes represent cells that must be placed in different partitions. Thus the interleaver partitioning in sections can be stated as a graph coloring problem [35]. In Fig. 14.11, a simple example of interleaver partitioning is reported. On the left the interleaver access graph is shown, where the "in order" and the "scrambled" sequences are represented with solid and dashed lines respectively. On the right the obtained partitioning is reported: in this case only two memories (RAM1 and RAM2) are needed.

A simple technique to design a collision-free interleaver, with assigned block size and number of parallel SISOs, applies special transformations to a starting row-column interleaver [36] [33]: these transformations are a sequence of cyclic shifts along the columns of the starting matrix and their effect is to evenly distribute the accesses among available memories, so obtaining a collision-free interleaver. In order to reduce the regularity of the obtained permutation law and to increase its spreading properties, both intra-row and inter-row permutations have been suggested as additional transformations that do not affect the property of avoiding collisions, but at the same time improve decoder performance. The authors also show how the proposed

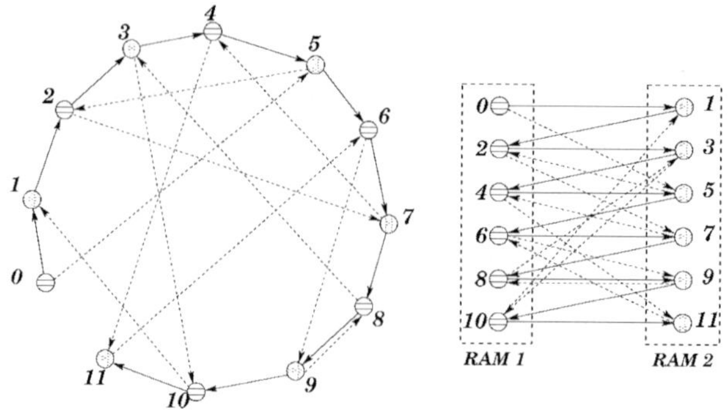

Fig. 14.11. A simple example of interleaver partitioning.

technique can be applied for all possible combinations of block size and degree of parallelism of the decoder architecture.

In [33] a scalable and parametric architecture implementing parallel PCCC decoders is presented: the overlapping windows technique is adopted in conjunction with collision-free interleavers (obtained through column cyclic shifts) and results are provided in terms of occupied area, dissipated power and offered throughput for several block sizes. The proposed solution has been shown to achieve the same performance as the UMTS interleaver [33]. Additionally it has the advantage of generating simple and highly scalable decoder architectures; on the other hand, comparisons in terms of performance with other interleavers, such as S-random interleavers, are not provided.

The generation of a collision-free S-random interleaver is described in [37] and [38], together with its implementation in the specific case of a rate 1/3 turbo code and block size equal to 4096. The proposed technique basically inserts collision avoidance constraints in a procedure for the generation of S-random interleavers and the resulting penalty in terms of code performance is limited.

The combination of spatial and temporal permutations is another technique that has been recently explored to obtain dividable interleavers resulting in good performance. In [39] a bit reversal permutation is concatenated with optimized shuffle patterns capable of avoiding low-weight error events. In [40] a two stage interleaver is proposed able to cover a wide range of interleaver sizes: the first stage implements a spatial permutation by means of a cross-bar unit, while the second one applies a temporal permutation to each parallel SISO; the most interesting feature of this solution is perhaps that it offers trade-offs between latency of the obtained architecture and performance of the code. A further proposal recently published [34] again combines spatial and temporal permutations with specific optimizations aimed at obtaining a high spread on single error events; the main limitation of the this solution deals with the application of the tail-biting technique to the case of short sub-blocks.

14.4.2 Design of Collision-Free Architectures

Instead of constraining the interleaver design to be supported in a parallel decoder, the collision problem can be faced by designing proper architectures capable of masking the conflicts in the memory access. The indubitable advantage of this approach is that it is viable in principle for any permutation law and, as a consequence, it is standard compliant.

Two main solutions can be mentioned for this interleaver-independent approach. The first solution is based on the simple observation that, when an interleaver is mapped to more parallel memories, frequent conflicts in the access to a specific RAM are generated for a random interleaver, but each memory is on average accessed once per cycle. Rather than modifying the permutation in order to eliminate the conflicts, one can simply manage the conflicts by buffering data that cannot be immediately written to (or read from) the memory. A number of papers have been published on this approach, proposing different buffering architectures [41] [42] [43] [44], which in fact work as routers. Basically, for a parallel architecture with M SISOs and M interleaving memories, a ring-shaped network is allocated, consisting of M RIBB (Ring Interleaver Bottleneck Breaker) cells placed between each pair SISO unit - memory: each cell is connected to the previous and to the following ones in a ring structure and include a local buffer. A LLR generated by a SISO is accompanied with a tag identifying the destination RAM; according to this tag, the LLR is send along the ring and, when received by the cell connected to the destination RAM, it is either written into the memory or, in case a conflict occurs, stored locally for later writing. A RIBB cell also needs buffers at the left and right ring outputs to accommodate simultaneous circulating data. As authors show, the required buffer length decreases with M for the local buffer, but the size increases for right and left buffers; due to their impact on required area and dissipated energy, the large size sometimes reached for these buffers is the major limitation of the proposed solution.

With the purpose of combating the growth of buffer size, the RIBB scheme has been generalized in a subsequent work [43], allowing for more connections between one cell and the others. The new scheme, named GIBB, for Generalized Interleaver Bottleneck Breaker, makes use of more complex network topologies that reduce both the latency in the transport of LLRs and the length of the buffers; on the other hand, in this case the area required for the communication structure is expected to increase significantly with respect to the RIBB case.

The second approach recently proposed to obtain parallel interleavers with no constraints on the permutation law is described in [45] [46]: the work shows that, given an arbitrary permutation law and a degree of parallelism M, it is always possible to find a proper mapping that associates each interleaver input to a specific memory bank with no collision. A two step algorithm is proposed that starts with a preliminary partial mapping and iteratively completes the mapping with a procedure called "annealing". The described approach results

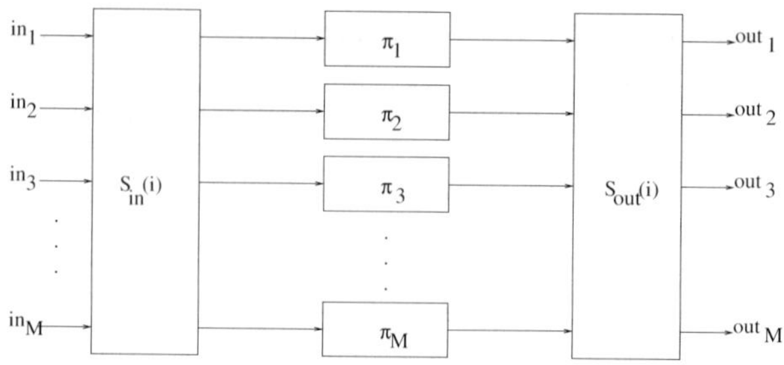

Fig. 14.12. Implementation of a parallel interleaver

in the implementation architecture shown in Fig. 14.12, consisting of a three layers of processing: an input nonblocking switch, which applies a time-varying permutation to the M inputs received at each cycle; a bank on M memories acting as independent interleavers of size N/M, where N is the block size; a final switch implementing a second time-varying permutation.

In this approach, the natural order of memory access cannot be grantied for neither of the two steps of a decoding iteration; this implies that two mappings for the two half iterations have to be stored, so doubling the memory requirements with respect to schemes that make use of a single scrambling sequence. Although this penalty can be severe for large block sizes, this approach has the advantage of allowing to implement standard compliant turbo decoders.

14.5 Energy Aware Techniques

The adoption of turbo codes as the Forward Error Correction technique in important standards for wireless communications imposes the search for IC solutions capable of reducing the energy computation. This is a hard task for complex systems such as turbo decoders, that tend to have a larger power consumption than other decoders [47] because of three main reasons: (1) the hardware complexity is larger than for other decoders, such as Viterbi decoders for example; (2) the decoding process is iterative and this implies that the clock frequency must be kept high with respect the the decoding throughput; (3) large memories are included in the decoder architecture, such as the interleaver memories and the input-output buffers.

In order to reduce as much as possible the dissipated energy per decoded bit and per iteration, one has to face the problem optimizing the implementation at different levels: algorithm, architecture and circuit.

At the algorithm level, a popular technique for controlling the energy dissipation is to introduce in the decoding process some methods for stopping the iterations after a given reliability has been reached in the decoded outputs. A very simple idea is to employ a CRC (cyclic redundancy checking) scheme concatenated with the turbo decoder: when the frame is completely decoded and no errors are signaled by the CRC, no redundant iteration is performed.

The algorithm for stopping the iterations can be adaptive:

(a) Initialize the number of iterations S to 1 and identify a maximum number of supported iterations, $S < S_{MAX}$.
(b) Decode the frame, increment S and see from the concatenated CRC if there is any bit error in the frame.
(c) Steps (a) and (b) are repeated until $S = S_{MAX}$ or no errors are detected in the current frame.

When the job is finished, the circuit is shut down before the next data frame arrives. As the energy dissipation tends to grow linearly with iteration, the percentage of power saved with this approach is roughly equal to the average reduction achieved in the number of iterations, which can be as large as 75% at high SNRs. The effect of this algorithm on a rate 1/3 turbo decoder with block length 736 has been reported in [48].

The iteration number can also be controlled by means of a decision-aided stop criterion, somehow related to the QoS requirements. A known method makes use of output log-likelihood ratios that are compared to a threshold previously set on the basis of the target bit error rate [49]: if LLRs magnitudes of all bits in a block are under the threshold, decisions are not considered reliable enough are new iterations are scheduled, while a passed threshold indicates that a sufficient confidence in the decoded outputs has been achieved and the block decoding is stopped. Also this technique gives percentages of energy reduction as high as 50%. Simpler and less precise techniques to control the iteration number include the evaluation of the number of 1's accumulated from the output of the decoded block at each iteration [13]: a necessary condition for having identical decoded bits from current and previous iterations is that the two accumulated values are equal.

It is well known that in complex digital systems very large margins of power reduction are offered by architecture-level optimizations, which basically consist of design exploration aimed at finding the best area-energy trade-off for a given application.

From the analysis of the main contributions to power consumption in a turbo decoder, many authors founded that most of power is dissipated in the memories; more precisely, for a sliding window approach, around 90% of energy consumption in a SISO module is due to the branch and path metric memories; moreover a large part of the energy consumption is typically due to the interleaver memories and to the input/output buffers. Of course the relevance of the memory contribution to the global power dissipation tends to reduce when high-speed, parallel or pipelined architectures are considered, because of the increased amount of logic and clock frequency.

In systems including memories, large percentages of reduction in energy dissipation are achievable by RAM partitioning. In fact splitting a memory in multiple RAMs allows relaxing the constraints on access times and the gained

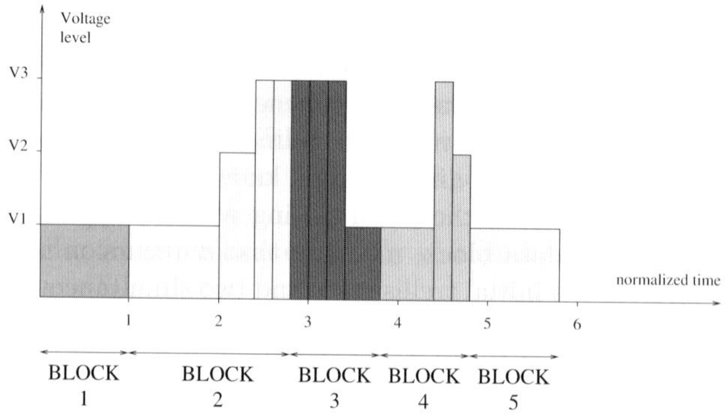

Fig. 14.13. Profile of supply voltage assignments in different iterations

The energy consumption for the kth processing iteration of block n is $E_{nk} = C_{eff}V_{nk}^2$, where C_{eff} is the equivalent total switching capacitance of the decoder and V_{nk} is the assigned supply voltage.

According to Eq. 14.10, the time delay in a digital circuit can be approximately expressed as

$$t_{nk} = \frac{\theta V_{nk}}{(V_{nk} - V_{th})^{\eta}}$$

where θ and η are process dependent constants and V_{th} is the threshold voltage.

The problem is to find a set of voltages V_{nk} to be assigned to each iteration applied to a sequence of N data blocks, so that the total energy consumption is minimum and a given constraint on the decoding delay (T_{tot}). The total energy for the decoding of the N blocks can be expressed as

$$E_{tot}(V_{nk}; n = 1..N, k = 1..K_n) = \sum_{n=1}^{N}\sum_{k=1}^{K_n} E_{nk}$$

where a variable number of iterations K_n can be used for each block n; the delay constraint is stated as

$$\sum_{n=1}^{N}\sum_{k=1}^{K_n} t_{nk} \leq T_{tot}$$

The proposed optimization algorithm starts with a low voltage (*as-slow-as-possible* approach) and increases the supply voltage from one iteration to the other when required, so obtaining profiles of supply voltage such as that provided in Fig. 14.13: here the decoding of each block is started assigning the lowest voltage level; the decoding process then continues with supply voltages set according to the available time: when new iterations are started for the the same block, the supply voltage is increased in order to speed up the decoding. If the previous block required a large number of iterations and as a consequence there is the need of saving decoding time while processing the current block, a high supply voltage is assigned for the new block (see the third and fourth blocks in the Figure).

In order to better exploit the potential energy advantage coming from the supply voltage scaling, the SNR estimation has been suggested as a method to determine in advance the required number of iteration and so optimally assign V_{nk} to the iterations of each block. It can be proved that the lowest energy consumption is achieved when the circuit is running at constant speed all the time and this implies that the best voltage assignment is found by distributing uniformly the global workload.

The actual effectiveness of this approach depends on the SNR and on the number of available voltage levels: reported results show that with 5 levels the energy consumption can be reduced by percentages ranging from 10% at low SNR up to 80% at high SNR.

At the publication time, the lowest values of dissipated power in actually implemented turbo decoders are large than 1 nJ per bit and per decoding iteration. Example of recent published data are in [29] and [51]. The first work presents a 0.18 μm CMOS turbo decoder for 3GPP-HSDPA (High Speed Downlink Packet Access) where power dissipation is reduced combining architectural techniques, such as clamping of extrinsic information to save memory, and iteration control by means of efficient stopping criteria. The second work achieves better power performance (1.45 nJ/ bit per decoding iteration) by means of several algorithm and architecture level optimization: reported data refer to a UMTS-like decoder, implemented in 0.18 μm CMOS technology and supporting 75 Mbps.

14.6 Standards & Products

After years from their introduction, iterative decoding for forward error correction have been adopted for a number of communication standards. The most widely known is probably the 3GPP UMTS standard for wireless cellular communication that, in addition to a traditional convolutional code, specifies the use of a “turbo” code for forward error correction: the selected solution is a parallel concatenation scheme, with two 8-state convolutional encoders and a wide selection of interleavers, up to a maximum size of 5120 bits; the global code rate is 1/3 and the highest bit rate is 2 Mb/s.

Several commercial implementations have been developed in the last years for decoders compliant with the 3GPP standard, both in the form of proprietary “hard IP” cores, integrated in complete transceiver components by manufacturing companies, and in the form of “soft IP” cores, made available as synthesizable HDL (Hardware Description Language) source code: some examples of available IP cores are given in [52], [53], [54], [55], [56].

VLSI design figures, such as complexity, power dissipation and clock frequency, are sometimes available for these hard and soft cores. An example of a complete 3GPP channel decoder known to the author and implemented by a major microelectronic company using a 0.25 μm CMOS technology is characterized by a total gate count of 75,000 and an operating frequency of 60 MHz.

The ASIC approach is not the only solution for the implementation of turbo decoders. Different platforms can be selected for the development of high performance decoders, according to the application and cost constraints. However processing speed, latency, energy consumption and complexity are not the only specifications for modern platforms, but a rapidly increasing role is played by two additional features, namely scalability and flexibility. Scalability is the capability of a platform to adapt to different choices for system level parameters, such as for example the mother convolutional codes or the size of the processed block; throughput, latency and power consumption typically change with these parameters and additional implementation complexity is paid to support scalability, however the decoder architecture is not changed or reconfigured when adapting to a different set of parameters.

On the other hand, the term flexibility is used to indicate the possibility to update an implementation platform in order to support a completely different decoder that does not simply requires a change in some parameters; as an example, a decoder that can be configured for different concatenation schemes (parallel as well as serial) would be flexible.

The most flexible and versatile solution is for sure offered by software programmable platforms such as microprocessors and DSP (Digital Signal Processor). Especially for the case of wireless cellular communications, the use of DSPs to support several processing functions, including the forward error correction, has become very popular and dedicated hardware units have been added to the execution units of some recent DSP, with the purpose of achieving a sufficient speed-up in the processing to support decoding throughputs as high as 2 Mb/s. The turbo-decoder coprocessor TMS320C64X DSP is a typical example of this approach [57].

A high degree of flexibility is also offered by the use of embedded processors specifically targeted to the decoding application: this approach, also known as ASIP (Application Specific Instruction set Processor), is well suited to high performance applications, giving the designer the possibility to exactly tune the structure of the processor to the specific tasks to be executed; at the same time, an ASIP, being software programmed, is almost as flexible as a commercial DSP. ASIP implemented turbo decoders compliant to the 3GPP standard have been recently demonstrated [58] [59].

A different kind of flexibility is provided by FPGA based solutions. Here the large amount of internal resources are exploited to achieve higher processing capabilities with respect to DSP implementations, while still offering flexibility through device reconfiguration. The programming process for an FPGA consists in the uploading of a bit stream containing the information for the internal configuration of logic blocks, interconnects and memories. For most devices, the reconfiguration process takes a long time and implies that the

Table 14.2. VLSI design figures for some ASIC implementations

	[63]	[29]	[64]	[65]
technology process	0.13 μm 1.2 V	0.18 μm 1.8 V	0.25 μm 2.5 V	0.18 μm 1.8 V
core area	10 mm^2	14.5 mm^2	8.9 mm^2	7.16 mm^2
clock frequency	350 MHz	145 MHz	135 MHz	160 MHz
throughput	350 Mb/s	24 Mb/s	5.48 Mb/s	75.6 Mb/s
power figure of merit	1.4 nJ/bit/it.	10 nJ/bit/it.	6.98 nJ/bit/it.	1.45 nJ/bit/it.
gate count		410k	34,400	373k
memory		0.45 Mb	201 kb	36 kb
max block size	2048	5114	20730	432
nr. of iterations	5	16	6	
architecture	5 stages	serial, radix-4 SISO	serial	7 concurrent units
application	generic	3GPP-HSDPA	W-CDMA and CDMA2000	generic

hardware previously mapped to the FPGA is stopped; these two main difficulties are overcome in recent devices that support partial and dynamic reconfiguration. Another major limitation to the adoption of FPGA platforms for wireless communications comes from the high power dissipation of field programmable devices, especially in static conditions.

In spite of these problems, the very short development time connected with the use of FPGAs, coupled to their good performance in terms of processing speed and flexibility determined an impressive growth in the development of FPGA based systems. A number of implementations addressed various turbo decoders achieving medium throughput figures [60], [61], [62], [56].

Iterative concatenated convolutional codes have also been specified by the digital video broadcasting standard (DVB-RCS) that requires decoding throughput significantly higher that wireless cellular communications (68 Mb/s); even higher precessing speed would be required for other applications, such as high performance storage and optical communications. In order to support throughput ranging from tens of Mb/s up to several Gb/s, high parallelism architectures have to be implemented and the custom ASIC is the most suited platform for these applications. Implemented ASIC turbo decoders therefore offer high performance and reduced complexity at the cost of almost no flexibility in the obtained result. Among the commercially available ASIC solutions, a number of cores are offered in [56] for DVB-RCS (60 Mb/s) and other high speed applications, such as disk storage (320 Mb/s) and Echostar/Broadcom video broadcast (110 Mb/s). Several ASIC turbo decoders are available in the scientific literature; Table 14.2 summarizes some of the recently published IC implementations, with available figures and architecture details. The four decoders have been published in [63] [29] [64] [65].

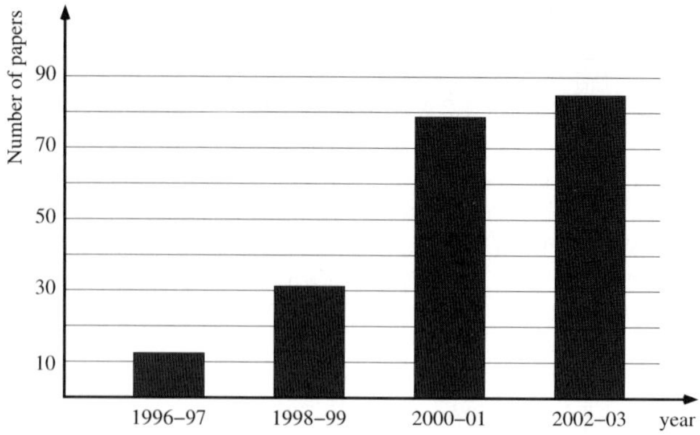

Fig. 14.14. Published IEEE/IEE papers on the implementation of turbo decoders

14.7 Concluding Remarks

The interest of the international community on the implementation of turbo codes has grown very fast after the historical paper published by Berrou, Glavieux and Thitimajshima, and it is now reaching the age of maturity, although many researchers are still active in this field.

A simple search for key words such as "turbo", "implementation" and "architecture" on the IEEE web engine provides evidence for this trend, showing that the number of published works on the implementation of turbo decoders has been continuously increasing for 10 years (see Fig. 14.14).

"Turbo decoders" represent one of the most fascinating cases of study for architecture designers and a formidable example of how the close cooperation between theory experts and VLSI architects is the best via to efficiently translate an idea into a practical application. The attempt to obtain cheaper, faster or more energy efficient decoders has driven the designer efforts for several years, giving up to the commercial diffusion of many implementations. Nevertheless the search for new architectures is still active, primarily directed by the need of more than 1 Gb/s throughput that is typical of some application fields and by the increasing request for scalable or flexible solutions, able to efficiently support multiple codes with different figures of throughput, latency, code rate, etc.

References

1. S. Benedetto, D. Divsalar, G. Montorsi, F. Pollara (1998) "Soft-input soft-output modules for the construction and distributed iterative decoding of code networks", *European transactions on telecommunications*, vol. ETT 9, March-April 1998

2. O. Y. Takeshita and D. J. Costello (2000) "New deterministic interleaver designs for turbo-codes", *IEEE Trans Info. theory*, IT-46, Sept. 2000, pp. 1988–2000
3. L.R. Bahl, J. Cocke, F. Jelinek, and J. Raviv (1974) "Optimal Decoding of Linear Codes for Minimizing Symbol Error Rate", *IEEE Transactions on Information Theory*, March 1974, pp.284–287.
4. S. Benedetto, D. Divsalar, G. Montorsi, F. Pollara (1996) "Soft-output decoding algorithms for continuous decoding of parallel concatenated convolutional codes", *Proceedings of ICC'96*, Dallas, Texas, June 1996.
5. www.etek.chalmers.se/groups/arithdb
6. S. Benedetto, D. Divsalar, G. Montorsi, F. Pollara (1996) "Soft input soft output MAP module to decode parallel and serial concatenated codes", in *TDA Progr. Rep. 42-127*, Jet Propulsion Lab., Pasadena, CA, pp. 1–20, 1996.
7. S. A. Barbulescu (1996) *Iterative decoding of turbo codes and other concatenated codes*, Ph.D. dissertation, Univ. South Australia, pp. 23-24, 1996.
8. S. S. Pietrobon (1996) "Efficient implementation of continuous MAP decoders and a synchronization technique for turbo decoders", in *Proc. Int. Symp. Inform. Theory Appl.*, Victoria, B.C., Canada, 1996, pp. 586–589.
9. P. Robertson, E. Villebrun, P. Hoeher (1995) "A comparison of optimal and sub-optimal MAP decoding algorithms operating in the log domain" *Proc. ICC95*, pp. 1009–1013, 1995.
10. V. C. Gaudet, P. G. Gulak (2003) "A 13.3-Mb/s 0.35 μm CMOS analog turbo decoder IC with a configurable interleaver", *Solid-State Circuits, IEEE Journal of* , Volume: 38 , Issue: 11 , Nov. 2003 pp. 2010–2015
11. H. A. Loeliger, F. Tarkoy, F. Lustenberger, M. Helfenstein (1999) "Decoding in analog VLSI" *Communications Magazine, IEEE* , Volume: 37 , Issue: 4 , April 1999 pp. 99–101
12. H. A. Loeliger (2001) "Probability propagation and decoding in analog VLSI", *IEEE Trans. Info. Theory*, Vol.47, Feb. 2001, pp. 837-843
13. Z. Wang, H. Suzuki, and K.K. Parhi (1999) "VLSI Implementation Issues of Turbo Decoder Design for Wireless Applications", *IEEE Workshop on Signal Processing Systems*, 20-22 October 1999, pp. 503–512
14. G. Montorsi, and S. Benedetto (2001) "Design of Fixed-Point Iterative Decoders for Concatenated Codes with Interleavers" *IEEE Jurnal on Selected Areas in Communications*, Vol. 19, No. 5, May 2001, pp.871-882
15. A.P. Hekstra (1989) "An alternative to metric rescaling in Viterbi decoders", *IEEE Trans. Commun.*, Vol. 37, No. 11, pp. 1220-1222, 1989
16. P.J. Black, T.H. Meng (1992) "A 140-Mb/s, 32-State, Radix-4 Viterbi Decoder" *IEEE Journal of Solid-State Circuits* vol. 27, no. 12, December 1992, pp.1877-1885.
17. A.J. Viterbi (1998) "An intuitive justification of the MAP decoder for convolutional codes", *IEEE Journal on Selected Areas in Communications*, Vol. 16, No. 2, February 1998.
18. C. Schurgers, F. Catthoor, M. Engels (2001) "Memory Optimization of MAP Turbo Decoder ALgorithms", *IEEE Trans. on VLSI Systems*, Vol. 9, No. 2, April 2001, pp. 305–312
19. M. Mansour, N. R. Shanbhag (2003) "VLSI Architectures for SISO-APP Decoders", *IEEE Tras. on VLSI Systems*, Vol. 11, No. 4, August 2003
20. K. K. Parhi (1999) *VLSI Signal Processing Systems: Design and Implementation*, Jhon Wiley & Sons, 1999

21. H. Diwid, H. Meyr (1995) "Real-time algorithms and VLSI architectures for soft output MAP convolutional decoding," *Proc. Personal, Indoor, and Mobile Radio Communications, PIMRC'95*, vol. 1, 1995, pp. 193–197
22. C. Schurgers, F. Catthoor, M. Engels (1999) "Energy efficient data transfer and storage organization for a MAP turbo decoder module," *Proc. of Low Power Electronics and Design Symposium*, 16-17 Aug. 1999, pp. 76–81
23. A. Worm, H. Lamm, N. When (2000) "A High Speed MAP architecture with optimized memory size and power consumption," *Proc. IEEE Workshop Signal processing Systems*, SiPS 2000, pp. 265-274
24. F. Viglione, G. Masera, G. Piccinini, M. Ruo Roch, M. Zamboni (2000) "A 50 Mbit/s Iterative Turbo-Decoder," *Proc. of DATE 2000 Conference*, pp. 176–180, March 2000.
25. Z. Wang, Z. Chi, K. K. Parhi (2002) "Area-efficient high-speed decoding schemes for turbo decoders," *IEEE Trans. on VLSI Systems*, Vol. 10, No. 6, December 2002, pp. 902–912
26. Y. Wu, W. J. Ebel, B. D. Woerner (2000) "Forward computation of backward path metrics for MAP decoders," *IEEE VTC*, pp. 2257–2261, 2000
27. I. Atluri, T. Arslan (2003) "Low power VLSI implementation of the MAP decoder for turbo codes through forward recursive calculation of reverse state metrics," *IEEE Int. SOC Conf.* 17-20 Sept. 2003, pp. 408–411
28. J. Kwak, S. M. Park, K. Lee (2003) "Reverse tracing of forward state metric in log-MAP and max-log-MAP decoders," *Int. Symp. on Circuits and Systems*, 25-28 May 2003
29. M. Bickerstaff, L. davis, C. Thomas, D. Garret, C. Nicol (2003) "A 24 Mb/s radix-4 LogMAP turbo decoder for 3GPP-HSDPA mobile wireless," *International Solid-State Circuits Conference, ISSCC*, February 11, 2003, Session 8, paper 8.5
30. G. Masera, G. Piccinini, M. Ruo Roch, M. Zamboni (1999) "VLSI architectures for turbo codes", *IEEE Trans. on VLSI Systems*, Vol.7, No.3, September 1999, pp.369-379
31. E. Yeo, P. Pakzad, B. Nikolic, V. Anantharam (2001) "VLSI architectures for iterative decoders in magnetic recording channels," *IEEE Trans. Magn.*, vol. 37, pp. 748–755, Mar. 2001
32. A. Hunt, S. Crozier, M. Richards, K. Gracie (1999) "Performance degradation as a function of overlap depth when using sub-block processing in the decoding of turbo codes," *Proc. of IMSC'99*, 1999, Ottawa, Canada, pp. 276–280
33. B. Bougard, A. Giulietti, L. Van der Perre, F. Catthoor (2002) "A class of power efficient VLSI architectures for high speed turbo-decoding," *Global Telecommunications Conference*, 2002, GLOBECOM 2002, Vol. 1 , pp. 549–553
34. D. Gnaedig, E. Boutillon, M. Jezequel, V.C. Gaudet, P.G. Gulak (2003) "Multiple Slice Turbo Codes," *Proceedings of the 3rd International Symposium on Turbo Codes and Related Topics*, pp 343-346, Brest, France, Sept. 2003
35. G. Masera, M. Mazza, G. Piccinini, F. Viglione, M. Zamboni (2002) "Architectural Strategies for Low-Power VLSI Turbo-Decoders," *IEEE Trans. on VLSI*, vol. 10, No. 3, June 2002, pp. 279-285
36. A. Giulietti, L. Van der Perre, M. Strum (2002) "Parallel turbo coding interleavers: avoiding collisions in accesses to storage elements", Electronics Letters, Vol. 38, Iss. 5, Feb. 2002, pp. 232-234
37. J. Kwak, K. Lee (2002) "Design of dividable interleaver for parallel decoding in turbo codes", *Electronics Letters*, Vol. 38, Iss. 22, Oct. 2002, pp.1362-1364

38. J. Kwak, S. Min Park, S. Yoon, K. Lee (2003) "Implementation of a parallel turbo decoder with dividable interleaver," *Int. Symp. on Circuits and Systems*, 25-28 May 2003
39. A. Nimbalker, T.K. Blankenship, B. Classon, T.E. Fuja, D.J. Costello Jr. (2003) "Inter-Window Shuffle Interleavers for High Throughput Turbo Decoding," *Proceedings of the 3rd International Symposium on Turbo Codes and Related Topics*, pp 355-358, Brest, France, Sept. 2003
40. R. Dobkin, M. Peleg, R. Ginosar (2003) "Parallel VLSI architectures and Parallel Interleaving Design for Low- Latency MAP Turbo Decoders," *Technical Report CCIT-TR436*, Electrical Engineering, Technion-Israel Institute of Technology, July 2003
41. M.J. Thul, F. Gilbert, N. Wehn (2002) "Optimized concurrent interleaving architecture for high-throughput turbodecoding," *9th Int. Conf. On Electronics, Circuits and Systems 2002*, vol. 3, pp. 1099-1102
42. F. Gilbert, M.J. Thul, N. Wehn (2002) "Communication centric architectures for turbo-decoding on embedded multiprocessors," *Conference and Exhibition on Design*, Automation and Test in Europe 2003, pp. 356–361
43. M.J. Thul, F. Gilbert, N. Wehn (2003) "Concurrent Interleaving architectures for high-throughput channel coding," *Proceedings of ICASSP 2003*, Vol. 2, pp. 613–616
44. F. Speziali, J. Zory (2004) "Scalable and area efficient concurrent interleaver for high throughput turbo-decoders," *Euromicro Symposium on Digital System Design*, Aug. 31 - Sept. 3, 2004 pp. 334–341
45. A. Tarable, G. Montorsi, S. Benedetto (2003) "Mapping interleaving laws to parallel Turbo decoder architectures," *Proceedings of the 3rd International Symposium on Turbo Codes and Related Topics*, pp. 153–156, Brest, France, Sept. 2003
46. A. Tarable, S. Benedetto (2004) "Mapping interleaving laws to parallel Turbo decoder architectures," *IEEE Comm. Letters*, Vol. 8, No. 3, March 2004, pp. 162–164
47. A. P. Worthen, S. Hong, R. Gupta, W. E. Stark (1999) "Performance optimization of VLSI transceiver for low-energy communications systems," *Military Communication Conference*, MILCOM 1999, Vol. 2, 31 Oct.-3 Nov. 1999, pp. 1434-1438
48. O.Y. Leung, C. Y. Tsui, R. S. Cheng (2001) "Reducing power consumption of turbo decoder using adaptive iteration with variable supply voltage," *IEEE Trans. on VLSI Systems*, Vol. 9, No. 1, Feb. 2001, pp. 34-40
49. C. Schurgers, L. Van der Perre, M. Engels, H. De Man (1999) "Adaptive Turbo Decoding for Indoor Wireless Communication," *IEEE Wireless Communications and Networking Conference (WCNC)*, 21-24 September 1999, pp. 1498–1502
50. R. Gonzales, B. M. Gordon, M. A. Horowitz (1997) "Supply and Threshold Voltage Scaling for Low Power CMOS," *IEEE Journal of Solid-State Circuits*, vol. 32, no. 8, pp. 1210–1216, August 1997.
51. B. Bougard, A. Giulietti, C. Desset, L. Van der Perre, F. Catthoor (2003) "A Low-Power High Speed Parallel Concatenated Turbo-decoding Architecture," *3rd Int. Symp. on Turbo Codes and Related Topics*, Brest, France, 2003, pp. 511–514
52. www.latticesim.com/products

53. www.xilinx.com/products/logicore/alliance
54. www.altera.com/products/ip
55. www.us.design-reuse.com/sip
56. www.icoding.com/products.htm
57. www.tij.co.jp/jsc/docs/dsps/support/dowload/c6000
58. A. La Rosa, C. Passerone, F. Gregoretti, L. Lavagno (2004) "Implementation of a UMTS turbo-decoder on a dynamically reconfigurable platform", *Design, Automation and Test in Europe*, Volume: 2 , 16-20 Feb. 2004 pp. 1218–1223 Vol.2
59. P. Salz, O. Schliebusch, D. Kammler, G. Ascheid, R. Leupers, H. Meyr (2004) *Specification and Implementation of an Application Specific Instruction Set Processor (ASIP) for Turbo Decoding,* Diploma Thesis, Institute for Integrated Signal Processing Systems, RWTH Aachen"
60. S. Sharma, S. Attri, F.C. Chauhan (2003) "A simplified and efficient implementation of FPGA-based turbo decoder" *Proceedings of the 2003 IEEE International Conference on Performance, Computing, and Communications Conference*, 9-11 April 2003 pp. 207–213
61. Xiao-Jun Zeng, Zhi-Liang Hong (2002) "Design and implementation of a turbo decoder for 3G W-CDMA systems" *Consumer Electronics, IEEE Transactions on* , Volume: 48 , Issue: 2 , May 2002 pp. 284–291
62. J. Steensma, C. Dick (2001) "FPGA implementation of a 3GPP turbo codec," *Thirty-Fifth Asilomar Conference on Signals, Systems and Computers,* Volume: 1 , 4-7 Nov. 2001, pp. 61–65 vol.1
63. P. Urard et al. (2004) "A Generic 350 Mb/s Turbo-Codec Based on a 16-states SISO Decoder," *International Solid-State Circuits Conference, ISSCC*, February 18, 2004, Session 23, paper 23.4
64. Myoung-Cheol Shin, In-Cheol Park (2003) "A Programmable Turbo Decoder for Multiple 3G Wireless Standards," *International Solid-State Circuits Conference, ISSCC*, February 11, 2003, Session 8, paper 8.7
65. B. Bougard et al. (2003) "A Scalable 8.7 nJ/bit 75.6 Mb/s Parallel Concatenated Convolutional (Turbo-) CODEC," *International Solid-State Circuits Conference, ISSCC*, February 11, 2003, Session 8, paper 8.6

Index